A COURSE IN NUMBER THI

A Course in Number Theory

H. E. ROSE

School of Mathematics, University of Bristol

CLARENDON PRESS · OXFORD

1988

Oxford University Press, Walton Street, Oxford OX2 6DP

Oxford New York Toronto
Delhi Bombay Calcutta Madras Karachi
Petaling Jaya Singapore Hong Kong Tokyo
Nairobi Dar es Salaam Cape Town
Melbourne Auckland

and associated companies in
Beirut Berlin Ibadan Nicosia

Oxford is a trade mark of Oxford University Press

Published in the United States
by Oxford University Press, New York

British Library Cataloguing in Publication Data

Rose, H. E.
A course in number theory.
Bibliography: p.
Includes indexes.
1. Numbers, Theory of. I. Title.
QA241.R665 1988 512'.7 87-7316
ISBN 0-19-853262-8
ISBN 0-19-853261-X (pbk.)

Library of Congress Cataloging in Publication Data

(Data Available)

Typeset and printed in Northern Ireland
at The Universities Press (Belfast) Ltd.

PREFACE

The central concern of number theory is the properties of the integers and rational numbers. Questions relating to algebraic number fields, individual real or complex numbers, integer matrices, points on algebraic curves, and similar entities are also considered. The subject has a very long history, as long as mathematics itself, and is as actively researched today as at any time in the past. For example, Faltings's remarkable solution of the Mordell conjecture has appeared during the writing of this book (see Chapter 15).

Number theory is not an organized theory in the usual sense but a vast collection of individual topics and results, with some coherent subtheories and a long list of unsolved problems. Some of these topics are highly specialized, for example giving a single solution to a Diophantine equation, whilst others have wide applicability, the Euclidean algorithm being an example. It is possible to treat number theory as applied mathematics. Problems are solved and the theory is developed using concepts and theorems from the whole range of mathematical disciplines, although group and field theory, linear algebra, and both real and complex analysis are the most commonly used. On the other hand, number theory is quite definitely a branch of pure mathematics; a problem or a result is studied entirely for its own sake with no application envisaged. The fact that many results are useful elsewhere is incidental.

This book concentrates on the central areas of number theory. Topics relying heavily on complex analysis and algebraic number theory have been excluded. This is partly to keep to a manageable length but mainly because these areas are best treated separately on their own merits. The choice of topics is a personal one, but it is hoped that a number of lecture and study courses will be constructed around a selection of the 15 chapters presented below.

Eight main subject areas are considered. Except for partition theory, each of these is allocated two chapters—a division which gives this book its main structure. The first chapter of a pair deals with the basic material whilst the second extends this and, in some cases, treats more specialized or difficult topics. Four of these more difficult topics are included because they contain important results but, in the opinion of the author, are best explored in the quiet of one's own study and away from the (undergraduate) lecture theatre; they are the Gelfond–Schneider theorem (Chapter

8), the class number formula for quadratic forms (Chapter 10), the Prime Number theorem (Chapter 13), and the Mordell–Weil theorem (Chapter 15).

Many exercises and problems have been provided with hints for their solution in the Appendix. *They form an integral part of this book.* The reader should try a good selection whilst reading a chapter, and *read* all of them because they contain a number of technical or subsidiary results not otherwise given. A star (☆), or in some extreme cases two stars, indicates that a problem is difficult. It is a good idea to have a pencil and paper handy when reading, both to try problems and to go over arguments in the text.

The first two chapters deal with divisibility and multiplicative functions, they form the basis for the whole book. Congruences are considered in the second pair of chapters; included is an introduction to p-adic numbers and a detailed discussion of quadratic congruences and Gauss's famous law of quadratic reciprocity. Chapter 5 presents some algebraic topics and primitive roots both of which have applications later in the book. The representations of integers as sums of squares is dealt with in Chapter 6 together with Gauss sums whose applications include the estimation of the number of solutions that some congruences possess.

The more advanced part of this book begins with two chapters on the number theory of real numbers. This includes results on the approximation of reals by rational numbers (Diophantine approximation), continued fractions and transcendence. Chapters 9 and 10 treat the arithmetic theory of quadratic forms and give the three squares theorem and a discussion of composition (of binary forms) which enables a closer examination of the theory to be undertaken. This latter topic is not often discussed in an elementary text but is included as it illustrates some useful algebraic techniques. Between this chapter and the final four chapters we have a change of pace; we discuss partition theory which considers different ways of representing a positive integer as a sum of smaller integers: this subject has a surprisingly rich and diverse development. Prime number theory is considered in Chapters 12 and 13. This is a major topic with an extensive literature. Proofs of Dirichlet's theorem on the existence of primes in arithmetic progressions, and the Prime Number theorem which estimates the density of the primes, are given together with some elementary results and a discussion of the Riemann hypothesis. In the final pair of chapters we study Diophantine equations, that is the solution of equations in integers or rational numbers. Some typical equations are discussed as well as the elementary theory of rational points on elliptic curves which uses a few basic ideas from algebraic geometry. This final topic is one of considerable current research interest. The book ends with some tables listing prime numbers, reduced quadratic forms, and so on, and a section giving answers, hints, or sketch solutions to problems.

Apart from the first five introductory chapters, Chapter 6, the four remaining pairs of chapters, and the chapter on partitions are largely independent of one another and can be studied in isolation. The work on Gauss sums in Chapter 6 is required in Chapter 10, Pell's equation (Chapter 7) and Legendre's Theorem (Chapter 14) are used in Chapter 9, and the material on quadratic forms given in Chapters 9 and 10 provides a starting point for the final pair of chapters.

Some historical comments are given throughout the text but no attempt has been made to treat this aspect in detail. Considerable attention has been paid to this recently and a number of books and papers have appeared giving the latest historical research; see Weil (1983), although Dickson's famous three-volume history (1919–22), Ore (1948), Schlarlau and Opolka (1984), and the historical notes in Hardy and Wright (1954), Le Veque (1977), and Ireland and Rosen (1980) also contain valuable information.

Prerequisites

The mathematical background needed for this book is covered in most first-year honours undergraduate courses for mathematicians given in the United Kingdom. A knowledge of matrices and basic linear algebra, group theory up to the homomorphism theorem, the elements of field theory, and some calculus and real analysis is assumed. Except briefly in the last sections of Chapters 8 and 12, no complex analysis is used. Of course the main requirement is an interest in the subject and the ability to 'think clearly'!

We shall use standard mathematical notation throughout. Curly brackets $\{\ldots\}$ will be used to denote sets including sets of solutions of congruences and Diophantine equations. In most contexts lower-case Roman letters denote integer or rational number variables or constants, and lower-case Greek letters are used either as integer exponents or, later, as real number variables or constants. τ, σ, ϕ, μ, and ζ denote the number theoretic functions defined in Chapters 1 and 2. We use the big 'O', little 'o' and $\sim$ notation:

$$f(x) = O(g(x)) \text{ stands for } \lim_{x\to\infty} |f(x)|/g(x) \text{ is bounded,}$$

$$f(x) = o(g(x)) \text{ stands for } \lim_{x\to\infty} f(x)/g(x) = 0,$$

$$f(x) \sim g(x) \text{ stands for } \lim_{x\to\infty} f(x)/g(x) = 1.$$

Finally the standard number systems will be denoted by $\mathbb{Z}$ (integers), $\mathbb{Q}$ (rational numbers), $\mathbb{R}$ (real numbers), and $\mathbb{C}$ (complex numbers).

Finally we mention four results of a slightly more specialized nature we shall use.

(i) The Vandermonde determinant. If $a_1, \ldots, a_n$ are real numbers then

$$\det\begin{pmatrix} 1 & 1 & \cdots & 1 \\ a_1 & a_2 & \cdots & a_n \\ \cdot & \cdot & \cdots & \cdot \\ a_1^{n-1} & a_2^{n-1} & \cdots & a_n^{n-1} \end{pmatrix} = \prod_{1\le i<j\le n} (a_j - a_i).$$

(ii) The symmetric function theorem. An n-variable polynomial f symmetric in its variables $x_1, \ldots, x_n$ and having degree r in each is equal to a polynomial of total degree r with integer coefficients in the elementary symmetric functions

$$\sum x_i, \sum x_i x_j, \ldots, \prod x_i$$

and the coefficients of f.

(iii) Euler's integral theorem. There is a constant γ $(=0.56\ldots)$, called the *Euler constant,* such that, if $x \ge 1$,

$$\int_1^x \frac{\mathrm{d}t}{t} = \ln x + \gamma + O\left(\frac{1}{x}\right),$$

where ln denotes the natural logarithm function.

(iv) The series

$$\sum_{n=1}^{\infty} \frac{1}{n^2} = \frac{\pi^2}{6}.$$

Proofs of these results can be found in many elementary algebra or analysis texts although derivations of (iii) and (iv) will be given in Chapter 12.

Acknowledgements

It is a pleasure to acknowledge the considerable assistance that I have received during the writing of this book. My greatest debt is to John Cremona, Ken Falconer, and Bob Odoni; they each read versions of the whole manuscript and made many helpful suggestions. Bryan Birch, J. W. S. Cassels, Harold Diamond, and the referees gave advice on some of the sections; and George Chauvet, Philip Drazin, and Philip Rose wrote some of the computer programs used to generate the tables. Finally I would like to thank the staff of the Clarendon Press, Oxford, and The Universities Press, Belfast, for the quality of their work.

Bristol H. E. R.
February 1987

CONTENTS

1

DIVISIBILITY

The simplest results in number theory concern the division of numbers. Everyone is familiar with this notion, for example when a week is divided into days, hours, and minutes. We shall begin with this topic to emphasize its central and fundamental role. Divisibility has been studied for at least three thousand years. From before the time of Pythagoras, the Greeks considered questions about even and odd numbers, perfect and amicable numbers, and the primes, amongst many others; even today a few of these are still unanswered. On the other hand some of their ideas have helped to lay the foundation of modern number theory. Their motivation was both practical and theoretical, the latter having a mystical or religious origin. Our motivation is 'study for its own sake' even though many results have important applications in other branches of mathematics and science.

1.1 The Euclidean algorithm and unique factorization

The set of integers

$$\ldots, -2, -1, 0, 1, 2, 3, \ldots$$

with the usual arithmetical operations is an integral domain denoted by $\mathbb{Z}$. This means that it is an Abelian group under addition $+$, with 0 as the neutral element, and with subtraction $-$ as the inverse operation. It is also closed under multiplication where the product of a and b is denoted by $a \cdot b$, although in most cases we suppress the dot and write ab. This operation is associative, commutative, obeys the usual distributive laws, and has 1 as its identity. Elements of the set $P = \{1, 2, \ldots\}$ are called *positive*, whilst elements of the set $N = \{-1, -2, \ldots\}$ are called *negative*. (We sometimes use the term *natural number* when referring to either zero or a positive integer.) $\mathbb{Z}$ is also linearly ordered by the relation $\leq$ defined by

$$a \leq b \quad \text{if and only if} \quad a = b \text{ or there is } c \in P \text{ such that } a + c = b.$$

Note that

$$a \leq b \quad \text{if and only if} \quad a + c \leq b + c,$$

and

$$a \le b \text{ and } 0 \le c \quad \text{implies} \quad ac \le bc.$$

As usual $a < b$ stands for $a \le b$ and $a \ne b$. Also note that P is well ordered, that is each subset of P has a least element.

The integral domain $\mathbb{Z}$ is not closed under division but it does have a number of divisibility properties which depend upon

THEOREM 1.1 *The division algorithm.*
Given a, $b \in \mathbb{Z}$, with $b > 0$, there exist unique integers c and d with the properties

$$a = bc + d \quad \text{and} \quad 0 \le d < b.$$

Proof. The set $\{a - bx : x \in \mathbb{Z}\}$ contains a least natural number y. If we let $d = y$ and $c = (a - y)/b$, then c, $d \in \mathbb{Z}$, $a = bc + d$, and $0 \le d < b$. To prove uniqueness assume that $a = bc + d = bc' + d'$ with $c \ge c'$ and $0 \le d$, $d' < b$, this gives

$$b(c - c') = d' - d.$$

If $c - c' > 0$, then $c - c'$ is at least one and so $d' - d \ge b$ which is impossible as both d and d' are less than b and non-negative; hence $c = c'$. This implies $d = d'$, and so the proof is complete. □

DEFINITION Divisibility.
If a, $b \in \mathbb{Z}$ then a divides b, denoted by $a|b$, if there exists $c \in \mathbb{Z}$ such that $b = ac$. We write $a \nmid b$ when a does not divide b. We have immediately from this definition

(i) for every integer a, $a|a$, $a|0$, $1|a$ and $-1|a$,
(ii) $0|a$ if and only if $a = 0$,
(iii) if $c > 0$ and $a|c$, then $c \ge a$,
(iv) if $a|b$ and $b|c$, then $a|c$,
(v) if $a|b$ and $a|c$, then for all x and y, $a|bx + cy$.

THEOREM 1.2 *The Euclidean algorithm.*
Suppose a and b are integers which are not both zero, then there exists a unique integer c satisfying the conditions

$$c > 0,\ c|a,\ c|b, \text{ and if } d|a \text{ and } d|b, \text{ then } d|c. \tag{1}$$

Proof. Let c be the least positive integer of the form

$$c = ax + by$$

where x, $y \in \mathbb{Z}$, we claim that c satisfies the conditions (1). Clearly $c > 0$ and if d divides both a and b, then $d|c$. We show that $c|a$, the derivation of the remaining condition is similar. By the division algorithm, integers t

and u exist satisfying $a = ct + u$ where $0 \le u < c$. Hence

$$a = ct + u = axt + byt + u,$$

and, rearranging, we have

$$a(1 - xt) + b(-yt) = u.$$

It follows that $u = 0$ by the definition of c as $u < c$, and so $c \mid a$. Finally suppose (1) also holds with c' replacing c. We have immediately $c' \mid c$ and $c \mid c'$, thus $c = c'$ and the proof is complete. □

We sketch another proof of this important result. Suppose $a > b > 0$, the argument is similar in the remaining cases. Using the division algorithm we have in turn

$$\begin{array}{ll} a = bq_1 + r_1 & 0 < r_1 < b \\ b = r_1 q_2 + r_2 & 0 < r_2 < r_1 \\ \dots & \dots \\ r_{t-2} = r_{t-1} q_t + r_t & 0 < r_t < r_{t-1} \\ r_{t-1} = r_t q_{t+1}, & \end{array} \tag{2}$$

applying Theorem 1.1 until we obtain a zero remainder. This will always occur as each r_i is non-negative and strictly less than its predecessor. Let r_t be the least non-zero remainder, using each equation of (2) in turn it is a simple matter to check that we can take r_t for c in (1). Begin by noting that $r_t \mid r_{t-1}$ and so the penultimate equation gives $r_t \mid r_{t-2}$.

Definition Greatest common divisor.
Let $a, b \in \mathbb{Z}$, not both zero, then the unique number c given by Theorem 1.2 is called the *greatest common divisor*, or GCD, of a and b. It is denoted by (a, b) and is the largest integer dividing both a and b. To complete the definition we set $(0, 0) = 0$.

The GCD of $a_1, \dots, a_n$, denoted by $(a_1, \dots, a_n)$, is given by $(a_1, \dots, a_n) = (a_1, (a_2, \dots, (a_{n-1}, a_n) \dots))$, see Problem 4.

The properties below follow easily from the definition, the reader should check this. We let $|a|$ denote the absolute value of a.

(i) $(a, b) = (b, a)$ and $(a, 0) = |a|$.
(ii) if $a \mid c$ and $b \mid c$ then $ab \mid c(a, b)$,
(iii) if $a \mid b$ then $(a, b) = a$ and $(a, c) \mid b$ for all $c \in \mathbb{Z}$,
(iv) if $(a, b) = d$ then a/d and b/d are integers and $(a/d, b/d) = 1$.

Example 1 Calculate $(533, 117)$.

Using the equations (2) on the previous page we have

$$533 = 117 \cdot 4 + 65$$
$$117 = 65 \cdot 1 + 52$$
$$65 = 52 \cdot 1 + 13$$
$$52 = 13 \cdot 4.$$

Hence $(533, 117) = 13$.

The Euclidean algorithm is an important construction in number theory with many applications. The first concerns solutions of the equation

$$ax + by = n$$

in integers x and y, a linear Diophantine equation. We use the term Diophantine (after the Greek mathematician Diophantus, see the introduction to Chapter 14) when considering solutions of algebraic equations which lie in a particular number class, for example the integers $\mathbb{Z}$ or the rational numbers $\mathbb{Q}$.

THEOREM 1.3 *The equation*

$$ax + by = n \tag{3}$$

has integer solutions x and y if and only if $(a, b)|n$.

Proof. Let $n \neq 0$, the result is trivial otherwise. If $(a, b) \nmid n$ there can be no solution, hence we may assume that $(a, b)|n$. In the proof of Theorem 1.2 we showed that integers x_1 and y_1 exist satisfying

$$ax_1 + by_1 = (a, b),$$

and so $x_0 = nx_1/(a, b)$, $y_0 = ny_1/(a, b)$ is an integer solution of equation (3). Note that, as $n \neq 0$, we have $(a, b) \neq 0$. □

The equations (2) given in the second proof of Theorem 1.2 provide an efficient method for solving (3). As $(a, b) = r_t$, the penultimate equation of (2) gives

$$r_{t-2} - r_{t-1}q_t = (a, b),$$

and combining this with the equation for r_{t-3} we have

$$-r_{t-3}q_t + r_{t-2}(1 + q_t q_{t-1}) = (a, b).$$

Repeating this process, using each equation of (2) in turn, we have finally

$$aq_1^* + bq_2^* = (a, b)$$

where q_1^* and q_2^* stand for polynomial expressions in $q_1, \ldots, q_t$. As in the proof of Theorem 1.3, this provides a solution of equation (3).

EXAMPLE 2 Find a solution of the equation $533x + 117y = 65$.

In Example 1 above we showed that $(533, 117) = 13$ and so, as $13|65$, this equation is soluble. Using the equations in the solution of Example 1 we have, following the construction described above,

$$\begin{aligned} 13 &= 65 - 52 = 65 - (117 - 65) = 65 \cdot 2 - 117 \\ &= (533 - 117 \cdot 4)2 - 117 = 533 \cdot 2 - 117 \cdot 9. \end{aligned}$$

Hence $x = 10$ and $y = -45$ is a solution.

Further numerical examples are given in the problem section.

We shall extend Theorem 1.3 to give the general solution of equation (3). For this we need

LEMMA 1.4 *If $a|bc$ and $(a, b) = 1$ then $a|c$.*

Proof. As $(a, b) = 1$, we can find integers x and y to satisfy $ax + by = 1$ using Theorem 1.3. Hence $acx + bcy = c$, but $a|bc$ and so a divides the left-hand side of this equation. It follows that $a|c$. □

THEOREM 1.5 *Suppose $n \neq 0$ and $(a, b)|n$. The general solution of the equation*

$$ax + by = n \tag{3}$$

has the form

$$\{x_0 + tb/(a, b), y_0 - ta/(a, b)\}$$

where $t \in \mathbb{Z}$ and $\{x_0, y_0\}$ is any solution of (3).

Proof. By Theorem 1.3, a solution $\{x_0, y_0\}$ of (3) exists. So clearly $\{x_0 + tb/(a, b),\ y_0 - ta/(a, b)\}$ is also a solution for each $t \in \mathbb{Z}$. Now assume that $\{x_1, y_1\}$ and $\{x_1 + u, y_1 - v\}$ are solutions of (3) with $0 < u < b/(a, b)$, i.e. assume that not all solutions of (3) are of the form stated in the theorem. It follows that

$$ax_1 + by_1 = a(x_1 + u) + b(y_1 - v),$$

and so $au = bv$ and

$$(a/(a, b))u = (b/(a, b))v.$$

Hence $(b/(a, b))|u$ by Lemma 1.4 and (iv) on p. 3, and so $u \geq b/(a, b)$ as $b \neq 0$. This contradicts the inequality defining u, thus our assumption is false and the proof is complete. □

If we refer back to Example 2, Theorem 1.5 shows that the general solution of the equation $533x + 117y = 65$ is $\{10 + 9t,\ -45 - 41t\}$.

Using the GCD we can introduce the *least common multiple* (LCM) of two non-zero integers a and b by $\mathrm{LCM}(a, b) = |ab|/(a, b)$.

THEOREM 1.6 *Let a and b be non-zero integers.*

(i) *a and b both divide* $\mathrm{LCM}(a, b)$.

(ii) *If $a|m$ and $b|m$ then* $\mathrm{LCM}(a, b)|m$.

Proof. (i) As (a, b) divides $|b|$, we have $|b|/(a, b)$ is an integer and so $a|\mathrm{LCM}(a, b)$. Similarly $b|\mathrm{LCM}(a, b)$.

(ii) Let

$$d = (a, b) \quad \text{with} \quad a = a_1 d,\ b = b_1 d \quad \text{and} \quad (a_1, b_1) = 1.$$

Now we can write $m = ra_1 d = sb_1 d$ for some integers r and s. This gives $a_1|sb_1$; and so $a_1|s$, by Lemma 1.4, and $s = a_1 t$ for some integer t. Thus we have

$$m = a_1 t b_1 d = t\frac{ab}{d} = \pm t\mathrm{LCM}(a, b)$$

and (ii) follows. □

The Euclidean algorithm also has some connections with the theory of continued fractions. Let α be a real number, let q_1 be the largest integer less than or equal to α, and write

$$\alpha = q_1 + \frac{1}{\alpha_1}.$$

Note $\alpha_1 > 1$ unless $\alpha \in \mathbb{Z}$. Now let q_2 be the largest integer less than or equal to α_1 and write

$$\alpha_1 = q_2 + \frac{1}{\alpha_2},$$

again $\alpha_2 > 1$ unless $\alpha_1 \in \mathbb{Z}$. Repeating this process we have the *continued fraction* expansion of α, namely:

$$\alpha = q_1 + \cfrac{1}{q_2 + \cfrac{1}{q_3 + \ddots \; q_n + \cfrac{1}{q_{n+1} + \ddots}}}$$

with a finite number of entries if and only if α is rational. If $\alpha = a/b$, with $a, b \in \mathbb{Z}$, then the numbers $q_1, q_2, \ldots$ are given by the Euclidean algorithm equations (2). We shall return to this topic in Chapter 7 where we shall show that continued fractions give efficient rational approximations to real numbers.

Unique factorization

We come now to the unique factorization theorem for $\mathbb{Z}$ which is proved using Lemma 1.4. We begin with a

DEFINITION An integer p, $p > 1$, is called *prime* if it has no divisors lying strictly between 1 and p. An integer greater than 1 and not prime is called *composite*. Further we say a is *coprime* to b, or sometimes a and b are *relatively prime*, when a and b have no common prime factors; that is when $(a, b) = 1$.

Note that if $a \in \mathbb{Z}$ and $a < -1$, then $a = (-1)b$ where b is prime or composite.

Except in Chapter 11, we shall use the symbol p (with suffices etc. in some cases) to denote primes throughout this book.

LEMMA 1.7 *Every integer greater than* 1 *is a prime or a product of primes.*

Proof. By contradiction. Assume that n is the least integer which is neither a prime nor a product of primes, by definition n is composite. So $n = ab$ where both a and b are less than n, and hence both a and b are either prime or a product of primes. But then n is also a product of primes contradicting our assumption. □

THEOREM 1.8 (Euclid) *There are infinitely many prime numbers.*

Proof. Assume the contrary and let $p_1, \ldots, p_k$ be a list of all primes. By Lemma 1.7 the number

$$z = p_1 p_2 \ldots p_k + 1$$

is a product of primes $p_1' p_2' \ldots p_j'$. Using the assumption, we have $p_1' = p_t$ for some t where $1 \le t \le k$. But this is impossible because p_t does not divide z (it has remainder 1), and so $p_1' \notin \{p_1, \ldots, p_k\}$. This contradiction proves the theorem. □

Further proofs of this theorem are given in Problem 18.

THEOREM 1.9 *Unique factorization theorem.*
Every integer greater than 1 *can be expressed as a product of prime factors in a unique way except for the order of the factors.*

Proof. Using Lemma 1.7 suppose we have two prime decompositions

$$\begin{aligned} a &= p_1^{\alpha_1} \ldots p_j^{\alpha_j} = P \\ &= q_1^{\beta_1} \ldots q_k^{\beta_k} = Q \end{aligned}$$

where each p_i and q_i is prime. For all i $(1 \le i \le j)$, we have $p_i | Q$, and so, by Lemma 1.4, $p_i | q_t^{\beta_t}$ for some t $(1 \le t \le k)$. As p_i and q_t are both prime it follows that $p_i = q_t$. Hence each prime occurring in P also occurs in Q and vice versa. This shows that $j = k$ and the decomposition Q has the form

$$p_1^{\gamma_1} \dots p_j^{\gamma_j}.$$

Secondly suppose $\gamma_i > \alpha_i$ for some i. Dividing both decompositions by $p_i^{\alpha_i}$, we obtain decompositions of $a/p_i^{\alpha_i}$ one of which involves p_i whilst the other does not. This is impossible by the argument above; hence, by symmetry, $\alpha_i = \gamma_i$ for all i and the theorem is proved. □

Using this theorem we can give formulas for GCD and LCM as follows.

THEOREM 1.10 (i) *Let*

$$m = \prod_{i=1}^{k} p_i^{\alpha_i} \quad \textit{and} \quad n = \prod_{i=1}^{k} p_i^{\beta_i}$$

where some of the α and β may be zero. Then

$$(m, n) = \prod_{i=1}^{k} p_i^{\min(\alpha_i, \beta_i)} \quad \textit{and} \quad \mathrm{LCM}(a, b) = \prod_{i=1}^{k} p_i^{\max(\alpha_i, \beta_i)}.$$

(ii) $(m_1, \mathrm{LCM}(m_2, m_3)) = \mathrm{LCM}((m_1, m_2), (m_1, m_3))$.

Proof. (i) Exercise. (ii) Use (i) and the result (proved by cases): $\min(x, \max(y, z)) = \max(\min(x, y), \min(x, z))$. □

Unique factorization is so familiar to every mathematician that it is difficult to realize its true significance. Many algebraic number fields do not possess this property. For example in the integral domain $\mathbb{Z}[\sqrt{-5}] = \{a + b\sqrt{-5} : a, b \in \mathbb{Z}\}$ we have

$$6 = 2 \cdot 3 = (1 - \sqrt{-5})(1 + \sqrt{-5}),$$

and it is an easy matter to check that 2, 3, $1 - \sqrt{-5}$ and $1 + \sqrt{-5}$ are irreducible or 'prime', see Chapter 5 for details.

We shall often use unique factorization for $\mathbb{Z}$ in the following pages, usually without explicitly referring to it. One application concerns the Riemann zeta function ζ which is defined by

$$\zeta(s) = \sum_{n=1}^{\infty} \frac{1}{n^s},$$

where s is a complex variable whose real part is greater than 1. Using unique factorization, Euler showed that

$$\zeta(s) = \prod_{\text{all primes}} \left(1 - \frac{1}{p^s}\right)^{-1}.$$

For a proof and further discussion of this important result see Section 3 of Chapter 12.

1.2 Prime numbers

The main prime number results will be considered in Chapters 12 and 13, here we discuss some formulas for the primes and computer factorization of large numbers.

Many attempts have been made to find simple formulas which would map the integers into the primes. For example

$$x^2 + x + 41$$

takes prime values for $-40 \le x < 40$. Of course formulas taking prime values infinitely often are the most useful. Until recently no polynomial formulas of this type existed, but the work of the recursion theorists has altered this. In the last decade a number of polynomial prime-generating formulas have been described using the result that a set or relation which can be represented on a computer, however complex, can also be represented by a collection of polynomial equations. The degree k and the number of variables m of these polynomial formulas vary slightly; in one version $k = 16$ and $m = 26$ (see Riesel, 1985 p. 42) and, at the present time, all known formulas have $m \ge 12$. But in every version the polynomial involved is defined over $\mathbb{Z}$ and maps $\mathbb{N}^m$ *onto* the primes, where $\mathbb{N}$ denotes the natural numbers. (Note that these formulas also take non-prime values when some of the arguments are negative.) Unfortunately as both k and m are relatively large, these polynomial formulas are only of theoretical interest.

We shall consider some of the early attempts to find prime number formulas. Although these attempts failed in their primary purpose the formulas are not without interest.

LEMMA 2.1 *Let $a > 1$ and $c > 1$. The integer $a^c - 1$ is composite if $a > 2$ or if c is composite.*

Proof. If $c = de$ we have $a^e - 1 | a^c - 1$ by writing

$$a^c - 1 = (a^e - 1)(a^{e(d-1)} + a^{e(d-2)} + \ldots + 1).$$

Hence $a^c - 1$ is composite if $a > 2$ or if c is not prime (let both d and e be greater than 1). □

The numbers $M_p = 2^p - 1$ (p a prime) are known as *Mersenne numbers* after the seventeenth-century French mathematician. M_p is prime if

$$p = 2, 3, 5, 7, 13, 17, 19, 31, 61, 89, 107, 127, 521, 607,$$
$$1279, 2203, 2281, 3217, 4253, 4423, 9689, 9941, 11213,$$
$$19937, 21701, 23209, 44497, 86243, 132049, 216091,$$

and M_p is composite for all other primes less than 100000. It seems likely that further Mersenne primes will be discovered and this suggests that M_p is prime for infinitely many values of p, but this appears to be very difficult to prove. Incidently at the present time (late 1985) M_{216091} is the largest explicitly known prime! (See also Lemma 1.6, Chapter 4.)

LEMMA 2.2 (i) *Let $a > 1$ and $c > 1$. The integer $a^c + 1$ is composite if a is odd or if c has an odd factor.*
(ii) *If $F_n = 2^{2^n} + 1$, then $(F_m, F_n) = 1$ whenever $m \neq n$.*

Proof. (i) If $a = 2j + 1$ then $a^c + 1 = (2j + 1)^c + 1$ which is even and larger than 2. Secondly if $c = k(2m + 1)$, $m > 0$, we have

$$a^c + 1 = (a^k + 1)(a^{2km} - a^{k(2m-1)} + \ldots + 1)$$

and so (i) follows as $a^k + 1 > 1$.
For (ii) we use a similar argument to show that, if $n > m$,

$$2^{2^m} + 1 | 2^{2^n} - 1,$$

and so $F_m | F_n - 2$. Hence F_m and F_n cannot have a common factor greater than 2, the result follows as they are both odd. □

The numbers $F_n = 2^{2^n} + 1$ are called *Fermat numbers*. Using the lemma above, Fermat conjectured that F_n is always prime. This is true if $n < 5$ but we have

THEOREM 2.3 (Euler) *F_5 is composite.*

Proof. (One of the cheekiest in mathematics!) We show $641 | F_5$. Clearly $641 = 2^4 + 5^4 = 5 \cdot 2^7 + 1$, so $2^4 = 641 - 5^4$ and

$$\begin{aligned} 2^{32} &= 2^4 \cdot 2^{28} = 641 \cdot 2^{28} - (5 \cdot 2^7)^4 = 641 \cdot 2^{28} - (641 - 1)^4 \\ &= 641k - 1. \end{aligned}$$

This gives the result. Incidently the only other factor is 6700417, a prime number. □

No other Fermat prime is known, F_n has been shown to be composite for $5 \le n \le 19$ and a number of other values up to 23471 usually without explicitly exhibiting the factors, although it is known that

$$274177 | F_6$$
$$5\,96495891\,27497217 | F_7$$
$$12389263\,61552897 | F_8.$$

At one time it was thought that Fermat's conjecture could be replaced by: the numbers

$$2 + 1, 2^2 + 1, 2^{2^2} + 1, \ldots$$

are all prime. This is also false as F_{16} is composite, it is divisible by

$1575 \cdot 2^{19} + 1$ (see Problem 11, Chapter 3). Further numerical details are given in Riesel (1985).

It is of interest to note that these numbers arise in other connections. First a regular polygon with n sides can be constructed using only a ruler and a compass if and only if n is the product of a power of 2 and distinct Fermat primes; see Problem 16, Chapter 6. Secondly the Mersenne primes have connections with perfect numbers, we shall discuss this in the next chapter.

Many formulas of a more complicated nature have been given, amongst these are:

$$\prod_{y=2}^{n-1} \prod_{z=1}^{n/y} (n - yz) > 0$$

and

$$\lim_{r\to\infty} \lim_{s\to\infty} \lim_{t\to\infty} \sum_{u=0}^{s} \left[1 - \left(\cos \frac{(u!)^r \pi}{n}\right)^{2t}\right] = n$$

where each proposition is equivalent to the statement 'n is prime', the second is due to Hardy. Unfortunately, as with the polynomial formulas mentioned earlier, these are too complicated for meaningful results to follow.

Computer prime recognition and factorization

It is a straightforward, if rather tedious, task to show that a 5-digit integer is prime or composite and to find its factors; but it is a very different matter if a 50-digit integer is to be tackled. Highly sophisticated computer techniques are required if the factorization is to be completed in a reasonable time, and for a typical 500-digit integer it would take more than a lifetime to find its factors even using the fastest computers available today. This fact has been used recently to construct virtually uncrackable codes; see Riesel (1985) for a discussion of this and the other topics mentioned in this subsection.

If a computer is to be used to factorize large integers, the first task is to introduce a method for representing long strings of digits in the program; let us assume that this has been done. Lists of all primes less than some fixed number can be produced easily using the 'Sieve of Eratosthenes' (see p. 215) provided the computer memory can cope with the volume of data involved. Hence prime recognition for relatively small integers (say less than a million) can be achieved by simple division, but larger integers require different techniques. We shall describe a few of these now; another method which has some connection with the Riemann hypothesis is discussed briefly on p. 229.

In Chapter 3 we prove that, given an integer n, if another integer a exists with the properties $(a, n) = 1$ and

$$n \nmid a^{n-1} - 1$$

then n is composite. The converse holds in many, but not all, cases. We call an integer n *pseudoprime* with respect to a if $(a, n) = 1$ and $n \mid a^{n-1} - 1$. Note that a prime p is pseudoprime with respect to any a satisfying $(a, p) = 1$. So for example 347 is pseudoprime with respect to 2 whilst 341 is not, see Problem 13, Chapter 3. Extremely efficient methods exist for checking whether or not an integer is pseudoprime with respect to some fixed integer a. Also there are only 1770 primes less than $25 \cdot 10^9$ which are not pseudoprime with respect to 2, 3, 5 or 7. Thus once these 1770 exceptions are coded into the program, a good prime recognition test will be given by this construction.

Another method, due to Fermat, can be effective in some cases. Given an integer n, positive integers a and b are sought satisfying

$$n = a^2 - b^2$$

when, of course, n is composite having the factors $a + b$ and $a - b$. Clearly $a > \sqrt{n}$, so as a ranges over the integers larger than $\sqrt{n}$, $a^2 - n$ is checked to see if it is square. For example suppose $n = 13199$, then $\sqrt{n} = 114.8\ldots$ and so the integers $115^2 - 13199 = 26$, $116^2 - 13199 = 257, \ldots$ are checked for squareness. In this case we have $132^2 - 13199 = 4225 = 65^2$ and so $13199 = 67 \cdot 197$. There are many versions of this method, see Problem 22.

In general it is more time consuming to factorize a number than to decide whether it is prime or not. Hence in factorization problems it is best to check for primeness first. A great variety of factorization methods have been discussed, for example the Fermat method described above can be used to factorize integers. Another method involves the Euclidean algorithm as follows. Calculate $P_1, \ldots, P_{10}$ where P_{n+1} is the product of all primes lying between $n \cdot 100$ and $(n + 1) \cdot 100$, so for example

$$P_1 = 23055\ 67963945\ 51842475\ 31021473\ 31756070$$

and the remaining numbers have similar orders of magnitude. Given an integer n to be factorized, first check that it is composite. Secondly calculate

$$(n, P_1), (n, P_2), \ldots, (n, P_{10})$$

using the Euclidean algorithm, this will give all prime factors of n less than 1000. As Euclidean algorithm calculations can be completed very quickly, this gives an efficient program for factorizing most integers less than 10^8. Many other methods are given in Riesel's book, one of these is described in Problem 11, Chapter 3. Recently a new method has been discovered by Lenstra (1986) using elliptic curve theory (Chapter 15), in the great majority of cases it is extremely efficient.

1.3 Problems 1

1. Given $a, b \in \mathbb{Z}$, $b > 0$, show that there exist unique integers $n, c_0, \ldots, c_n$ such that

$$a = c_0 b^n + c_1 b^{n-1} + \ldots + c_n$$

where each c_i satisfies $0 \leq c_i < b$.

2. Let $n > 1$ and $S_n = \frac{1}{2} + \frac{1}{3} + \ldots + \frac{1}{n}$, show that S_n cannot be an integer.

3. Solve in integers (where possible)
   (i) $3x - 5y = 7$,
   (ii) $21x - 35y = 24$,
   (iii) $97x + 127y = 1$,
   (iv) $91x - 143y = 13$.

4. Given $a_1, \ldots, a_n$, let c be the least positive integer which can be expressed in the form $a_1x_1 + \ldots + a_nx_n$. Show that $c = (a_1, \ldots, a_n)$. Deduce

$$a_1x_1 + \ldots + a_nx_n = b$$

is soluble if and only if $(a_1, \ldots, a_n) | b$.

5. Let n be a positive integer, prove that n can be written as a sum of (at least two) consecutive positive integers if and only if $n \neq 2^k$ for some k.

6. Let a and b be positive integers with $(a, b) = 1$. Show that $ax + by = n$ is soluble in non-negative integers x and y whenever $n \geq (a-1)(b-1)$.

7. Use Theorem 1.5 and induction on k to show that, if $(a, b) = 1$, integers t and u can be found to solve the equation

$$at^2 + bt + c = a^k u.$$

8. Show that
   (i) if $(m, n) = 1$ then $(m + n, m - n) = 1$ or 2,
   (ii) $(m, m + n) | n$,
   (iii) if $(a, m) = 1 = (a, n)$ then $(a, mn) = 1$.

9. Prove the following, where n is a positive integer,
   (i) if n is odd then $8 | n^2 - 1$,
   (ii) if $3 \nmid n$ and n is odd then $24 | n^2 - 1$,
   (iii) $30 | n^5 - n$,
   (iv) $42 | n^7 - n$,
   (v) $504 | (n^3 - 1)n^3(n^3 + 1)$.

10. Given a set X with n elements and a finite number of properties (of elements of X) $P_1, P_2, \ldots, P_r$, let

$$X_{ij\ldots k} = \{x : x \in X \text{ and } x \text{ has the properties } P_i, P_j, \ldots \text{ and } P_k\}$$

and suppose $X_{ij\ldots k}$ has $n_{ij\ldots k}$ elements. Using a counting argument and the binomial theorem show that if m is the number of elements of X which possess none of the properties $P_1, \ldots, P_r$ then

$$m = n - n_1 - \ldots - n_r + n_{12} + n_{13} + \ldots + n_{r-1,r} - n_{123} - \ldots.$$

11. Suppose $a_1, \ldots, a_r$ are positive integers, show that
    (i) $\max(a_1, \ldots, a_r) = a_1 + \ldots + a_r - \min(a_1, a_2) - \ldots - \min(a_{r-1}, a_r)$ $+ \min(a_1, a_2, a_3) + \ldots \pm \min(a_1, \ldots, a_r)$.
    (ii) $\text{LCM}(a_1, \ldots, a_r) = a_1 a_2 \ldots a_r (a_1, a_2)^{-1} \ldots (a_{r-1}, a_r)^{-1} (a_1, a_2, a_3)$ $(a_1, a_2, a_4) \ldots (a_1, a_2, \ldots, a_r)^{(-1)^{r+1}}$.

12. Show that the general solution in integers of the equation

$$a^2 + b^2 = c^2$$

can be written in the form $a = k(m^2 - n^2)$, $b = 2kmn$ and $c = k(m^2 + n^2)$ where $(m, n) = 1$.

13. Show that the only solutions of the equation

$$(n-1)! = n^k - 1$$

are $n = 2$, $k = 1$; $n = 3$, $k = 1$ or $n = 5$, $k = 2$.

14. (i) Let $\text{ord}_p\, n$ denote the exponent of the prime p in the unique factorization of n. Show that

$$\text{ord}_p\, n! = [n/p] + [n/p^2] + \ldots,$$

where [.] denotes the integer part.

(ii) Use (i) to show that the binomial coefficients $\binom{m}{n}$ are integers.

15. Prove that $[2x] + [2y] \geq [x] + [y] + [x + y]$ for real numbers x and y. Deduce that, if m and n are positive integers,

$$\frac{(2m!)(2n!)}{m!\, n!\, (m+n)!}$$

is an integer.

16. Show that there exist arbitrarily long sequences of consecutive composite integers.

17. Prove that there are infinitely many primes of the form $4n + 3$ and $6n + 5$.

18. Give new proofs of Euclid's result (Theorem 1.8) using the propositions
    (i) $\zeta(1)$ is divergent,
    (ii) $\zeta(2) = \pi^2/6$ is irrational,
    (iii) one of our results involving the Fermat numbers,
    (iv) $(n!)^{1/n} \leq \prod_{p|n} p^{1/(p-1)}$ [Hint. Use Problem 14 to show that $\text{ord}_p\, n! \leq n/(p-1)$.]

Further proofs will be given in Chapter 12.

19. The Fibonacci sequence $\{u_n\}$ is given by: $u_0=0$, $u_1=1$, and $u_{n+2}=u_{n+1}+u_n$; the first few terms are 0, 1, 1, 2, 3, 5, 8, 13, ... Show that

(i) if $x=(1+\sqrt{5})/2$ and $y=(1-\sqrt{5})/2$ then $u_n\sqrt{5}=x^n-y^n$ (x and y are the roots of the equation $x^2-x-1=0$),

(ii) $(u_n, u_{n+1})=1$,

(iii) $u_{m+n}=u_{n-1}u_m+u_nu_{m+1}$,

(iv) if $r>0$, $u_n|u_{nr}$,

(v) if $(m, n)=d$ then $(u_m, u_n)=u_d$.

20. An integer d is called a *fundamental discriminant* if $4|d$ or $4|d-1$, it has no odd prime squared factor, and if $d=4d^*$ then $4|d^*-2$ or $4|d^*-3$. We shall use these discriminants in Chapter 10.

(i) Show that every integer d satisfying $4|d$ or $4|d-1$ can be expressed in the form d_0e^2 where d_0 is a fundamental discriminant.

(ii) Further show that every fundamental discriminant can be expressed as a product of distinct members of the set $\{(-1)^{(p-1)/2}p, -8, -4, 8\}$ where p is an odd prime.

21. The *Farey series* F_n of order n is the increasing sequence of all irreducible fractions a/b lying between 0 and 1 whose denominators do not exceed n, so $0\le a\le b\le n$ and $(a, b)=1$.

(i) Write down the Farey series of order 4 and 7.

In the next three parts assume that c/d immediately succeeds a/b in the series F_n. Show that:

(ii) $b+d>n$,

(iii) $b\ne d$ unless $n<2$,

(iv) $bc-ad=1$.

[Hint. To prove (iv) consider the general solution of $bx-ay=1$, show that y can be chosen so that $0\le n-b<y\le n$, and then $x/y\in\mathsf{F}_n$ and $a/b<x/y$. Deduce $x/y=c/d$ by considering $1/by$.]

(v) Show further that if a/b, c/d, and e/f are consecutive terms in F_n then $c/d=(a+e)/(b+f)$.

(vi) Use (iv) to find the two terms which succeed 3/7 in F_{11}.

22. Fix k and suppose n can be represented in two distinct ways as

$$n=a^2+kb^2=c^2+kd^2.$$

Use this and the Euclidean algorithm to find two factors of n. Now apply this to extend Fermat's method for factorizing large integers and consider the case $n=34889$ with $k=10$.

2

MULTIPLICATIVE FUNCTIONS

The function method is a powerful tool in some branches of number theory. Suppose we are given a property R of the positive integers, and suppose we can define a function r which represents R, then we can apply algebraic or analytic techniques to r in the hope of characterizing R. For instance a positive integer is a square if and only if it has an odd number of divisors, and we can derive this property using the function τ which counts the number of divisors of an integer. In this chapter we shall discuss a class of functions, called the multiplicative functions, which preserve products in certain cases; they arise naturally from divisibility.

Two important multiplicative functions will be treated in detail. They are (i) the Möbius function μ which was introduced in the last century to solve a problem concerning the Riemann zeta function (see Lemma 2.4), and (ii) the Euler function ϕ introduced by Euler to generalize a congruence result of Fermat (see Chapter 3). They have many applications later in this book. The average order of some of these functions will be considered in the second section.

A number of proofs involve double sums and the interchange of the order of summation, or the replacement of a sum of products by a product of sums or vice versa. If these arguments seem confusing they will become clearer if the reader chooses a small round number like 24 or 30, writes out the sums and products in full, and checks the claimed equality term by term. In this chapter m and n denote positive integer constants or variables.

2.1 The Möbius and Euler functions

We begin with a

DEFINITION (i) A function that maps the positive integers into a set which is closed under a product is called a *number theoretic* function.

(ii) A number theoretic function f is called *multiplicative* if

$$(m, n) = 1 \quad \text{implies} \quad f(mn) = f(m)f(n).$$

(iii) A number theoretic function g is called *completely multiplicative* if

$$g(mn) = g(m)g(n)$$

without restriction.

The range of a multiplicative function may be finite or infinite and may involve complex numbers or matrices in some cases. The divisor function τ mentioned in the Introduction is multiplicative, but the function π, where $\pi(n)$ equals the number of primes less than or equal to n, does not have this property [for instance $\pi(2)\pi(3) \neq \pi(6)$] and this explains one of the difficulties in the evaluation of π. Note that if f is multiplicative and not identically zero, then $f(1) = 1$ and

$$n = p_1^{\alpha_1} \dots p_k^{\alpha_k} \quad \text{implies} \quad f(n) = f(p_1^{\alpha_1}) \dots f(p_k^{\alpha_k}). \tag{1}$$

Some results follow which will provide important examples of multiplicative functions. We let $\sum_{d|n} f(d)$ denote the sum of the values of a function f as the argument d ranges over the positive divisors of n, so for example

$$\sum_{d|12} f(d) = f(1) + f(2) + f(3) + f(4) + f(6) + f(12).$$

Note that

$$\sum_{d|n} f(d) = \sum_{d|n} f(n/d) \tag{2}$$

as the second sum is simply the first sum written in reverse order.

THEOREM 1.1 *If f is a multiplicative function and F is given by*

$$F(n) = \sum_{d|n} f(d)$$

then F is also multiplicative.

Proof. Let $(m, n) = 1$. A divisor d of mn can be uniquely expressed as $d = d_1 d_2$ where $d_1 | m$ and $d_2 | n$, and then $(d_1, d_2) = 1$. Now $(m, n) = 1$ implies

$$\begin{aligned} F(mn) &= \sum_{d|mn} f(d) \\ &= \sum_{d_1|m\, d_2|n} f(d_1 d_2) && \text{by above} \\ &= \sum_{d_1|m\, d_2|n} f(d_1) f(d_2) && \text{as } f \text{ is multiplicative} \\ &= \sum_{d_1|m} f(d_1) \sum_{d_2|n} f(d_2) && \text{by writing the sum of products above as a product of sums} \\ &= F(m)F(n) && \text{by definition.} \end{aligned}$$

□

DEFINITION The functions τ (called the *divisor function*) and σ are given by

$$\tau(n) = \text{the number of positive divisors of } n$$
$$\sigma(n) = \text{the sum of the positive divisors of } n.$$

LEMMA 1.2 *τ and σ are multiplicative.*

Proof. This follows immediately from Theorem 1.1 as the constant function 1, where $1(n) = 1$, and the identity function i, where $i(n) = n$, are both clearly multiplicative. □

Using this lemma we can give formulas for τ and σ. If p is a prime then the divisors of p^α are $1, p, p^2, \ldots, p^\alpha$ and so we have by inspection

$$\tau(p^\alpha) = 1 + \alpha \quad \text{and} \quad \sigma(p^\alpha) = 1 + p + p^2 + \ldots + p^\alpha = \frac{p^{\alpha+1} - 1}{p - 1}.$$

Hence if $n = p_1^{\alpha_1} \ldots p_k^{\alpha_k}$, we have by (1) on the previous page,

$$\tau(n) = \prod_{i=1}^{k} (1 + \alpha_i) \quad \text{and} \quad \sigma(n) = \prod_{i=1}^{k} \frac{p_i^{\alpha_i+1} - 1}{p_i - 1}.$$

EXAMPLE Calculate $\tau(300)$ and $\sigma(300)$.

As $300 = 2^2 \cdot 3 \cdot 5^2$, we have $\tau(300) = 3 \cdot 2 \cdot 3 = 18$ and $\sigma(300) = 7 \cdot 4 \cdot 31 = 868$.

We shall return to these formulas in the next section to find their average values for large n.

A very old problem first considered by the Greeks over 2000 years ago concerns 'perfect numbers'.

DEFINITION A positive integer n is called *perfect* if it equals the sum of its divisors except itself, that is if $\sigma(n) = 2n$.

The only perfect numbers less than 10^6 are 6, 28, 496 and 8128. The result below characterizes the even perfect numbers. It is not known if there are infinitely many perfect numbers or if any odd perfect numbers exist.

LEMMA 1.3 *The integer n is even and perfect if and only if it has the form $2^{p-1}(2^p - 1)$ where both p and $2^p - 1$ are prime, that is exactly one perfect number is associated with each Mersenne prime.*

Proof. If $n = 2^{p-1}(2^p - 1)$ and $2^p - 1$ is prime (note, this implies p is prime by Lemma 2.1, Chapter 1) then the divisors of n are $1, 2, 2^2, \ldots, 2^{p-1}, (2^p - 1), 2(2^p - 1), \ldots,$ and $2^{p-1}(2^p - 1)$. As $1 + 2 + \ldots + 2^{p-1} = 2^p - 1$, it follows that n is perfect.

Conversely suppose n is even and perfect, write $n = 2^{k-1}n'$ where n' is odd and $k \geq 2$. Now

$$\begin{aligned} \sigma(n) &= \sigma(2^{k-1})\sigma(n') && \text{as } \sigma \text{ is multiplicative} \\ &= (2^k - 1)\sigma(n') \\ &= 2n && \text{by hypothesis} \\ &= 2^k n', \end{aligned}$$

so $2^k - 1 \mid n'$ and we have $n' = (2^k - 1)n''$ with $\sigma(n') = 2^k n''$. As n' and n'' both divide n' and

$$n' + n'' = (2^k - 1)n'' + n'' = 2^k n'' = \sigma(n'),$$

it follows that n' and n'' are the only divisors of n', so $n'' = 1$ and n' is prime. Hence $n' = 2^k - 1$ and $n = 2^{k-1}(2^k - 1)$; note k is prime using Lemma 2.1, Chapter 1, again. □

The Möbius function and inversion formula

The Möbius function μ has a deceptively simple definition; despite this it has many applications in number theory. Using it we prove the Möbius inversion formula which gives a kind of 'Laplace transform' for our sums over divisors.

Definition The Möbius function μ is the number theoretic function given by

$$\begin{aligned} \mu(1) &= 1 \\ \mu(n) &= 0 \quad \text{if } n \text{ has a squared factor} \\ \mu(p_1 p_2 \ldots p_k) &= (-1)^k \text{ where } p_1, \ldots, p_k \text{ are distinct primes.} \end{aligned}$$

Lemma 1.4 (i) *μ is multiplicative.*

(ii)
$$\sum_{d \mid n} \mu(d) = \begin{cases} 1 & \text{if } n = 1 \\ 0 & \text{if } n > 1. \end{cases}$$

Proof. (i) This follows immediately by considering integers with an even or odd number of distinct prime factors separately.

(ii) Define the function m by $m(n) = \sum_{d \mid n} \mu(d)$, we have

$$m(p^\alpha) = 1$$

if $\alpha = 0$, and

$$\begin{aligned} m(p^\alpha) &= \mu(1) + \mu(p) + \mu(p^2) + \ldots + \mu(p^\alpha) \\ &= 1 - 1 + 0 + \ldots + 0 = 0 \end{aligned}$$

if $\alpha > 0$.
The result follows using (1) on p. 17 and Theorem 1.1. □

THEOREM 1.5 *Möbius inversion formula.*
If f is a number theoretic function and F is defined by

$$F(n) = \sum_{d|n} f(d)$$

then

$$f(n) = \sum_{d|n} F(d)\mu\left(\frac{n}{d}\right) = \sum_{d|n} F\left(\frac{n}{d}\right)\mu(d). \qquad (*)$$

Proof. The second equality in $(*)$ follows by (2) on p. 17. Consider

$$\sum_{d|n} \mu(d)F\left(\frac{n}{d}\right) = \sum_{d_1d_2=n} \mu(d_1)F(d_2)$$

this sum is taken over all pairs d_1, d_2 such that $d_1d_2 = n$

$$= \sum_{d_1d_2=n} \left[\mu(d_1) \sum_{d|d_2} f(d)\right]$$

by definition

$$= \sum_{d_1d|n} \mu(d_1)f(d)$$

by multiplying out the terms in the square brackets above, this new sum is taken over all pairs d_1, d such that $d_1d|n$

$$= \sum_{d|n} f(d) \sum_{d_1|n/d} \mu(d_1)$$

by collecting together all multiples of $f(d)$ when $d|n$

$$= f(n)$$

by Lemma 1.4(ii), the inner sum above is zero unless $d = n$. □

Our first application of this result gives the converse of Theorem 1.1.

THEOREM 1.6 *If F is a multiplicative function and satisfies*

$$F(n) = \sum_{d|n} f(d)$$

for some number theoretic function f, then f is also multiplicative.

Proof. As Theorem 1.5 holds whether or not f is multiplicative, we have using that result

$$f(mn) = \sum_{d_1|m\, d_2|n} F(d_1d_2)\mu\left(\frac{mn}{d_1d_2}\right)$$

where d_1d_2 is a divisor of mn. Hence as both F and μ are multiplicative we have $(m, n) = 1$ implies

$$f(mn) = \sum_{d_1|m\, d_2|n} F(d_1)F(d_2)\mu\left(\frac{m}{d_1}\right)\mu\left(\frac{n}{d_2}\right)$$

$$= \sum_{d_1|m} F(d_1)\mu\left(\frac{m}{d_1}\right) \sum_{d_2|n} F(d_2)\mu\left(\frac{n}{d_2}\right)$$

by writing the sum of products above as a product of sums so separating m and n

$$= f(m)f(n)$$

by Theorem 1.5. □

The Euler function

This function, denoted by ϕ, is one of the most important in number theory; it was introduced by Euler in 1749 to solve a congruence problem, see Section 1 of the next chapter.

DEFINITION The Euler function ϕ is given by

$$\phi(n) = \text{the number of integers } a \text{ satisfying } 1 \le a \le n \text{ and } (a, n) = 1,$$

that is, if $n > 1$, $\phi(n)$ is the number of positive integers less than n which are coprime to n.

We shall show that this function is multiplicative, and hence find a formula for it using Theorem 1.5. To begin with we prove

LEMMA 1.7 $\sum_{d|n} \phi(d) = n.$

Proof. For a proposition P let $\text{card}\{a : P(a)\}$ denote the number of integers a such that $P(a)$ holds. Now

$$\sum_{d|n} \text{card}\{a : 1 \le a \le n \text{ and } (a, n) = d\} = n \qquad (*)$$

as each integer a, $1 \le a \le n$, is counted exactly once in this sum because each a and n have a unique GCD d which is a divisor of n. Further if $d|n$ then

$$\begin{aligned}\phi(n/d) &= \text{card}\{a : 1 \le a \le n/d \text{ and } (a, n/d) = 1\} \quad \text{by definition} \\ &= \text{card}\{a : 1 \le a \le n \text{ and } (a, n) = d\}.\end{aligned}$$

This holds as the mapping $a \to ad$ is a bijection between the two sets. The result follows by (2) on p. 17 and (*). □

THEOREM 1.8 (i) *The function ϕ is multiplicative.*

(ii) *If $n = p_1^{\alpha_1} \dots p_k^{\alpha_k}$ then*

$$\phi(n) = \prod_{p_i|n} p_i^{\alpha_i} - p_i^{\alpha_{i-1}} = n \prod_{p_i|n} \left(1 - \frac{1}{p_i}\right).$$

(iii) $\phi(n) = n \sum_{d|n} \frac{\mu(d)}{d}.$

Proof. (i) Use Theorem 1.6 and Lemma 1.7.

(ii) If p is a prime then we have $\phi(p^\alpha) = p^\alpha - p^{\alpha-1}$ (if $\alpha > 0$), as each integer having a factor in common with p^α is a multiple of p and there are $p^{\alpha-1}$ such factors. Now use (1) on p. 17.

(iii) This follows by applying the Möbius inversion formula to Lemma 1.7. □

A second proof of (i) above is provided by Corollary 1.11, Chapter 3.

As a final example of multiplicativity we consider briefly the *Liouville function* λ.

DEFINITION Let $\nu(n)$ denote the number of (not necessarily distinct) prime factors of n, so for example $\nu(24)=4$. The number theoretic function λ is given by

$$\lambda(n)=(-1)^{\nu(n)}.$$

LEMMA 1.9 (i) *λ is completely multiplicative.*

(ii)
$$\sum_{d\mid n}\lambda(d)=\begin{cases}1 & \text{if } n \text{ is a square}\\ 0 & \text{otherwise.}\end{cases}$$

Proof. (i) This follows immediately as $\nu(mn)=\nu(m)+\nu(n)$. (ii) The function N given by $N(n)=\sum_{d\mid n}\lambda(d)$ is multiplicative by (i) and Theorem 1.1. But if p is a prime we have

$$N(p^{\alpha})=1-1+1-\ldots=\begin{cases}1 & \text{if } \alpha \text{ is even}\\ 0 & \text{if } \alpha \text{ is odd,}\end{cases}$$

the result follows using (1) on p. 17. □

We shall return to this topic in Chapter 5 and introduce a class of completely multiplicative functions, called characters, which we use to codify arithmetic progressions.

2.2 Average order

Although this topic is not of central importance, the methods employed will be used extensively later in this book.

The formulas for the multiplicative functions given above allow us to calculate particular values of these functions easily. The values can vary erratically, and so we shall look for new formulas giving the average values. These will enable us to give simple estimates for some of these functions which are valid for most large arguments.

Consider, for example, the divisor function τ. We have for all $n>1$

$$2\leq\tau(n)\leq n,$$

but the maximum value of $\tau(i)$, $1\leq i\leq n$, grows at a much slower rate than n itself. So for instance $\tau(n)\leq 448$ if $n\leq 10^7$, and the erratic nature is illustrated by the fact that if $n=8648640$, $\tau(n-1)=4$, $\tau(n)=448$, and $\tau(n+1)=8$. Hence it is reasonable to consider the average value of τ, that is

$$\frac{1}{n}\sum_{i\leq n}\tau(i).$$

We shall show below (Theorem 2.3) that this is approximately $\ln n$, and so $\ln n$ is close to $\tau(n)$ in most cases when n is large. A better approximate value is $(\ln n)^{\ln 2} - 2^{\ln \ln n}$, see Problem 13, Chapter 12. (This can also be deduced by analysing the error term in Theorem 2.3.) As a consequence we see that integers with a large number of factors are rare. For example if n is of the order of 8600000, as above, then $\tau(n)$ is approximately 6.8, that is n is most likely to have two or three prime factors. More details are given in Hardy and Wright (1954, Chapter 22).

We shall discuss the average values of τ and ϕ, the first because of its historical importance and the second as it will have applications later. The average value of μ will be considered in Chapter 13, it has connections with the prime number theorem. We begin with some notation and a useful identity.

DEFINITION If x is a real number then $[x]$ denotes the largest integer less than or equal to x, $[x]$ is called the *integer part* of x.

We have immediately

(i) $x - 1 < [x] \le x < [x] + 1$,

(ii) $[x] + [y] \le [x + y]$,

(iii) the number of integers n satisfying the inequality $x < n \le y$ is $[y] - [x]$.

We shall use the convention: if x is a real variable and $x \ge 1$, then $\sum_{i \le x} f(i)$ denotes the sum $\sum_{i=1}^{[x]} f(i)$.

Let f and g be number theoretic functions, we have the following identities.

$$\sum_{i \le n} f(i) \sum_{d \mid i} g(d) = \sum_{d \le n} g(d) \sum_{j \le n/d} f(dj)$$

and, if f is completely multiplicative, both of these sums equal

$$\sum_{d \le n} f(d) g(d) \sum_{j \le n/d} f(j).$$

We call these identities the *divisor sum identities*, or DSI, they have important applications. To derive the first identity multiply out the left-hand side

$$f(1)g(1) + f(2)(g(1) + g(2)) + f(3)(g(1) + g(3)) + \ldots,$$

and collect together the terms involving $g(1), g(2), \ldots$ separately

$$g(1)(f(1) + f(2) + \ldots) + g(2)(f(2) + f(4) + \ldots) \\ + g(3)(f(3) + \ldots) + \ldots.$$

Now note that there are $[n/d]$ integers of the form dj lying between 1 and n. The second identity follows as f is completely multiplicative.

Lemma 2.1 *Let f be a number theoretic function and let F be given by*

$$F(n) = \sum_{d|n} f(d),$$

then

$$\sum_{i\le n} F(i) = \sum_{i\le n} \left[\frac{n}{i}\right] f(i).$$

Proof. We have

$$\sum_{i\le n} F(i) = \sum_{i\le n} \sum_{d|i} f(d) = \sum_{d\le n} f(d) \sum_{j\le n/d} 1 = \sum_{d\le n} \left[\frac{n}{d}\right] f(d),$$

using DSI. □

An extension of this to be used in the following theorem is

Lemma 2.2 *Let f and F be as given in the lemma above, and let* $F^*(x) = \sum_{i\le x} f(i)$ *then, if* $1 \le t \le n$ *where* $t \in \mathbb{Z}$,

$$\sum_{i\le n} F(i) = \sum_{i\le t} \left[\frac{n}{i}\right] f(i) - \left[\frac{n}{t}\right] F^*(t) + \sum_{i\le n/t} F^*\left(\frac{n}{i}\right).$$

Proof. As $f(i) = F^*(i) - F^*(i-1)$ (if $i \in \mathbb{Z}$) we have by Lemma 2.1

$$\begin{aligned}\sum_{i\le n} F(i) &= \sum_{i\le t} \left[\frac{n}{i}\right] f(i) + \sum_{i=t+1}^{n} \left\{\left[\frac{n}{i}\right](F^*(i) - F^*(i-1))\right\} \\ &= \sum_{i\le t} \left[\frac{n}{i}\right] f(i) - \left[\frac{n}{t}\right] F^*(t) + \sum_{i=t}^{n} \left\{F^*(i)\left(\left[\frac{n}{i}\right] - \left[\frac{n}{i+1}\right]\right)\right\}\end{aligned} \qquad (*)$$

as $\left[\frac{n}{n+1}\right] = 0$. (This process is called 'partial summation', see Lemma 2.2, Chapter 12.)

By (iii) on p. 23 $\left[\frac{n}{i}\right] - \left[\frac{n}{i+1}\right]$ is the number of integers u which satisfy

$$\frac{n}{i+1} < u \le \frac{n}{i}.$$

Rearranging these inequalities we have $\frac{n}{u} - 1 < i \le \frac{n}{u}$ and so $i = \left[\frac{n}{u}\right]$, that is i is unique. Hence

$$F^*(i)\left(\left[\frac{n}{i}\right] - \left[\frac{n}{i+1}\right]\right) = \sum_{n/(i+1) < u \le n/i} F^*\left(\frac{n}{u}\right),$$

and so

$$\sum_{i=t}^{n} F^*(i)\left(\left[\frac{n}{i}\right]-\left[\frac{n}{i+1}\right]\right)=\sum_{i=t}^{n}\sum_{n/(i+1)<u\le n/i} F^*\left(\frac{n}{u}\right)=\sum_{u\le n/t} F^*\left(\frac{n}{u}\right)$$

and the result follows using the equation (*). □

THEOREM 2.3 $\sum_{i\le n}\tau(i)=n\ln n+(2\gamma-1)n+O(\sqrt{n})$, *where* γ *denotes Euler's constant, see p.* viii.

Proof. Using Lemma 2.2 with $f=1$, $F=\tau$ and $t=[\sqrt{n}]$ we have, as $F^*(x)=[x]$,

$$\sum_{i\le n}\tau(i)=\sum_{i\le\sqrt{n}}\left[\frac{n}{i}\right]-\left[\frac{n}{[\sqrt{n}]}\right][\sqrt{n}]+\sum_{i\le n/[\sqrt{n}]}\left[\frac{n}{i}\right]$$

$$=2\sum_{i\le\sqrt{n}}\left[\frac{n}{i}\right]-n+O(\sqrt{n})$$ as the second term above lies between $n-\sqrt{n}$ and n

$$=2n\sum_{i\le\sqrt{n}}\frac{1}{i}+O(\sqrt{n})-n+O(\sqrt{n})$$ as the difference between this sum and the one above is at most $2\sqrt{n}$

$$=2n\left(\ln\sqrt{n}+\gamma+O\left(\frac{1}{\sqrt{n}}\right)\right)-n+O(\sqrt{n})$$ by (iii) on p. viii

$$=n\ln n+n(2\gamma-1)+O(\sqrt{n}).$$ □

The Dirichlet divisor problem is: improve the error term in Theorem 2.3, it lies between $O(n^{1/3})$ and $O(n^{1/4})$. See N40 in LeVeque (1974) and Guy (1984).

We come to the Euler function ϕ, we need

LEMMA 2.4 $\sum_{n=1}^{\infty}\frac{\mu(n)}{n^2}=\frac{6}{\pi^2}$.

Proof. The zeta function ζ was defined in Section 1 of Chapter 1, we shall use the result $\zeta(2)=\pi^2/6$ (see p. viii). The lemma follows using

$$s>1 \quad \text{implies} \quad \frac{1}{\zeta(s)}=\sum_{n=1}^{\infty}\frac{\mu(n)}{n^s}. \qquad (**)$$

To prove (**) we note that all the series involved are absolutely convergent when $s>1$, and so we have

$$\sum_{m=1}^{\infty}\frac{1}{m^s}\sum_{n=1}^{\infty}\frac{\mu(n)}{n^s}=\sum_{m,n=1}^{\infty}\frac{\mu(n)}{(mn)^s}=\sum_{i=1}^{\infty}\left\{\frac{1}{i^s}\sum_{d\mid i}\mu(d)\right\}=1$$

by collecting together terms with the same denominator in the double sum and using Lemmas 2.1 and 1.4. □

The main result for ϕ is

THEOREM 2.5 $\sum_{i\le n} \phi(i) = \dfrac{3n^2}{\pi^2} + O(n \ln n)$.

Proof. By Theorem 1.8 we have

$$\sum_{i\le n} \phi(i) = \sum_{i\le n} \left\{ i \sum_{d|i} \frac{\mu(d)}{d} \right\} = \sum_{d\le n} \mu(d) \sum_{i\le n/d} i \qquad \text{using DSI (see p. 23)}$$

$$= \sum_{d\le n} \left\{ \mu(d) \frac{1}{2} \left(\left[\frac{n}{d}\right]^2 + \left[\frac{n}{d}\right] \right) \right\} \qquad \text{summing the inner sum above}$$

$$= \frac{1}{2} \sum_{d\le n} \left\{ \mu(d) \left(\left(\frac{n}{d}\right)^2 + O\left(\frac{n}{d}\right) \right) \right\} \qquad \text{as } ((n/d)-1)^2 < [n/d]^2 \le (n/d)^2$$

$$= \frac{n^2}{2} \left\{ \sum_{d=1}^{\infty} \frac{\mu(d)}{d^2} - \sum_{d=n+1}^{\infty} \frac{\mu(d)}{d^2} \right\} + O(n \ln n) \qquad \text{using (iii) on p. viii}$$

$$= \frac{3n^2}{\pi^2} + n^2 O\left(\int_{n+1}^{\infty} \frac{\mathrm{d}x}{x^2} \right) + O(n \ln n) \qquad \text{by Lemma 2.4}$$

$$= \frac{3n^2}{\pi^2} + O(n \ln n).$$

□

COROLLARY 2.6 The probability that two positive integers are coprime is $6/\pi^2$.

Proof. The number of pairs of positive integers $\{r, s\}$ satisfying

$$1 \le r \le s \le n$$

is $n(n+1)/2$, and $\sum_{i\le n} \phi(i)$ is the number of such pairs which are also coprime by the definition of ϕ. We define the probability of two positive integers being coprime to be

$$\lim_{n\to\infty} \frac{\sum_{i\le n} \phi(i)}{n(n+1)/2},$$

and this limit equals $6/\pi^2$ by Theorem 2.5. □

2.3 Problems 2

1. Establish the following properties of the divisor function.

   (i) $\sum_{d|n} \tau^3(d) = \left(\sum_{d|n} \tau(d) \right)^2$ (see Problem 21, Chapter 12);

(ii) $\prod_{d|n} d = n^{\tau(n)/2}$;

(iii) the number of ordered pairs of positive integers whose least common multiple is n equals $\tau(n^2)$;

(iv) the relation $\tau(n) = O(\ln^r n)$ is false for all fixed powers r;

(v) the relation $\tau(n) = O(n^s)$ is true for all constant positive s.

2. Prove the following properties of the Möbius function.

(i) $\sum_{d^2|n} \mu(d) = |\mu(n)|$;

(ii) $\sum_{i \le n} \mu(i)\left[\frac{n}{i}\right] = 1$ and so $\left|\sum_{n \le x} \frac{\mu(n)}{n}\right| \le 1$;

(iii) $\sum_{\substack{d|n \\ (d,n/d)=1}} 1 = \sum_{d|n} |\mu(d)| = 2^{\nu(n)}$, and so $\sum_{\substack{d|n \\ \mu(d)=1}} |\mu(d)| = 2^{\nu(n)-1}$

where $\nu(n)$ is the number of distinct prime factors of n.

3. We have $\zeta(s) = \sum_{n=1}^{\infty} \frac{1}{n^s}$. Show that, giving conditions on s,

(i) $(\zeta(s))^2 = \sum_{n=1}^{\infty} \frac{\tau(n)}{n^s}$;

(ii) $\zeta(s)\zeta(s-1) = \sum_{n=1}^{\infty} \frac{\sigma(n)}{n^s}$;

(iii) $\frac{\zeta(s-1)}{\zeta(s)} = \sum_{n=1}^{\infty} \frac{\phi(n)}{n^s}$.

4. Show that if $\sigma(n)$ is odd then n is a square or twice a square.

5. By considering divisors of the form $2^c e$, where e is an odd divisor of n, show that the sum of the odd divisors of n is

$$-\sum_{d|n} (-1)^{n/d} d.$$

Deduce that, if n is even,

$$2\sigma(n/2) - \sigma(n) = \sum_{d|n} (-1)^{n/d} d.$$

6. (i) A pair of positive integers $\{a, b\}$ is called *amicable* if $\sigma(a) = \sigma(b) = a + b$. Find an amicable pair with both components less than 300.

(ii) A integer n is called *k-perfect* if $\sigma(n) = kn$. Find a 3-perfect integer less than 150 and show that

$$2^{15} \cdot 3^7 \cdot 5 \cdot 7 \cdot 11 \cdot 17 \cdot 41 \cdot 43 \cdot 257$$

is 5-perfect. [For more information on these and similar topics see Guy (1981).]

7. Show that

$$\sum_{i \le n} \sigma(i) = \frac{\pi^2 n^2}{12} + O(n \ln n).$$

8. Let $\psi_k(n)$ denote the number of ordered sets of k positive integers (equal or not), none of which exceed n, and whose GCD is prime to n. Prove that

(i) $\sum_{d|n} \psi_k(d) = n^k$;

(ii) ψ_k is multiplicative;

(iii) $\psi_k(n) = n^k \prod_{p|n} \left(1 - \frac{1}{p^k}\right)$.

9. Establish the following properties of Euler's function where $m, n \geq 1$.

(i) $\phi(mn)\phi((m, n)) = (m, n)\phi(m)\phi(n)$;

(ii) the sum of the integers $t: 1 \leq t \leq n$ and $(n, t) = 1$ is $\frac{1}{2}n\phi(n)$, provided $n > 1$;

(iii) $\sum_{i \leq n} \phi(i)\left[\frac{n}{i}\right] = \frac{n(n+1)}{2}$;

(iv) $\frac{n}{\phi(n)} = \sum_{d|n} \frac{\mu^2(d)}{\phi(d)}$;

(v) $\sum_{n \leq x} \frac{1}{\phi(n)} \sim \ln x \sum' \frac{1}{n\phi(n)}$;

where $\sum'$ denotes a sum over square-free n such that $1 \leq n \leq x$;

(vi) if $n > 1$, $n^2 < 2\phi(n)\sigma(n) < 2n^2$.

10. Let $\det(f(i, j))$ denote the determinant of the $n \times n$ matrix whose i, j entry is $f(i, j)$. Show that

$$\det((i, j)) = \prod_{k=1}^{n} \phi(k).$$

☆11. Let D_n be the proposition: all integers t satisfying $2 \leq t \leq n$ and $(n, t) = 1$ are prime. Let

$$\delta_n = \text{card}\{i : i \in \mathbb{Z}, 2 \leq i \leq \sqrt{n}, \text{ and } (n, i) = 1\}.$$

Show that

(i) D_{30} holds;

(ii) if $\delta_n > 0$ then D_n is false;

(iii) $\delta_n = \sum_{d|n} \mu(d)[(\sqrt{n})/d] - 1$;

(iv) using Problem 1(iii),

$$\delta_n > \frac{\phi(n)}{\sqrt{n}} - (2^{\nu(n)-1} + 1);$$

(v) working numerically on the prime factorization of n deduce that D_n is false for all $n > 30$.

12. Let λ be the Liouville function, show that if $x \geq 1$

$$\sum_{n \leq x} \lambda(n)\left[\frac{x}{n}\right] = [\sqrt{x}],$$

and deduce

$$\left|\sum_{n\le x}\frac{\lambda(n)}{n}\right|<2.$$

13. Let $\Lambda(n)=\begin{cases}\ln p & \text{if } n \text{ is a power of a prime } p\\ 0 & \text{otherwise.}\end{cases}$

This is called the *Mangoldt function,* see Chapter 13.
Show that

(i) $\ln n=\sum_{d|n}\Lambda(d)$;

(ii) $\sum_{d|n}\mu(d)\ln d=-\Lambda(n)$.

14. If $f(s)=\sum a_n n^{-s}$ and $g(s)=\sum b_n n^{-s}$ show that $g(s)=\zeta(s)f(s)$ if and only if

$$\sum a_n\left(\frac{x^n}{1-x^n}\right)=\sum b_n x^n \quad \text{with} \quad 0\le x<1.$$

Deduce (see Problem 3)

(i) $\displaystyle\sum_{n=1}^{\infty}\frac{\mu(n)x^n}{1-x^n}=x$;

(ii) $\displaystyle\sum_{n=1}^{\infty}\tau(n)x^n=\frac{x}{1-x}+\frac{x^2}{1-x^2}+\frac{x^3}{1-x^3}\cdots$;

(iii) $\displaystyle\sum_{n=1}^{\infty}\frac{\phi(n)x^n}{1-x^n}=\frac{x}{(1-x)^2}$.

15. Let f, g and h be number-theoretic functions and let $f*g$ denote the sum $f*g(n)=\sum_{d|n}f(d)g(n/d)$. $f*g$ is sometimes called the *convolution* of f and g. Show that this operation is (i) commutative, (ii) associative, and (iii) if L denotes $\ln n$ (and so $Lf(n)=f(n)\ln n$) then $L(f*g)=(Lf)*g+f*(Lg)$.

16. Show that the number of terms in the Farey series F_n, defined in Problem 21, Chapter 1, is approximately $3n^2/\pi^2$.

3

CONGRUENCE THEORY

By universal agreement C. F. Gauss (1777–1855) was a mathematical genius of the highest order, to be considered alongside such giants as Newton and Einstein. He made many discoveries both inside and outside mathematics but in his time he was probably best known as an astronomer, he was director of the Göttingen Observatory for many years. Number theory was one of his great passions, he called it the 'Queen of Mathematics'. In 1801, at the age of 24, he published one of the classics of mathematical literature, *Disquisitiones Arithmeticae,* which contained a wealth of new ideas and theorems and which laid the foundation for modern number theory. A translation has been published recently (Gauss 1966) and the reader should try to study it. Many of the results presented below first appeared in Gauss's classic, particularly those concerning quadratic reciprocity, primitive roots and quadratic form theory.

Gauss's book begins with the definition of congruence and modulus: if m divides $a - b$ then a is congruent to b modulo m. This deceptively simple definition underlies a major discovery, that is, in modern terminology, congruences are equivalence relations. Also he chose a remarkably good notation which we still use today and which is suggestive of deeper meanings. The ideas associated with this definition have helped in the development of many branches of mathematics, for example with the notions of subgroup and factor group in group theory. In number theory these ideas led to questions concerning the solubility of polynomial congruences which in turn introduced the important concepts of reciprocity and the 'local–global' dichotomy. The general theory is beyond the scope of this book; we shall treat the linear and quadratic cases including Gauss's famous law of quadratic reciprocity, but even here some problems remain. In this chapter we give the basic definitions, deal with the linear case completely and consider a few general results and ideas including a discussion of local and global properties of equations. In Chapter 4 we shall study the quadratic case in detail and in Chapter 5 we shall introduce some of the connections with group theory associated with primitive roots.

Congruences are used in everyday language regularly; 'Today is Tuesday' or 'It is 3.15 p.m.' are examples. The periodic nature of dates and time are described exactly using congruences. We are not interested (and do not know) how many days have passed since the beginning of time but this does not matter. 'Today is Tuesday' or

'Today is the second day of the week' is a congruence relation modulo 7 which describes the day for us. There are many other instances in common use.

3.1 Definitions, linear congruences, and Euler's theorem

We begin with the basic definitions and lemmas due to Gauss.

DEFINITION Let a, b and m be integers with m positive. The expression

$$a \equiv b \pmod{m},$$

called a *congruence,* stands for the proposition $m|b-a$. m is called the *modulus* and if the relation is true, b is called a *residue* of a and vice versa. We shall sometimes write $a \equiv b(m)$ for this relation.

The basic congruence properties are given by

LEMMA 1.1 (i) *For fixed* m, *the proposition* $a \equiv b \pmod{m}$ *is an equivalence relation,*
(ii) *if* $a \equiv b(m)$ *and* $c \equiv d(m)$ *then* $a + c \equiv b + d(m)$ *and* $ac \equiv bd(m)$,
(iii) *if* $a \equiv b(m)$ *and* $d|m$ *then* $a \equiv b(d)$,
(iv) *if* $a \equiv b(m)$, $a = a'd$, $b = b'd$ *and* $(d, m) = 1$ *then* $a' \equiv b'(m)$.

Proof. (i) We have $m|a-a$, if $m|b-a$ then $m|a-b$, and if $m|b-a$ and $m|c-b$ then $m|c-a$; (i) follows. The remaining parts are left as an exercise for the reader. □

Some further properties are given in Problem 1.

DEFINITION We denote by $\bar{a}_m$ the set of integers c satisfying $c \equiv a(\bmod m)$, $\bar{a}_m$ is called the *congruence class* modulo m of a. The set $\bar{a}_m$ is the equivalence class of the congruence relation modulo m containing a. When it is clear which modulus is being used we write $\bar{a}$ for $\bar{a}_m$. As an example note that the even integers and the odd integers are congruence classes modulo 2.

We have immediately (with proofs as exercises)

LEMMA 1.2 (i) $\bar{a}_m = \bar{b}_m$ *if and only if* $a \equiv b \pmod{m}$;
(ii) *if* $\bar{a}_m \cap \bar{b}_m \neq \varnothing$ *then* $\bar{a}_m = \bar{b}_m$;
(iii) *there are* m *congruences classes modulo* m.

The algebraic structure of the set of congruence classes is given by

THEOREM 1.3 *Fix the modulus m. Let the operations $\oplus$ and $\otimes$ be given by*

$$\bar{a} \oplus \bar{b} = \bar{c} \quad \text{if and only if} \quad a + b \equiv c \pmod{m}$$
$$\bar{a} \otimes \bar{b} = \bar{c} \quad \text{if and only if} \quad ab \equiv c \pmod{m},$$

then the set of congruences classes with these operations forms a commutative ring with identity which we denote by $\mathbb{Z}/m\mathbb{Z}$ using the standard notation for a factor ring.

Proof. By Lemma 1.1(ii) we see that $\oplus$ and $\otimes$ are well defined, the remaining properties follow using the corresponding ones for $\mathbb{Z}$. □

We shall write + for $\oplus$ and use concatenation for $\otimes$ from now on. Lemma 1.2 and Theorem 1.3 allow us to transform propositions about $\mathbb{Z}$ involving congruences into propositions about $\mathbb{Z}/m\mathbb{Z}$ and vice versa; we shall do this freely using the structure most suitable for the problem in hand.

DEFINITION A *complete system of residues* modulo m is any set of m integers $\{c_1, \ldots, c_m\}$ such that $c_i \not\equiv c_j \pmod{m}$ if $i \neq j$.

Note. If a satisfies $0 \leq a < m$ then there is a unique j such that $c_j \equiv a \pmod{m}$; also $\{\bar{c}_i : i = 1, \ldots, m\} = \mathbb{Z}/m\mathbb{Z}$. As an example we have $\{3, 4, 5, 6, 7\}$, $\{3, -4, 9, -8, 15\}$ and $\{0, 2, 4, 8, 16\}$ are all complete systems of residues (mod 5).

Linear congruences

Theorems on the solution of polynomial congruences are central to number theory, and have many consequences including applications to Diophantine problems over $\mathbb{Z}$ or $\mathbb{Q}$. We shall consider some general results in the next few sections and treat the linear case now.

Let f be a polynomial with integer coefficients. Consider the congruence

$$f(x) \equiv 0 \pmod{m}.$$

We always use the symbols a, b or c (with or without suffices etc.) to denote coefficients and x or y to denote the variable of the polynomial f. By the number of solutions of this congruence we mean the number of solutions x such that $0 \leq x \leq m-1$ or, equivalently, the number of solutions of the corresponding equation in $\mathbb{Z}/m\mathbb{Z}$.

THEOREM 1.4 *Let a, b, and m be integers with m positive. The congruence*

$$ax \equiv b \pmod{m}$$

is soluble if and only if $(a, m)|b$. *If* x_0 *is a solution, there are exactly* (a, m) *solutions given by* $\{x_0 + tm/(a, m)\}$ *where* $t = 0, 1, \ldots, (a, m) - 1$.

Proof. This is a restatement of Theorems 1.3 and 1.5 of Chapter 1. □

COROLLARY 1.5 *If* $(a, m) = 1$ *then the congruence*

$$ax \equiv b \pmod m$$

has exactly one solution.

This corollary has many applications, the first concerns the units of $\mathbb{Z}/m\mathbb{Z}$. In general an element of $\mathbb{Z}/m\mathbb{Z}$ does not have a multiplicative inverse, for instance (by Theorem 1.4) the congruence $3x \equiv 1 \pmod 6$ has no solution. Corollary 1.5 gives a necessary and sufficient condition for an inverse to exist: $\bar{a}$ has an inverse in $\mathbb{Z}/m\mathbb{Z}$, that is $\bar{a}$ is a unit of $\mathbb{Z}/m\mathbb{Z}$, if and only if $(a, m) = 1$. There are $\phi(m)$ units in $\mathbb{Z}/m\mathbb{Z}$ where ϕ is the Euler function defined in Section 1 of Chapter 2.

DEFINITION (i) The set of units of $\mathbb{Z}/m\mathbb{Z}$ is denoted by $(\mathbb{Z}/m\mathbb{Z})^*$, this is the set of elements of $\mathbb{Z}/m\mathbb{Z}$ having multiplicative inverses.

(ii) Each set of $\phi(m)$ integers $\{c_1, \ldots, c_{\phi(m)}\}$ whose elements satisfy

$$c_i \not\equiv c_j \pmod m \quad \text{if} \quad i \neq j, \quad \text{and}$$
$$(c_i, m) = 1 \quad \text{for} \quad 1 \leq i \leq \phi(m)$$

is called a *reduced system of residues* modulo m.

Clearly $\{\bar{c}_1, \ldots, \bar{c}_{\phi(m)}\} = (\mathbb{Z}/m\mathbb{Z})^*$ and, if $m > 1$, $(\mathbb{Z}/m\mathbb{Z})^*$ is a proper subset of $\mathbb{Z}/m\mathbb{Z}$. To illustrate the relationship between $\mathbb{Z}/m\mathbb{Z}$ and $(\mathbb{Z}/m\mathbb{Z})^*$ we consider two examples. First let $m = 12$. We have $\{\bar{0}, \bar{1}, \ldots, \overline{11}\} = \mathbb{Z}/12\mathbb{Z}$ and $\{\bar{1}, \bar{5}, \bar{7}, \overline{11}\} = (\mathbb{Z}/12\mathbb{Z})^*$, and an easy calculation shows that $(\mathbb{Z}/12\mathbb{Z})^*$ is a group under multiplication (use $5^2 \equiv 7^2 \equiv 11^2 \equiv 1$ and $5 \cdot 7 \equiv 11 \pmod{12}$, etc.). But the non-zero elements of $\mathbb{Z}/12\mathbb{Z}$ do not form a group under multiplication as this set is not closed (for instance $\bar{3} \cdot \bar{4} = \bar{0}$), and so $\mathbb{Z}/12\mathbb{Z}$ is not a field. Secondly let $m = 5$. Here the two systems fit together for we have $(\mathbb{Z}/5\mathbb{Z})^* = \mathbb{Z}/5\mathbb{Z} - \{\bar{0}\}$ and $\mathbb{Z}/5\mathbb{Z}$ is a finite field as the reader can check easily. These examples suggest

THEOREM 1.6 *Let m be an integer greater than* 1, *then* $(\mathbb{Z}/m\mathbb{Z})^*$ *is an Abelian group under multiplication modulo m with* $\phi(m)$ *elements.*

Proof. For closure use Problem 8(iii) of Chapter 1 and for the inverse property use Corollary 1.5. The remaining group axioms are easily verified. □

THEOREM 1.7 *Let p be a prime, then* $\mathbb{Z}/p\mathbb{Z}$ *is a field with p elements.*

Proof. Use Theorems 1.4 and 1.6 noting that $(\mathbb{Z}/p\mathbb{Z})^* = \mathbb{Z}/p\mathbb{Z} - \{\bar{0}\}$. □

The connection between congruence theory and finite fields is a fruitful one with many consequences (see p. 84), but these are mainly beyond the scope of an undergraduate text. For example the development of general reciprocity theory is best carried out in an abstract framework based on finite fields. Good accounts appear in Ireland and Rosen (1982) and Serre (1973).

Chinese remainder theorem

Let us consider the simultaneous solution of several linear congruences. The main result is usually called the Chinese remainder theorem because a version with two congruences was discovered by Sun Tse, a Chinese mathematician working in the first century AD.

Suppose we wish to find the common solutions of

$$a_i x \equiv b_i \pmod{m_i} \quad \text{for} \quad i = 1, \ldots, k. \qquad (*)$$

We begin by replacing this set of congruences with an equivalent set in which each leading coefficient is 1. Clearly each congruence in $(*)$ must be soluble separately, so we have, for $i = 1, \ldots, k$, $(a_i, m_i) | b_i$. Also there exist integers c_i with the property $a_i c_i \equiv (a_i, m_i) \pmod{m_i}$ by Theorem 1.4. Hence $(*)$ is equivalent to

$$x \equiv b_i^* c_i \pmod{m_i^*} \quad \text{for} \quad i = 1, \ldots, k,$$

where b_i^* and m_i^* satisfy $b_i = b_i^*(a_i, m_i)$ and $m_i = m_i^*(a_i, m_i)$. When solving simultaneous congruences this process should always be carried out first.

THEOREM 1.8 *Chinese remainder theorem.*
Suppose the positive integers $m_1, \ldots, m_k$ are coprime in pairs, that is $(m_i, m_j) = 1$ for all i, j where $i \neq j$, then the set of congruences

$$x \equiv c_i \pmod{m_i} \quad \textit{for} \quad i = 1, \ldots, k$$

has a unique common solution modulo m where $m = m_1 m_2 \ldots m_k$.

Proof. We construct a solution as follows. For each $i = 1, \ldots, k$ let $m_i' = m/m_i$, note that $m_i' \in \mathbb{Z}$ and $(m_i, m_i') = 1$. Using this and Corollary 1.5 we can find integers m_i'' satisfying $m_i' m_i'' \equiv 1 \pmod{m_i}$. A solution x is given by

$$x = c_1 m_1' m_1'' + \ldots + c_k m_k' m_k'',$$

for clearly $m_i | m_j'$ if $i \neq j$ and so $x \equiv c_i m_i' m_i'' \equiv c_i \pmod{m_i}$.

Secondly assume that both x and y are common solutions to the congruences in the theorem. It follows that $x - y \equiv 0 \pmod{m_i}$ for all $i = 1, \ldots, k$. This proves the result as the moduli are coprime in pairs.

□

A slightly more general version of this result can be derived, see Problems 16 and 17. As an example of the method given in the theorem above we calculate the least positive integer x satisfying $x \equiv i+1 \pmod{p_i}$, $0 \le i \le 4$, where p_i is the ith prime. We have in turn

$$m = 2 \cdot 3 \cdot 5 \cdot 7 \cdot 11 = 2310,$$
$$m_0' = 1155, \quad m_1' = 770, \quad m_2' = 462, \quad m_3' = 330, \quad m_4' = 210,$$
$$m_0'' = 1, \quad m_1'' = 2, \quad m_2'' = 3, \quad m_3'' = 1 \quad \text{and} \quad m_4'' = 1,$$

hence

$$x = 1155 + 3080 + 4158 + 1320 + 1050 = 10763 \equiv 1523 \pmod{2310}.$$

The Chinese remainder theorem has an algebraic interpretation. Suppose $R_1, \ldots, R_k$ are commutative rings with identity. The *direct sum* $R_1 \oplus \ldots \oplus R_k$ is the set of k-tuples $(r_1, \ldots, r_k)$, where $r_i \in R_i$ for $i = 1, \ldots, k$, with addition and multiplication defined componentwise, that is

$$(r_1, \ldots, r_k) + (s_1, \ldots, s_k) = (r_1 + s_1, \ldots, r_k + s_k)$$
$$(r_1, \ldots, r_k) \cdot (s_1, \ldots, s_k) = (r_1 s_1, \ldots, r_k s_k).$$

The zero is $(0, 0, \ldots, 0)$ and the identity is $(1, 1, \ldots, 1)$. It is a simple matter to check that $R_1 \oplus \ldots \oplus R_k$ is also a commutative ring with identity.

LEMMA 1.9 *Let $\times$ denote the direct product of groups defined componentwise. Using the notation above we have*

$$(R_1 \oplus \ldots \oplus R_k)^* = R_1^* \times \ldots \times R_k^*,$$

where R^* denotes the group of units of the commutative ring R.

Proof. Suppose $u = (u_1, \ldots, u_k)$ is a unit in $R_1 \oplus \ldots \oplus R_k$, so there exists $v = (v_1, \ldots, v_k)$ in the direct sum with the property $(u_1, \ldots, u_k) \cdot (v_1, \ldots, v_k) = (1, 1, \ldots, 1)$, that is $u_i \in R_i^*$ for $i = 1, \ldots, k$. Conversely if $u_i \in R_i^*$ for all i then clearly $(u_1, \ldots, u_k)$ is a unit in the direct sum; this proves the lemma. □

The Chinese remainder theorem is equivalent to

THEOREM 1.10 *Let $m = m_1 m_2 \ldots m_k$ where $(m_i, m_j) = 1$ if $i \ne j$, then there is an isomorphism between $\mathbb{Z}/m\mathbb{Z}$ and $\mathbb{Z}/m_1\mathbb{Z} \oplus \ldots \oplus \mathbb{Z}/m_k\mathbb{Z}$.*

COROLLARY 1.11 *There is a group isomorphism between $(\mathbb{Z}/m\mathbb{Z})^*$ and $(\mathbb{Z}/m_1\mathbb{Z})^* \times \ldots \times (\mathbb{Z}/m_k\mathbb{Z})^*$.*

Proof. The corollary follows immediately from the theorem and Lemma 1.9. For the main result let θ_i, $i = 1, \ldots, k$, be the natural homomorphism mapping $\mathbb{Z}$ onto $\mathbb{Z}/m_i\mathbb{Z}$ given by: $\theta_i(n) = \bar{c}$ where c is an integer satisfying $c \equiv n \pmod{m_i}$. By Lemma 1.1(ii) this is a ring homomorphism. Now define $\theta: \mathbb{Z} \to \mathbb{Z}/m_1\mathbb{Z} \oplus \ldots \oplus \mathbb{Z}/m_k\mathbb{Z}$ by

$$\theta(n) = (\theta_1(n), \ldots, \theta_k(n)).$$

Clearly this is also a ring homomorphism. It is an onto mapping, for if $\bar{c}_i \in \mathbb{Z}/m_i\mathbb{Z}$ $(i = 1, \ldots, k)$ we can find, using the Chinese remainder theorem, $c \in \mathbb{Z}$ such that $c \equiv c_i \pmod{m_i}$ for all i. This gives $\bar{c}_i = \theta_i(c)$ and so θ maps c to the k-tuple $(\bar{c}_1, \ldots, \bar{c}_k)$. Finally if b belongs to the kernel K of θ then $\theta_i(b) = 0$, or $b \equiv 0 \pmod{m_i}$ for $i = 1, \ldots, k$. This shows that $K = m\mathbb{Z}$ and the result follows by the homomorphism theorem. □

Note that as the order of the group $(\mathbb{Z}/m_i\mathbb{Z})^*$ is $\phi(m_i)$, Corollary 1.11 provides another proof of the multiplicativity of the Euler function ϕ.

The theorems of Fermat and Euler

As with the results discussed above these famous theorems have both group theoretic and number theoretic derivations and applications. They are both instances of the proposition: if G is a finite group with identity e and order g, and if $a \in G$, then $a^g = e$.

THEOREM 1.12 (Euler) *If $(a, m) = 1$ then $a^{\phi(m)} \equiv 1 \pmod{m}$.*

COROLLARY 1.13 (Fermat's theorem) *If p is prime then $a^p \equiv a \pmod{p}$.*

The corollary follows easily from the main result as $a^p \equiv a \equiv 0 \pmod{p}$ if $p \mid a$. We shall give two proofs of Euler's theorem.

First proof. Let $c_1, \ldots, c_{\phi(m)}$ be a reduced residue system modulo m, then $ac_1, \ldots, ac_{\phi(m)}$ is also a reduced system $\pmod{m}$. For, as $(a, m) = 1$, $ac_i \equiv ac_j \pmod{m}$ implies $c_i \equiv c_j \pmod{m}$. Also, for all i, we have $(ac_i, m) = 1$ because $(c_i, m) = 1$. Hence

$$c_1 c_2 \ldots c_{\phi(m)} \equiv ac_1 ac_2 \ldots ac_{\phi(m)} \pmod{m}$$

which gives

$$a^{\phi(m)} \equiv 1 \pmod{m},$$

as $(c_i, m) = 1$ for $i = 1, \ldots, \phi(m)$.

Second proof. We require four easy lemmas.

(i) Let p be a prime, then the binomial coefficient $\binom{p}{i}$ is an integer divisible by p if $1 \le i \le p - 1$.

By Problem 14, Chapter 1, $\binom{p}{i} \in \mathbb{Z}$, and (i) follows because $p \mid p!$ but $p \nmid i!(p - i)!$.

(ii) Fermat's theorem. $a^p \equiv a \pmod{p}$.

By (i) we have $(b_1 + \ldots + b_a)^p \equiv b_1^p + \ldots + b_a^p \pmod p$ for any $b_i \in \mathbb{Z}$. Now let $b_1 = b_2 = \ldots = b_a = 1$.

(iii) If $n \equiv 1 \pmod{p^\alpha}$ then $n^p \equiv 1 \pmod{p^{\alpha+1}}$.

Write $n = 1 + p^\alpha t$, we have, using the binomial theorem,

$$\begin{aligned} n^p &= (1 + p^\alpha t)^p = 1 + p^{\alpha+1}u \qquad \text{for } u \in \mathbb{Z} \\ &\equiv 1 \pmod{p^{\alpha+1}}. \end{aligned}$$

(iv) If $p \nmid a$ and $\alpha \geq 1$, $a^{p^{\alpha-1}(p-1)} \equiv 1 \pmod{p^\alpha}$.

As $p \nmid a$ (ii) gives $a^{p-1} \equiv 1 \pmod p$. Now apply (iii) $\alpha - 1$ times with $n = a^{p-1}$.

Proof of Euler's theorem. Let $m = \prod_{i \leq k} p_i^{\alpha_i}$. By Theorem 1.8, Chapter 2, $\phi(m) = \prod_{i \leq k} p_i^{\alpha_i - 1}(p_i - 1)$. Using (iv) this implies that, for each i, $a^{\phi(m)} \equiv 1 \pmod{p_i^{\alpha_i}}$ and the result follows by either Problem 1 (i) or the Chinese remainder theorem. □

Example As $\phi(16) = 8$, we have $7^8 \equiv 1 \pmod{16}$. But note that the power 8 is not the smallest with this property for clearly $7^2 \equiv 1 \pmod{16}$.

Some consequences of these results are given in the problem section.

3.2 Non-linear congruences and the theorems of Lagrange and Chevalley

The solution of non-linear polynomial congruences with prime moduli is a major topic in number theory with far-reaching consequences and many unsolved problems. The quadratic case will be considered in the next chapter. We begin by giving a straightforward procedure, based on the results of the previous section, for solving congruences with composite moduli once the solutions in the prime case have been found.

Let f be a polynomial in one variable. Consider the congruence

$$f(x) \equiv 0 \pmod m, \tag{1}$$

where $m = \prod_{i=1}^{k} p_i^{\alpha_i}$. Using the Chinese remainder theorem (1.8) we have immediately: (1) is soluble if and only if each of the congruences

$$f(x) \equiv 0 \pmod{p_i^{\alpha_i}}, \quad i = 1, \ldots, k$$

is soluble. For if x_i is a solution of the ith congruence then we can find an x such that $x \equiv x_i \pmod{p_i^{\alpha_i}}$, $i = 1, \ldots, k$, and this x will solve (1).

In the second reduction suppose we have a solution x_0 of the congruence

$$f(x) \equiv 0 \pmod{p^\alpha}. \tag{2}$$

We use a version of Taylor's theorem to solve, where possible,

$$f(x) \equiv 0 \pmod{p^{\alpha+1}}. \tag{3}$$

We have

$$f(x_0 + tp^\alpha) = f(x_0) + tp^\alpha f'(x_0) + \ldots + \frac{(tp^\alpha)^n}{n!} f^{(n)}(x_0), \tag{4}$$

where n is the degree of f and t is an integer. Note that the derivatives in this expression can be defined in a purely algebraic manner and all the coefficients in (4) are integers. This is because, for all k, each coefficient in the polynomial $f^{(k)}$, the kth derivative of f, is divisible by $k!$ (see Problem 14, Chapter 1). Dividing by p^α, (4) gives: $x_0 + tp^\alpha$ is a solution of (3) if and only if

$$tf'(x_0) \equiv -f(x_0)/p^\alpha \pmod{p}. \tag{5}$$

By Theorem 1.4, if $p \nmid f'(x_0)$, (5) has a unique solution t and so the solution x_0 of (2) gives rise to a unique solution $x_0 + tp^\alpha$ of (3). But if $p \mid f'(x_0)$, $f(x_0 + tp^\alpha) \equiv f(x_0) \pmod{p^{\alpha+1}}$, hence either x_0 is also a solution of (3) in which case so is $x_0 + tp^\alpha$ for all t, or x_0 is not a solution of (3) in which case (3) has no solution x satisfying $x \equiv x_0 \pmod{p^{\alpha+1}}$. These three cases are illustrated in the example below.

Thus given all solutions to $f(x) \equiv 0 \pmod{p}$ for each prime p dividing m we can find all solutions of (1) by applying these techniques.

EXAMPLE Consider the congruence $f(x) \equiv 0 \pmod{27}$ where $f(x) = x^3 - 2x^2 + 3x + 9$.

Case 1. Solutions modulo 3: clearly 0 and 2 are the solutions of $f(x) \equiv 0 \pmod{3}$.

Case 2. Solutions modulo 9:

(i) $x = 0$. We have $f(0) = 9$ and $f'(0) = 3$, so 0 is a solution modulo 9, and both 3 and 6 are also solutions (mod 9).

(ii) $x = 2$. Here $f(2) = 15$, $f'(2) = 7$, and the congruence

$$7t \equiv -15/3 \pmod{3}$$

gives $t = 1$. Hence 5 is another solution modulo 9.

Case 3. Solutions modulo 27:

(i) $x = 0$. We have $3 \mid f'(0)$ but $27 \nmid f(0)$ and so there is no solution.

(ii) $x = 3$. As $3 \mid f'(3)$ and $f(3) = 27$, the solutions (mod 27) in this case are 3, 12 and 21.

(iii) $x=5$. Here $f(5)=99$, $f'(5)=58$, $3\nmid 58$ and the congruence

$$58t \equiv -11 \pmod 3)$$

gives $t=1$. Hence 14 is another solution (mod 27).

(iv) $x=6$. This is similar to case (i).

Hence the original congruence has solutions 3, 12, 14, and 21 modulo 27.

Lagrange's theorem

Let F be a field, then the set of one-variable polynomials over F, denoted by $F[x]$, is an integral domain with a division algorithm: if $f, g \in F[x]$, g not identically zero, then there exist $q, r \in F[x]$ such that $f(x) = g(x)q(x) + r(x)$ and $\deg(r) < \deg(g)$, or r is identically zero [$\deg(f)$ stands for the degree of the polynomial f]. Now consider the field $\mathbb{Z}/p\mathbb{Z}$ for some prime p; if f is a polynomial with integer coefficients we denote by $\bar{f}$ the polynomial in $\mathbb{Z}/p\mathbb{Z}[x]$ formed by replacing each coefficient a of f by $\bar{a}_p$, its congruence class (mod p). We shall use the division algorithm for $\mathbb{Z}/p\mathbb{Z}[x]$ to show that the number of roots of a polynomial equation $\bar{f}=\bar{0}$ in $\mathbb{Z}/p\mathbb{Z}$ is bounded by its degree. This is the analogue of the corresponding result for the real field, but note that the result is false for non-prime moduli as we saw in the example above. Note also that the propositions 'x_0 is a solution of the congruence $f(x) \equiv 0 \pmod p$' and '$\bar{x}_0$ is a root of the equation $\bar{f}(x) = \bar{0}$ in $\mathbb{Z}/p\mathbb{Z}$' are equivalent (see p. 32).

LEMMA 2.1 *The polynomial $\bar{f}$ can be written in the form $(x-\bar{x}_0)\bar{q}(x)$ in $\mathbb{Z}/p\mathbb{Z}[x]$ if and only if $\bar{x}_0$ is a root of the equation $\bar{f}(x)=\bar{0}$ in $\mathbb{Z}/p\mathbb{Z}$.*

Proof. Divide $x-\bar{x}_0$ into $\bar{f}$ and we have

$$\bar{f}(x) = (x-\bar{x}_0)\bar{q}(x) + \bar{r}(x)$$

where $\deg(\bar{r}) < 1$ or $\bar{r}$ is identically zero, i.e. $\bar{r}$ is a constant. But as $\bar{f}(\bar{x}_0) = \bar{0}$, this constant must be zero. The converse is obvious. □

THEOREM 2.2 (Lagrange) *If $\bar{f}$ is a one-variable polynomial of degree n over $\mathbb{Z}/p\mathbb{Z}$ then it cannot have more than n roots unless it is identically zero.*

Proof. By induction on the degree. Use Theorem 1.4 if the degree is 1. Suppose now the result holds for all polynomials of degree less than n and let $\bar{x}_0$ be a root of the nth degree polynomial $\bar{f}$. By the lemma above we have

$$\bar{f}(x) = (x-\bar{x}_0)\bar{q}(x)$$

for some polynomial $\bar{q}$ of degree $n-1$. Hence

$$\bar{f}(x)=\bar{0} \quad \text{if and only if} \quad x=\bar{x}_0 \quad \text{or} \quad \bar{q}(x)=\bar{0}$$

because $\mathbb{Z}/p\mathbb{Z}[x]$ is an integral domain. As $\bar{q}$ has no more than $n-1$ roots (by the inductive hypothesis) the result follows. □

We give here two applications of Lagrange's theorem; they have useful consequences later.

THEOREM 2.3 *If $d|p-1$ then the congruence $x^d \equiv 1 \pmod{p}$ has exactly d solutions.*

Proof. By Theorem 2.2 this congruence cannot have more than d solutions. For the converse let $de = p-1$ and

$$x^{p-1} - 1 = (x^d - 1)g(x)$$

where

$$g(x) = x^{d(e-1)} + x^{d(e-2)} + \ldots + x^d + 1.$$

Now by Lagrange's theorem again $g(x) \equiv 0 \pmod{p}$ cannot have more than $p-1-d$ solutions, so if $x^d \equiv 1 \pmod{p}$ has less than d solutions then $x^{p-1} \equiv 1 \pmod{p}$ has less than $p-1$ solutions. This contradicts Fermat's Theorem (1.13) and so the result follows. □

THEOREM 2.4 *If p is prime then $(p-1)! \equiv -1 \pmod{p}$.*

(This result is sometimes known as Wilson's theorem.)

Proof. If $p>2$ let h be given by

$$h(x) = x^{p-1} - 1 - (x-1)(x-2)\ldots(x-p+1).$$

By Fermat's theorem each x, $1 \le x \le p-1$, is a root of $h(x) \equiv 0 \pmod{p}$, but h has degree less than $p-1$, and so all its coefficients must be congruent to zero $\pmod{p}$ by Theorem 2.2. If p is odd the result follows by considering the constant coefficient of h and the result is obvious if $p=2$. □

Chevalley's theorem

In this subsection congruences in more than one variable will be considered. The reduction from composite to prime moduli given above for one-variable polynomials also applies in this case. In 1935 Artin conjectured that a polynomial congruence with prime modulus always has a non-trivial solution if the number of variables in the polynomial is greater than its degree. For example if f is a quadratic form in three variables then $f(x, y, z) \equiv 0 \pmod{p}$ always has at least one non-trivial solution. This conjecture was proved by Chevalley a year later, we shall give his proof now.

Throughout this subsection let p be a fixed prime, let f, g and h be polynomials defined over $\mathbb{Z}/p\mathbb{Z}$, and let $\deg f$ denote the (total) degree of f. We begin with some

DEFINITIONS (i) If f and g are n-variable polynomials, f is *equivalent* to g, written as $f \equiv g$, if for all sets $\{a_1, \ldots, a_n\}$ of elements of $\mathbb{Z}/p\mathbb{Z}$ we have

$$f(a_1, \ldots, a_n) \equiv g(a_1, \ldots, a_n) \pmod p.$$

(ii) f is *congruent* to g, written as $f \sim g$, if all coefficients of corresponding monomials of f and g are congruent modulo p.

(iii) f is called *reduced* if f has degree less than p in *each* of its variables.

Note that $f \sim g$ implies $f \equiv g$, but the converse may not hold; for example $x^p \equiv x$, but $x^p \not\sim x$. Note also that x^p is not reduced.

LEMMA 2.5 *Given a polynomial f we can find a reduced polynomial f' such that* $\deg f' \leq \deg f$ *and* $f' \equiv f$.

Proof. This follows from Fermat's theorem. Given a monomial x^k where $k \geq p$, if $k = p$ then $x^k \equiv x$, and if $k > p$ we have, using the division algorithm, $k = tp + u$ with $0 \leq u < p$, and so $x^k = x^{tp}x^u \equiv x^{t+u}$. If $t + u < p$ then x^{t+u} is reduced, if not we can repeat this process until we obtain a reduced monomial. The lemma follows if we apply this transformation to each $x_i^{k_i}$ in f. □

The following lemma is an extension of Lagrange's theorem.

LEMMA 2.6 *If f and g are reduced polynomials and $f \equiv g$ then $f \sim g$.*

Proof. It is sufficient to assume that g is the zero polynomial. Let n be the number of variables in f, the proof is by induction on n. The case $n = 1$ follows immediately from Theorem 2.2 as $f(x) \equiv 0 \pmod p$ has more roots than its degree. For the general case we write

$$f(x_0, \ldots, x_n) = g_0(x_1, \ldots, x_n)x_0^{p-1} + \ldots + g_{p-1}(x_1, \ldots, x_n).$$

Given $a_1, \ldots, a_n$ in $\mathbb{Z}/p\mathbb{Z}$, let $g_i(a_1, \ldots, a_n) = b_i$ for $i = 0, \ldots, p-1$, and we have

$$f(x_0, a_1, \ldots, a_n) = b_0x_0^{p-1} + b_1x_0^{p-2} + \ldots + b_{p-1} = h(x_0).$$

By the assumption on f, h has p roots but is of degree $p - 1$, hence by Theorem 2.2, again, we have for $i = 0, \ldots, p-1$

$$g_i(a_1, \ldots, a_n) \equiv 0 \pmod p.$$

As this argument applies for all n-tuples $\{a_1, \ldots, a_n\}$, this gives $g_i \equiv 0$ for all i. But the polynomials g_i are reduced and have degree less than n, hence $f \sim 0$ follows by induction. □

THEOREM 2.7 *Chevalley's theorem.*
Suppose f and g are n-variable polynomials with degree less than n.
(i) *If the congruence*

$$f(x_1, \dots, x_n) \equiv 0 \pmod p \qquad (*)$$

is soluble then it has at least two solutions.

(ii) *Suppose g is a homogeneous polynomial then the congruence* $g(x_1, \dots, x_n) \equiv 0 \pmod p$ *has a solution distinct from* $\{0, 0, \dots, 0\}$.

Proof. (ii) follows immediately from (i) as $\{0, 0, \dots, 0\}$ is always a solution in the homogeneous case. For (i) assume that $\deg f = r$ and $(*)$ has the unique solution $x_i \equiv a_i \pmod p$, $i = 1, \dots, n$. Let h be given by

$$h(x_1, \dots, x_n) = 1 - f(x_1, \dots, x_n)^{p-1}.$$

By Fermat's theorem (1.13)

$$h(x_1, \dots, x_n) = \begin{cases} 1 & \text{if } x_i \equiv a_i \pmod p \text{ for all } i \\ 0 & \text{otherwise.} \end{cases}$$

Let h' be the reduced polynomial equivalent to h, by Lemma 2.5 it takes the same values as h. Now consider the polynomial h^* given by

$$h^*(x_1, \dots, x_n) = \prod_{i=1}^{n} (1 - (x_i - a_i)^{p-1}),$$

it also takes the same values $(\text{mod}\, p)$ as h, that is $h^* \equiv h'$. But h^* is reduced, so $h^* \sim h'$ by Lemma 2.6. Further, by Lemma 2.5, $\deg h' \le \deg h = r(p-1)$. As $\deg h^* = n(p-1)$ and h^* is congruent to h', this is impossible if $r < n$ and so our uniqueness assumption is invalid. □

Warning has shown that Theorem 2.7 can be improved to show that the number of solutions of $(*)$ is divisible by p, see Problem 24.

3.3 Local versus global considerations

This section is of a more specialized nature and can be omitted on first reading; the p-adic material will only be used again in Chapter 14.

Let f be a polynomial with integer coefficients. Consider the propositions

(i) there exist integers $x_1, \dots, x_n$ such that $f(x_1, \dots, x_n) = 0$, and
(ii) for all primes p and positive integers m the congruence

$$f(x_1, \dots, x_n) \equiv 0 \pmod{p^m} \text{ is soluble.}$$

There are a number of connections between these propositions. Clearly if (ii) is false then so is (i); this can be used to show that certain Diophantine equations are insoluble. More importantly, in some circumstances the converse holds; in this case we say that the *Hasse principle* applies, see p. 46.

In the context above the terms 'global' and 'local' are used to describe the two situations. The solution referred to in (i) is called a 'global' solution, although the term can also be used for solutions in the rationals or an algebraic number field. The individual solutions referred to in (ii) are called 'local' solutions of the equation $f(x_1, \ldots, x_n) = 0$.

We begin by treating a few examples where (ii) fails; no deep theory is involved but we can show that some Diophantine equations are insoluble.

Consider the equation

$$x^2 + y^2 = 3z^2$$

and congruences modulo 4. Clearly $x^2 + y^2$ is congruent to 0, 1 or 2 (mod 4), and so there is no solution if z is odd. If z is even then x and y must also be even (as each side of the equation is then divisible by 4), and so we can divide both sides by 4 and begin again. Note that as the equation is homogeneous it is also insoluble in $\mathbb{Q}$. Secondly we mention an example of a more general nature. Let p be a prime such that $p \nmid abc$ then the equation

$$ax^3 + bpy^3 + cp^2z^3 = 0$$

has no solutions in $\mathbb{Z}$ (or $\mathbb{Q}$). For suppose $\{x_0, y_0, z_0\}$ is a solution with $(x_0, y_0, z_0) = 1$ (there is no loss of generality in assuming this). Clearly $p \mid x_0$ and division by p gives

$$ap^2(x_0/p)^3 + by_0^3 + cpz_0^3 = 0.$$

It follows that $p \mid y_0$ and repeating this argument we also have $p \mid z_0$ contrary to assumption. Further examples of this type are given in the problem section, and many similar equations are discussed in Chapter 2 of Mordell (1969).

The p-adic numbers

Before considering Hasse's principle we shall introduce the p-adic number fields in which analytic methods are used to characterize congruence properties, no proofs will be given.

Let p be a fixed prime throughout.

DEFINITION $|\cdot|_p$ The p-adic valuation.

Let $x \in \mathbb{Q}$, where $0 \neq x = r/s$ and $r, s \in \mathbb{Z}$, let t_1 and t_2 be the largest integers such that $p^{t_1} \mid r$ and $p^{t_2} \mid s$, and let $t = t_1 - t_2$. Now define $|\cdot|_p$, a mapping $\mathbb{Q} \to \mathbb{Q}$, by

$$|x|_p = \frac{1}{p^t} \quad \text{if} \quad x \neq 0 \quad \text{and} \quad |0|_p = 0.$$

We have immediately for all x and y in $\mathbb{Q}$

(i) $|x|_p \geq 0$, and $|x|_p = 0$ if and only if $x = 0$,
(ii) $|xy|_p = |x|_p\, |y|_p$,
(iii) $|x+y|_p \leq \max(|x|_p, |y|_p) \leq |x|_p + |y|_p$.

It is clear that a p-adic valuation behaves in a similar way to the usual absolute value function on the reals except that the topological structure is different. Note also that the triangle inequality holds in a much stronger sense than usual.

Limits and Cauchy sequences are defined in an analogous manner to those in the real field except that $|\cdot|_p$ replaces $|\cdot|$ throughout, namely: a sequence of rational numbers $\{a_m\}$ is called a *null* sequence if $\lim_{m\to\infty} |a_m|_p = 0$, two sequences $\{a_m\}$ and $\{b_m\}$ are called *equivalent* if $\{a_m - b_m\}$ is a null sequence, a sequence of rational numbers $\{a_m\}$ is called a *p-adic Cauchy sequence* if given $\varepsilon > 0$ there exists $n = n(\varepsilon)$ such that whenever k, $m \geq n$ we have $|a_k - a_m|_p < \varepsilon$, and a *p-adic number* is defined to be an equivalence class of p-adic Cauchy sequences, also denoted by $\{\ldots\}$. Addition and multiplication of sequences is defined componentwise (that is $\{a_m\} + \{b_m\} = \{a_m + b_m\}$ and $\{a_m\}\{b_m\} = \{a_m b_m\}$) and so we have:

THEOREM 3.1 (i) *The set of p-adic numbers with componentwise addition and multiplication forms a field, denoted by $\mathbb{Q}_p$, which contains an isomorphic copy of $\mathbb{Q}$ under the correspondence $a \leftrightarrow \{a, a, a, \ldots\}$ for $a \in \mathbb{Q}$.*

(ii) *If x is a non-zero element of $\mathbb{Q}_p$ then x can be represented uniquely by the sequence*

$$x = p^u\{d_0, d_0 + d_1 p, d_0 + d_1 p + d_2 p^2, \ldots\} \qquad (*)$$

where $d_i \in \mathbb{Z}$ and $0 \leq d_i < p$ for all i, $u \in \mathbb{Z}$ and $d_0 \neq 0$.

(iii) *If $x = p^u\{e_0, e_1, \ldots\}$ and $x' = p^{u'}\{e_0', e_1', \ldots\}$, where u, u', e_i, $e_i' \in \mathbb{Z}$ and each e_i and e_i' is coprime to p, then $x = x'$ in $\mathbb{Q}_p$ if and only if $u = u'$ and, for all i, $e_i \equiv e_i' \pmod{p^{i+1}}$.*

The proof of this result is straightforward, see the references given at the end of this chapter. Note that the representation $(*)$ of x is often abbreviated to

$$x = d_0 p^u + d_1 p^{u+1} + d_2 p^{u+2} + \ldots.$$

We can now make the following consequential definitions. The integer u in $(*)$ is called the *order* of x. The mapping $|\cdot|_p$ can be extended to a mapping $\mathbb{Q}_p \to \mathbb{Q}$: if $x \in \mathbb{Q}_p$ and u is the order of x then $|x|_p = p^{-u}$. If u is non-negative then x is called a *p-adic integer*; the set of p-adic integers, denoted by $\mathbb{Z}_p$, is an integral domain with quotient field $\mathbb{Q}_p$. Finally if $|x|_p = 1$ then x is called a *unit* of $\mathbb{Q}_p$.

Let us consider some examples in $\mathbb{Q}_3$. Each of the following belong to $\mathbb{Q}_3$

$$x_1 = \{4, 4, 4, 4, \ldots\},$$
$$x_2 = \{1, 4, 13, 40, 121, \ldots\},$$
$$x_3 = \{1, 1, 10, 10, 91, \ldots\},$$
$$x_4 = \{1/9, 4/9, 13/9, 40/9, 121/9, \ldots\},$$
$$x_5 = \{9, 63, 225, 711, 2169, \ldots\}.$$

The standard form $(*)$ for x_1 is $\{1, 4, 4, 4, \ldots\}$; and, for example, it is also represented by $\{4, 13, 31, 85, \ldots\}$ by Theorem 3.1(iii). The remaining 3-adic numbers are in standard form, for instance $u = 0$, $d_{2i} = 1$ and $d_{2i+1} = 0$ in x_3. The order of x_1, x_2 and x_3 is zero, and so they are units (see Example 2 below). The orders of x_4 and x_5 are -2 and 2, respectively, and it is a simple matter to check that x_5 is the inverse of x_4 in $\mathbb{Q}_3$, for instance $225(13/9) \equiv 1 \pmod{3^3}$.

One application of the p-adic method concerns the solution of congruences. The solutions of $f(x_1, \ldots, x_n) \equiv 0 \pmod{p^m}$ for $m = 1, 2, 3, \ldots$ can be represented by single n-tuples of elements of $\mathbb{Q}_p$. Let us consider two examples.

Example 1 For $m = 1, 2, \ldots,$ study the congruence

$$x^2 \equiv 2 \pmod{7^m}.$$

Using the methods of the previous section this congruence has the solutions

$$x_1 = 3 + 1 \cdot 7 + 2 \cdot 7^2 + 6 \cdot 7^3 + 1 \cdot 7^4 + 2 \cdot 7^5 + \ldots,$$
$$x_2 = 4 + 5 \cdot 7 + 4 \cdot 7^2 + 5 \cdot 7^4 + 4 \cdot 7^5 + \ldots.$$

Hence we say that the equation $x^2 - 2 = 0$ has the 7-adic solutions $\{3, 10, 108, 2166, 4567, \ldots\}$ and $\{4, 39, 235, 235, 12240, \ldots\}$.

Example 2 Show directly that $x_1 = \{4, 4, 4, \ldots\}$ is a unit in $\mathbb{Q}_3$.

To do this we must find $y = \{y_0, y_1, \ldots\}$ in $\mathbb{Q}_3$ such that $x_1 y = \{1, 1, \ldots\}$. Hence, by Theorem 3.1 (iii), we must solve

$$4y_i \equiv 1 \pmod{3^{i+1}},$$

for each i. A simple calculation shows that the first few terms are given by $y = \{1, 7, 7, 61, 61, 547, \ldots\}$. One way to see this is to note that, in $\mathbb{Q}_3$,

$$y = \frac{1}{4} = \frac{1}{1+3} = 1 - 3 + 3^2 - 3^3 + \ldots,$$

so $y = \{1, -2, 7, -20, 61, -182, \ldots\}$, and this is 3-adically equivalent to the representation above.

A famous conjecture (due to Artin) states: if f is a homogeneous polynomial in n variables and has degree r, where $n > r^2$, then $f = 0$ always has a non-trivial solution in $\mathbb{Q}_p$ for all primes p (the inequality is best possible, see Problem 28). This proposition is true for $r = 2$ and 3 but false for $r = 4$ (with $n = 18$). Using deep methods from mathematical logic Artin's conjecture has been proved for all r provided a fixed set of primes (depending on r but not on n or f) is excluded, see Ax and Kochen (1966).

Hasse's principle

As we pointed out earlier, and using our new notation, Hasse's principle states that, in certain circumstances, if a polynomial equation $f = 0$ has solutions in $\mathbb{Q}_p$ for all p and in $\mathbb{R}$ (the real numbers) then it also has a solution in $\mathbb{Q}$ and possibly in $\mathbb{Z}$. Stated differently the principle is: under certain circumstances, $f = 0$ is soluble in $\mathbb{Q}$ if and only if it is soluble in all completions (in the metric space sense) of $\mathbb{Q}$, as it can be shown that $\mathbb{R}$ and $\mathbb{Q}_p$, for $p = 2, 3, 5, \ldots$, are the only completions of $\mathbb{Q}$.

The exact circumstances in which Hasse's principle holds are not known. It is false, for example, for the equations $3x^3 + 4y^3 + 5z^3 = 0$ and $x^4 - 17y^4 = 2z^4$, but it holds for quadratics by

THEOREM 3.2 (Hasse, Minkowski) *Let f be a homogeneous quadratic polynomial with integer square-free coefficients. $f = 0$ has a non-trivial solution in $\mathbb{Q}$ (or equivalently in $\mathbb{Z}$) if and only if it is soluble (non-trivially) in $\mathbb{R}$ and $\mathbb{Q}_p$ for all primes p.*

Special cases of this theorem will be discussed later, see Theorem 3.3, Chapter 4 and Section 1 of Chapter 14. Full details and proofs of all the results discussed in this section are given in Borevich and Shafarevich (1966), Mordell (1969) and Serre (1973).

3.4 Problems 3

1. Deduce the congruence properties

(i) $a \equiv b\ (m_i)$, for $i = 1, \ldots, k$, implies $a \equiv b(m)$ where m is the least common multiple of $m_1, \ldots, m_k$;

(ii) if $a \equiv b(m)$ then $(a, m) = (b, m)$, $an \equiv bn(mn)$ and $a^k \equiv b^k(m)$ for $k \geq 0$;

(iii) if $a \equiv b(m)$, $a = a'd$, $b = b'd$ and $m = m'd$ then $a' \equiv b'(m')$;

(iv) if $a_i \equiv b_i\ (i = 1, 2, \ldots, n)$ and $x \equiv y(m)$ then $\sum_{i=1}^{n} a_i x^i \equiv \sum_{i=1}^{n} b_i y^i$ and $\prod_{i=1}^{n} a_i \equiv \prod_{i=1}^{n} b_i(m)$.

2. Solve where possible

(i) $91x \equiv 84 \pmod{143}$;

(ii) $91x \equiv 84 \pmod{147}$;

(iii) $12x + 16y \equiv 6 \pmod{30}$.

3. (i) Show that an integer n is divisible by 3 or 9 if the sum of its digits is divisible by 3 or 9, respectively.

(ii) Show also that n is divisible by 11 if the number formed by alternately adding and subtracting its digits in turn is divisible by 11.

4. (i) Show that the last digit of an even perfect number is 6 or 8.

(ii) Further show that the sum of the digits of an even perfect number, larger than 6, is congruent to one modulo 9.

5. Find the last digit of 7^{139} and 13^{2001}.

6. Prove that if $(a, m) = (a - 1, m) = 1$ then

$$1 + a + a^2 + \ldots + a^{\phi(m)-1} \equiv 0 \pmod m,$$

and deduce that every prime other than 2 or 5 divides infinitely many of the integers 1, 11, 111, 1111, ...

7. Let $n > 2$. If m is the number of solutions of the congruence $x^2 \equiv 1 \pmod n$, show that $2|m$. Further let $a_1, \ldots, a_{\phi(n)}$ be a reduced system of residues modulo n, prove that

$$a_1 a_2 \ldots a_{\phi(n)} \equiv (-1)^{m/2} \pmod n.$$

8. Show that if $(m, n) = 1$ then $m^{\phi(n)} + n^{\phi(m)} \equiv 1 \pmod{mn}$.

9. Let $f(x, y) = ax^2 + bxy + cy^2$ where $(a, b, c) = 1$ and let $d = b^2 - 4ac$.

(i) By considering the cases $p = 2$ and $p > 2$ separately, show that if $p^\alpha | d$ then there are $p^\alpha \phi(p^\alpha)$ sets of integers $\{x, y\}$ in a complete residue system modulo p^α such that $p \nmid f(x, y)$. [Hint. Consider $4af(x, y)$.]

(ii) Deduce that, if both x and y take values in a complete residue system modulo $|d|$, there are exactly $|d|\,\phi(|d|)$ sets of integers $\{x, y\}$ such that $(d, f(x, y)) = 1$.

10. Let $(a, p) = 1$, where p is a prime. Show that $a^{p-2}b$ is a solution of $ax \equiv b \pmod p$, and so solve $3x \equiv 17 \pmod{29}$.

11. (i) Let F_n be a Fermat number and p be a prime. Show that if $p | F_n$ then $p = 2^{n+1}k + 1$ for some integer k.

(ii) Let p be a prime and suppose each prime factor of $p - 1$ is not more than c. Show that if $(a, p) = 1$ then $a^{c!} \equiv 1 \pmod p$. Deduce that if $p | n$ then $(a^{c!} - 1, n) > 1$. As it often happens that the prime power factors of $p - 1$ are relatively small, this can be used to factorize large numbers as follows: choose a (small) and calculate $a_1 = a, a_2, \ldots$ and $b_1, b_2, \ldots$ using

$$a_k \equiv a_{k-1}^k \pmod n \qquad \text{and} \qquad b_k = (a_k - 1, n).$$

Note that $a_k \equiv a^{k!} \pmod n$ and b_k is a non-trivial factor of n if $b_k > 1$. This is called the *Pollard p − 1 method,* see Section 2, Chapter 1. Use it to factorize 56759 beginning with $a = 2$.

12. (i) Prove that if $a^p \equiv b^p \pmod p$ then $a^p \equiv b^p \pmod{p^2}$.

(ii) Show that if $(a, p) = 1$ then $p^k | a^{p^k} - a^{p^{k-1}}$, and so deduce that $(a, n) = 1$ implies $n \Big| \sum_{d|n} a^d \mu\left(\frac{n}{d}\right)$.

13. Let the number theoretic function χ be given by: $\chi(1)=\chi(2)=1$, $\chi(4)=2$, $\chi(2^{n+3})=\frac{1}{2}\phi(2^{n+3})$, $\chi(p^n)=\phi(p^n)$ for an odd prime p, and if $m=\prod_{i=1}^{k} p_i^{\alpha_i}$ then $\chi(m)=\text{LCM}\{\chi(p_1^{\alpha_1}),\ldots,\chi(p_k^{\alpha_k})\}$. $\chi(m)$ is known as the *universal exponent* of m. Show that $a^{\chi(m)}\equiv 1 \pmod m$ if $(a, m)=1$. Now consider (with $m>1$ and odd)

(i) m is prime;
(ii) $\phi(m)|m-1$;
(iii) $\chi(m)|m-1$;
(iv) $2^{m-1}\equiv 1 \pmod m$.

Show that (i) implies (ii), (ii) implies (iii), (iii) implies (iv), (iv) does not imply (iii) and (iii) does not imply (ii). The relationship between (i) and (ii) is unknown. You should consider the integers 341, 645, . . . ; 561, 1105, ... Numbers satisfying (iv) are sometimes known as *pseudoprime* (see p. 12), for many years up to 1819 it was thought that (iv) implied (i). Show that 341 is the smallest counter-example. Finally prove that if (iv) holds for $m=k$ then it also holds for $m=2^k-1$.

14. Solve the basket of eggs problem: find the smallest number of eggs such that one egg remains when eggs are removed 2, 3, 4, 5, or 6 at a time, but no eggs remain if they are removed 7 at a time. (Problems of this kind have a very long history, this one appears in an Indian manuscript of the seventh century AD.)

15. Solve the simultaneous congruences

(i) $x\equiv 3 \pmod 6$
$x\equiv 5 \pmod{35}$
$x\equiv 7 \pmod{143}$
$x\equiv 11 \pmod{323}$

(ii) $x\equiv 5 \pmod 6$
$7x\equiv 5 \pmod{12}$
$17x\equiv 19 \pmod{30}$.

16. Show that

$$x\equiv a \pmod m \quad \text{and} \quad x\equiv b \pmod n$$

have a common solution if and only if $(m, n)|b-a$, and the solution is unique modulo the least common multiple of m and n.

17. By considering the prime power representation of the GCD and LCM of the moduli, state and prove the result corresponding to Problem 16 when k congruences are involved.

18. Solve, where possible,

(i) $x^3-2x+3\equiv 0 \pmod{27}$;
(ii) $x^3-5x^2+3\equiv 0 \pmod{27}$;
(iii) $x^3-2x+4\equiv 0 \pmod{125}$.

19. Show that (i) $(m-1)!\equiv -1 \pmod m$ if and only if m is prime or $m=1$; (ii) if p is an odd prime and $0<k<p$ then (note $0!=1$)

$$(p-k)!\,(k-1)!\equiv(-1)^k \pmod p.$$

20. Suppose the congruence $f(x)\equiv 0 \pmod p$ has t distinct solutions $x_1,\ldots,x_t$ and each of these solutions satisfies $p\nmid f'(x_i)$ for $i=1,\ldots,t$. Deduce that, for $e\geq 1$, the congruence $f(x)\equiv 0 \pmod{p^e}$ also has t distinct solutions. Use this to

show that if p is prime and $d|p-1$ then the congruence

$$x^d \equiv 1 \pmod{p^e}$$

has exactly d solutions for each $e \geq 1$.

21. Let p be a prime. Show that $\mathbb{Z}/p\mathbb{Z}[x]$ is a Euclidean domain, that is a version of the Euclidean algorithm can be used to find the greatest common factor of two given polynomials over $\mathbb{Z}/p\mathbb{Z}$. Now suppose $f \in \mathbb{Z}/p\mathbb{Z}[x]$. Show that $f(x)=0$ has t distinct roots in $\mathbb{Z}/p\mathbb{Z}$ where $t = \deg(\mathrm{GCD}(f(x), x^p - x))$ in $\mathbb{Z}/p\mathbb{Z}[x]$.

22. A polynomial f with integer coefficients is called *primitive* if

$$f(x) = a_0 + a_1x + \ldots + a_nx^n \quad \text{and} \quad (a_0, a_1, \ldots, a_n) = 1.$$

Prove that the product of two primitive polynomials is primitive.

23. Let p be a prime greater than 3, and let a^* denote a solution of the congruence $ax \equiv 1 \pmod{p^2}$. If

$$\prod_{i=1}^{p-1}(x-i) = x^{p-1} + a_1x^{p-2} + \ldots + a_{p-1}, \qquad (*)$$

show that
(i) $p|a_i$ for $i = 1, \ldots, p-2$,
(ii) $1^* + 2^* + \ldots + (p-1)^* \equiv 0 \pmod{p^2}$.
[Hint. For (ii) put $x = p$ in $(*)$ and consider the coefficient a_{p-2}.] (ii) is known as Wolstenholme's theorem.

24. Show that the number of solutions of the polynomial congruence

$$f(x_1, \ldots, x_n) \equiv 0 \pmod{p}$$

is divisible by p provided the degree of f is less than n. This can be derived in a similar manner to Theorem 2.7 using the polynomial H^* given by

$$H^*(x_1, \ldots, x_n) = \sum_{i=1}^{s}\prod_{j=1}^{n}(1 - (x_j - a_{i,j})^{p-1})$$

where, for $i = 1, \ldots, s$, the solutions of the congruence are $\{a_{i,1}, \ldots, a_{i,n}\}$. Consider the terms of highest power in H^*.

25. Construct a binary quadratic form f and a ternary cubic form g satisfying

$$f(x, y) \equiv 0 \pmod{5} \text{ implies } x \equiv y \equiv 0 \pmod{5},$$
$$g(x, y, z) \equiv 0 \pmod{2} \text{ implies } x \equiv y \equiv z \equiv 0 \pmod{2}.$$

26. Let $a \equiv d \equiv 4 \pmod{9}$, $b \equiv 0 \pmod{3}$ and $c \equiv \pm 1 \pmod{3}$. Show that the equation

$$ax^3 + 3bx^2y + 3cxy^2 + dy^3 = z^3$$

has no non-trivial integer solution by considering congruences modulo 9, and showing that $3 \nmid xy$ and $z \equiv ax + by \pmod{3}$.

27. (i) Show that $x^3 \equiv 0, \pm 1 \pmod 7$.
(ii) Prove that the equation

$$(7a+1)x^3 + (7b+2)y^3 + (7c+4)z^3 + (7d+1)xyz = 0$$

has no non-trivial solution by considering congruences modulo 7 and treating the cases $z \equiv 0$ and $z \not\equiv 0 \pmod 7$ seperately.

28. Let f_i be homogeneous cubic polynomials such that, for $i = 1, 2, 3$, $f_i(x, y, z) \equiv 0 \pmod p$ implies $x \equiv y \equiv z \equiv 0 \pmod p$. Show that the equation in nine variables

$$f_1(x_1, y_1, z_1) + pf_2(x_2, y_2, z_2) + p^2 f_3(x_3, y_3, z_3) + p^{3^{\cdot}} \sum_{\substack{k,m,n=1 \\ k \neq m \neq n \neq k}}^{3} a_{kmn} x_k y_m z_n = 0$$

has only the trivial solution. [Note. This shows that the inequality in Artin's conjecture (see p. 46) is best possible.]

29. Show that $|x| \prod_{\text{all } p} |x|_p = 1$ for non-zero $x \in \mathbb{Q}$.

30. Verify that $\mathbb{Z}_p$ is an integral domain.

31. Prove that in a p-adic field

(i) the series $\sum_{i=1}^{\infty} a_i$ converges if and only if $|a_i|_p \to 0$ as $i \to \infty$,

(ii) the power series $\sum_{i=0}^{\infty} a_i x^i$ converges for each p-adic integer x provided $|a_i|_p \to 0$ as $i \to \infty$.

32. Find the multiplicative inverses of $x_2 = \{1, 4, 13, 40, 121, \ldots\}$ and $x_3 = \{1, 1, 10, 10, 91, \ldots\}$ in $\mathbb{Q}_3$.

4

QUADRATIC RESIDUES

Here we shall continue the development of congruence theory begun in the previous chapter; we shall discuss the quadratic case in detail and prove Gauss's famous law of quadratic reciprocity. This illustrates many of the ideas used in the general (nth degree) theory whilst being amenable to elementary treatment. Over the past hundred years many mathematicians have been involved in the development of the cubic, quartic, and higher degree theory culminating in the results associated with Artin's reciprocity law, a considerable generalization of Gauss's law. The general theory is beyond the scope of an undergraduate text as it builds upon deep results about algebraic number fields, a good introduction is given in Ireland and Rosen (1982) where the cubic and quartic cases are discussed in detail.

In the first proof of Euler's theorem (see p. 36) we derived a property of a product taken over a reduced system of residues to obtain the result, but we did not use any information about the individual members of that product. Similar arguments will play a vital role in this chapter and in the primitive root section of Chapter 5, they are an essential feature of quadratic residue theory. For example we can easily demonstrate that exactly half of the elements of the set $\{1, \ldots, p-1\}$ are quadratic non-residues modulo p but we can only give a very crude estimate of where the first one occurs; see Section 3.

A typical quadratic congruence can be reduced to the basic type

$$x^2 \equiv a \pmod{p}$$

where p is prime. The quadratic reciprocity law provides an algorithm for solving these congruences. It has applications in many other areas; some of these are given at the end of this chapter.

4.1 The Legendre symbol

A general quadratic congruence

$$ax^2 + bx + c \equiv 0 \pmod{m}$$

can be reduced to the basic type as follows. First using the methods of Section 2, Chapter 3, we may assume that the modulus is a prime p. Also as the case $p = 2$ is trivial we shall suppose *p is an odd prime throughout this chapter.*

Given the congruence

$$ay^2 + by + c \equiv 0 \pmod p,$$

where $p \nmid a$, we can find $a' \in \mathbb{Z}$ such that $2aa' \equiv 1 \pmod p$, and so, multiplying by $2a'$, this congruence has the form

$$y^2 + 2b'y + c' \equiv 0 \pmod p,$$

where $b' = aa'b$ and $c' = 2aa'c$. Completing the square we have

$$(y + b')^2 \equiv b'^2 - c' \pmod p,$$

and this is of the basic type if we let $x = y + b'$.

Definition Suppose $p \nmid a$. If the congruence

$$x^2 \equiv a \pmod p \tag{1}$$

is soluble then a is called a *quadratic residue* modulo p, and if it has no solution a is called a *quadratic non-residue* modulo p.

Lemma 1.1 *Exactly half of the integers a satisfying $1 \le a \le p - 1$ are quadratic residues modulo p.*

Proof. No two elements of the set $S = \{1^2, 2^2, \ldots, (p-1)/2)^2\}$ are congruent $(\text{mod}\, p)$, and so if a is the least positive residue $(\text{mod}\, p)$ of an element of S, it is a quadratic residue modulo p. Hence there are at least $(p-1)/2$ quadratic residues $(\text{mod}\, p)$.

Conversely note that if x_0 is a solution of (1) then $p - x_0$ also solves (1), and $x_0 \ne p - x_0$ as $p > 2$. Further, neither x_0 nor $p - x_0$ are solutions of another congruence of this type, for if $x_0^2 \equiv a$, and $x_0^2 \equiv b$ or $(p - x_0)^2 \equiv b \pmod p$, where $1 \le b \le p - 1$, then $b \equiv a \pmod p$; that is $b = a$. Hence the total number of soluble congruences of the type (1) is at most $(p-1)/2$. Combining this with the above completes the proof. □

Definition The Legendre symbol.
Let p be an odd prime. The values of $(\cdot/p)$, called the *Legendre symbol*, are given by

$$(a/p) = \begin{cases} +1 & \text{if } p \nmid a \text{ and } a \text{ is a quadratic residue } (\text{mod}\, p) \\ 0 & \text{if } p \mid a \\ -1 & \text{if } p \nmid a \text{ and } a \text{ is a quadratic non-residue } (\text{mod}\, p). \end{cases}$$

Note that the congruence (1) has $(a/p) + 1$ solutions, for if $(a/p) = 1$ then, by Lagrange's theorem, (1) has exactly two solutions, and if $(a/p) = 0$ then (1) has the unique solution $x = 0$.

THEOREM 1.2 *Euler's criterion.*
If p is an odd prime and $(a, p) = 1$ then

$$(a/p) = 1 \qquad \text{if and only if} \qquad a^{(p-1)/2} \equiv 1 \pmod{p},$$

or, equivalently,

$$(a/p) \equiv a^{(p-1)/2} \pmod{p}.$$

Proof. By Lemma 1.1 the set of quadratic residues (mod p) has $(p-1)/2$ elements and by Theorem 2.3, Chapter 3, the set of solutions of the congruence

$$x^{(p-1)/2} \equiv 1 \pmod{p} \qquad (*)$$

also has $(p-1)/2$ elements. These sets are identical. For if a is a quadratic residue, there is a b such that $b^2 \equiv a \pmod{p}$, and so $a^{(p-1)/2} \equiv b^{p-1} \equiv 1 \pmod{p}$ by Fermat's theorem (1.13, Chapter 3); that is a solves (*). The result follows as both sets have the same cardinality. □

Another proof of Euler's criterion and Lemma 1.1 is given in Problem 1.

An algorithm for evaluating Legendre symbols is given in the next few results, the difficult part—Gauss's reciprocity law—will be derived in Section 2.

THEOREM 1.3 (i) $(ab/p) = (a/p)(b/p)$;
(ii) *if* $a \equiv b \pmod{p}$ *then* $(a/p) = (b/p)$;
(iii) $(a^2/p) = 1$ *and so* $(1/p) = 1$;
(iv) $(-1/p) = (-1)^{(p-1)/2}$.

Proof. (i) This is trivial if p divides a or b. Otherwise we have by Theorem 1.2.

$$(a/p)(b/p) \equiv a^{(p-1)/2}b^{(p-1)/2} = (ab)^{(p-1)/2} \equiv (ab/p) \pmod{p},$$

and the result follows as $p > 2$ and $|(a/p)| \leq 1$. (ii) and (iii) are obvious and for (iv) use Theorem 1.2, again noting that $p > 2$. □

EXAMPLE Does $x^2 \equiv 63 \pmod{11}$ have a solution? We have

$$\begin{aligned} (63/11) &= (8/11) && \text{by (ii) as } 63 \equiv 8 \pmod{11} \\ &= (2/11) && \text{by (i) and (iii)} \\ &= -1 && \text{by trial and error.} \end{aligned}$$

The next two results will remove the trial and error element of the above and similar examples. We use the terminology: the *absolute least residue* of a (mod n) is the unique integer b satisfying $b \equiv a \pmod{n}$ and $-n/2 < b \leq n/2$.

THEOREM 1.4 *Gauss's lemma.*
Suppose $(a, p) = 1$. *If* t *is the number of elements of the set* $Q = \{a, 2a, \ldots, ((p-1)/2)a\}$ *whose absolute least residues modulo p are negative then*

$$(a/p) = (-1)^t.$$

If we apply Gauss's lemma to the example above we have $a = 2$, $p = 11$, $Q = \{2, 4, 6, 8, 10\}$ and the absolute least residues are 2, 4, -5, -3 and -1. Hence $t = 3$ and $(2/11) = -1$.

Proof. Let $r_1, r_2, \ldots$ be the positive absolute least residues of the elements of Q and $-s_1, -s_2, \ldots$ be the negative absolute least residues. We claim that no two members of the set $R = \{r_1, r_2, \ldots, s_1, s_2, \ldots\}$ are congruent modulo p. As $(a, p) = 1$, this is clearly so for the first and second halves of R separately. So suppose, for some i and j, we have $r_i \equiv s_j$, $m_1 a \equiv r_i$ and $m_2 a \equiv s_j \pmod p$. Then $a(m_1 + m_2) \equiv 0$ and so $m_1 + m_2 \equiv 0 \pmod p$, but this is impossible as $m_1, m_2 < p/2$. It follows that $R = \{1, 2, \ldots, (p-1)/2\}$ and we have (see p. 51)

$$a2a \ldots ((p-1)/2)a \equiv (-1)^t((p-1)/2)! \pmod p,$$

which gives

$$a^{(p-1)/2} \equiv (-1)^t \pmod p.$$

Now use Theorem 1.2. □

Our first application of this result is the evaluation of $(2/p)$.

THEOREM 1.5 2 *is a quadratic residue (quadratic non-residue) of primes of the form* $8k \pm 1$ ($8k \pm 3$, respectively) *or, equivalently,*

$$(2/p) = (-1)^{(p^2-1)/8}.$$

Proof. Apply Theorem 1.4; in this case $Q = \{2, 4, \ldots, p-1\}$ and t is the number of elements of Q greater than $p/2$, that is

$$t = (p-1)/2 - [p/4]$$

where the square brackets denote integer part. Hence

$$p = 8k + 1 \quad \text{gives} \quad t = 4k - [2k + \tfrac{1}{4}] \equiv 0 \pmod 2 \quad \text{and} \quad (a/p) = 1.$$

The remaining cases are similar. For the second part note that $16 \mid p^2 - 1$ if and only if p has the form $8k \pm 1$. □

An application of this result is

LEMMA 1.6 (i) *If* $k > 1$, $p = 4k + 3$, *and p is prime, then* $2p + 1$ *is also prime if and only if* $2^p \equiv 1 \pmod{2p+1}$.

(ii) *If* $2p + 1$ *is prime then* $2p + 1 \mid M_p$, *and* M_p *is composite, where* $M_p = 2^p - 1$ *is a Mersenne number.*

Proof. (i) Let $2p+1$ be prime. $2p+1 \equiv 7 \pmod 8$, so by the above result $(2/(2p+1)) = 1$ and Euler's criterion (Theorem 1.2) gives

$$2^p = 2^{((2p+1)-1)/2} \equiv 1 \pmod{2p+1}.$$

Conversely suppose $2^p \equiv 1 \pmod{2p+1}$ and $2p+1 = \prod_{i=1}^{k} p_i^{\alpha_i}$. Then $p \mid \phi(2p+1)$ by Euler's theorem (1.12, Chapter 3), that is

$$p \Big| \prod_{i=1}^{k} p_i^{\alpha_i - 1}(p_i - 1).$$

by Theorem 1.8, Chapter 2. Hence there is an i such that $p \mid p_i - 1$. Moreover $2p+1 = p_i n$ for some integer n. If $n > 1$ then $n > 2$, but then $p \nmid p_i - 1$ as $p > p_i - 1$; this gives $n = 1$ and $2p+1$ is prime.

(ii) This follows from (i) for, as $k > 1$, we see that $p > 3$ and so $M_p > 2p+1$. □

Using this lemma we have for instance $23 \mid M_{11}$, $47 \mid M_{23}$, and $167 \mid M_{83}$ and so these Mersenne numbers are composite.

4.2 Quadratic reciprocity

We come now to one of the great theorems of mathematics—Gauss's Law of Quadratic Reciprocity

THEOREM 2.1 *If p and q are distinct odd primes then*

$$(p/q)(q/p) = (-1)^{(p-1)(q-1)/4}$$

or, equivalently, $(p/q) = (q/p)$ *unless* $p \equiv q \equiv 3 \pmod 4$, *in which case* $(p/q) = -(q/p)$.

This remarkable result was first proved by Gauss (at the age of 19). Strong numerical evidence existed and both Euler and Legendre had produced incomplete proofs, but from a theoretical point of view there is no obvious reason for a connection to exist between the congruences, with different moduli,

$$x^2 \equiv p \pmod q \quad \text{and} \quad x^2 \equiv q \pmod p. \qquad (*)$$

As a consequence all proofs are non-trivial and further, although the result as stated refers to solutions of (*), it has profound consequences and generalizations in areas far removed from congruence theory.

A large number of proofs exist—over one hundred and fifty have been claimed—but many are only slight variants of others. Gauss's first proof used induction and a modern version of this appears in Venkov (1970). There are many proofs which use Gauss's lemma (Theorem 1.4) and either counting arguments or trigonometrical identities. We shall give one of each, some others can be found in the problem section. In Chapter 6 we shall reprove the result using Gauss sums, this proof does not rely on Gauss's lemma; perhaps the most elegant proof is a version of this one which uses some results from finite field theory; see, for example, Serre (1973). Other proofs use results from algebra, geometry, permutation theory, and even fluid dynamics—there is one (Lewy, 1946) which begins by considering the wave equation on a beach with constant angle $\pi p/2q$. Bachmann (1968) has given a detailed analysis of all proofs known before 1910, see also A14 and R40 in LeVeque (1974) and Guy (1984).

The number and variety of proofs is not accidental for the result is central to number theory, some would say central to mathematics as a whole. We shall mention here an algebraic and an analytic application, many others will be given later. A major problem in algebraic number theory concerns the splitting of ideals in extension fields. Kronecker's version of the Legendre symbol (see Section 3), whose behaviour is determined by the law of quadratic reciprocity, answers this completely for quadratic fields: the ideal (p) splits in $\mathbb{Q}(\sqrt{q})$ if and only if $(p/q)=1$. Further, the law can be used to study the behaviour of the series (using the Kronecker symbol)

$$\sum_{n=1}^{\infty} \frac{(d/n)}{n^s}$$

for complex s, and this has important consequences for the Riemann hypothesis; we shall discuss this in more detail in Chapters 10 and 13.

First proof of Theorem 2.1. This is a considerably amended version of Gauss's third proof. Let s denote the number of integers in the set $\{q, 2q, \ldots, (p-1)/2q\}$ with negative absolute least residue (mod p) and t the number of integers in the set $\{p, 2p, \ldots, (q-1)/2p\}$ with negative absolute least residue (mod q). By Gauss's lemma

$$(p/q)=(-1)^t \quad \text{and} \quad (q/p)=(-1)^s,$$

so $(p/q)=(q/p)$ if and only if $s+t$ is even. Hence we need:

$$s+t \quad \text{is odd if and only if} \quad p\equiv q\equiv 3 \pmod 4). \tag{1}$$

This is derived by counting the number of lattice points (points with integer co-ordinates) inside a hexagon (Figure 1) in two different ways; the hexagon picks out those points with negative absolute least residue. First we show that the hexagon has $s+t$ points, and secondly we show that it has an odd number of points if and only if $p\equiv q\equiv 3$ (mod 4).

Our hexagon H has vertex co-ordinates $A_0, \ldots, A_5$ where

$$A_0=(0,0) \qquad A_1=(\tfrac{1}{2},0) \qquad A_2=\left(\frac{p}{2}, \frac{q(p-1)}{2p}\right)$$

$$A_3=\left(\frac{p}{2}, \frac{q}{2}\right) \qquad A_4=\left(\frac{p(q-1)}{2q}, \frac{q}{2}\right) \qquad A_5=(0,\tfrac{1}{2}),$$

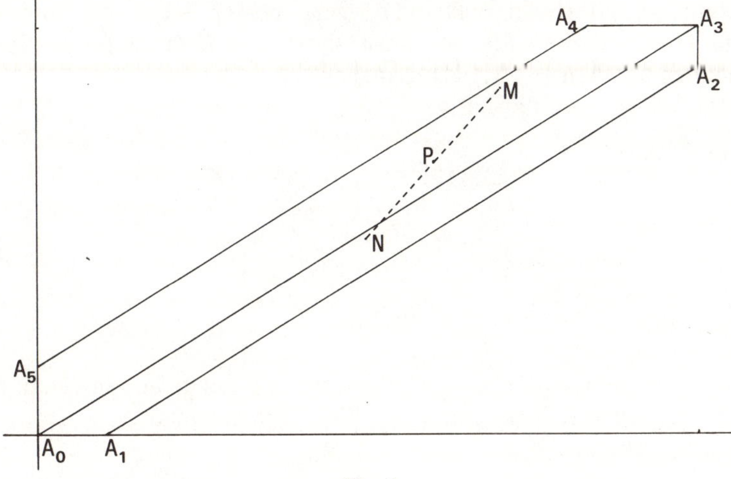

Fig. 1

and the lines A_1A_2 and A_4A_5 have equations

$$y = \frac{qx}{p} - \frac{q}{2p} \quad \text{and} \quad y = \frac{qx}{p} + \frac{1}{2}$$

respectively, see Figure 1. Hence the coordinates (x, y) of a point inside H satisfy

$$0 < x < \frac{p}{2}, \quad 0 < y < \frac{q}{2}, \quad y < \frac{qx}{p} + \frac{1}{2} \quad \text{and} \quad y > \frac{qx}{p} - \frac{q}{2p}. \tag{2}$$

Also, as p and q are distinct odd primes, there are no lattice points on the line A_0A_3.

Now if (i, j) is a lattice point in $A_0A_1A_2A_3$ we have $1 \le j \le (q-1)/2$ and

$$\frac{qi}{p} - \frac{q}{2p} < j < \frac{qi}{p}.$$

This gives

$$-\frac{q}{2} < jp - iq < 0. \tag{3}$$

$jp - iq$ is the absolute least residue $(\text{mod}\, q)$ of jp and, by (3), it is negative and so is one of the points counted by t. Conversely if jp has a negative absolute least residue $(\text{mod}\, q)$ then we can find an i such that (3) holds. Hence there are exactly t lattice points in $A_0A_1A_2A_3$. Similarly there are s lattice points in $A_3A_4A_5A_0$ and so there are $s + t$ lattice points in H.

Secondly we define a one-to-one correspondence on the lattice points in H: if the point (x_0, y_0) (M in Figure 1) is a lattice point in H, then so is the point (x_1, y_1) (N in Figure 1) where $x_1 = (p+1)/2 - x_0$ and $y_1 = (q+1)/2 - y_0$. To see this we must show that if (x_0, y_0) satisfies the inequalities (2) then (x_1, y_1) also satisfies (2).

As $0 < x_0 < p/2$ we have

$$\frac{p+1}{2} > \frac{p+1}{2} - x_0 > \frac{p+1}{2} - \frac{p}{2}$$

or

$$\frac{1}{2} < x_1 < \frac{p+1}{2}.$$

This is equivalent to $0 < x_1 < p/2$ as both x_1 and p are integers. The second inequality follows similarly. The third inequality for (x_0, y_0) is $y_0 < qx_0/p + 1/2$ and so

$$y_1 = \frac{q+1}{2} - y_0 > \frac{q+1}{2} - \frac{qx_0}{p} - \frac{1}{2}$$
$$= \frac{q}{2} - \frac{q}{p}\left(\frac{p+1}{2} - x_1\right) = \frac{qx_1}{p} - \frac{q}{2p}.$$

This is the fourth inequality for (x_1, y_1). By a similar argument we can show that the fourth inequality for (x_0, y_0) implies the third one for (x_1, y_1). Hence the correspondence is established as it is self-inverse.

The points (x_0, y_0) and (x_1, y_1) are distinct unless $x_0 = (p+1)/4$, $y_0 = (q+1)/4$ and these numbers are integers; that is, unless $p \equiv q \equiv 3 \pmod 4$. In this case there is one point (P in Figure 1) which corresponds to itself. Hence H has an odd number of points if and only if $p \equiv q \equiv 3 \pmod 4$. Thus (1) holds and the theorem is proved. □

Second proof of Theorem 2.1. This proof, due originally to Eisenstein, also uses Gauss's lemma and the trigonometrical identity: if m is an odd positive integer then

$$\frac{\sin mx}{\sin x} = (-4)^{(m-1)/2} \prod_{j=1}^{(m-1)/2} \left(\sin^2 x - \sin^2 \frac{2\pi j}{m}\right). \tag{4}$$

To show this we prove first, by induction on m, that $\sin mx/\sin x$ and $\cos mx/\cos x$ can be expressed as polynomials, of degree $(m-1)/2$, in $\sin^2 x$ with leading coefficient $(-4)^{(m-1)/2}$. This is trivial if $m = 1$, and the main induction step follows from the identities

$$\frac{\sin(m+2)x}{\sin x} = \frac{\sin mx}{\sin x}(1 - 2\sin^2 x) + \frac{\cos mx}{\cos x}(2 - 2\sin^2 x)$$

and

$$\frac{\cos(m+2)x}{\cos x} = \frac{\cos mx}{\cos x}(1 - 2\sin^2 x) - 2\frac{\sin mx}{\sin x}\sin^2 x.$$

Now $\sin mx/\sin x$ is zero when $mx = 2\pi i$, $i = 1, \ldots, m-1$, hence each of the terms

$$\sin x \pm \sin\frac{2\pi j}{m},$$

$j = 1, \ldots, (m-1)/2$, are factors of the polynomial expression for $\sin mx/\sin x$. Equation (4) follows by noting that there are exactly $m-1$ of these factors.

For the main part of Eisenstein's proof we let $S = \{1, \ldots, (q-1)/2\}$ and define e and s_i by: if $q \nmid i$ and $s \in S$,

$$e(i, s) = \begin{cases} 1 & \text{if the absolute least residue (mod } q\text{) of } si \text{ is positive} \\ -1 & \text{if the absolute least residue (mod } q\text{) of } si \text{ is negative,} \end{cases}$$

and s_i is the unique integer satisfying: $s_i \in S$ and

$$si \equiv e(i, s)s_i \pmod q.$$

Putting $i = p$ we have, as $\sin(-x) = -\sin x$,

$$\sin\frac{2\pi ps}{q} = e(p, s)\sin\frac{2\pi s_p}{q}. \tag{5}$$

Now, if $T = \{1, \ldots, (p-1)/2\}$,

$$\begin{aligned}
(p/q) &= \prod_{s \in S} e(p, s) && \text{by Gauss's lemma} \\
&= \prod_{s \in S} \sin\frac{2\pi ps}{q} \Big/ \sin\frac{2\pi s}{q} && \text{by (5) as the mapping } s \to s_p \text{ is a bijection} \\
&= \prod_{s \in S} (-4)^{(p-1)/2} \prod_{t \in T}\left(\sin^2\frac{2\pi s}{q} - \sin^2\frac{2\pi t}{p}\right) && \text{by (4) with } m = p \text{ and } x = \frac{2\pi s}{q} \\
&= (-4)^{(p-1)(q-1)/4} \prod_{s \in S\, t \in T}\left(\sin^2\frac{2\pi s}{q} - \sin^2\frac{2\pi t}{p}\right)
\end{aligned}$$

as S has $(q-1)/2$ members. Now interchange p and q and we have

$$(q/p) = (-4)^{(q-1)(p-1)/4} \prod_{s \in S\, t \in T}\left(\sin^2\frac{2\pi t}{p} - \sin^2\frac{2\pi s}{q}\right).$$

The right-hand sides of these two equations are identical except for

$(p-1)(q-1)/4$ minus signs and so

$$(q/p)=(-1)^{(p-1)(q-1)/4}(p/q). \qquad \square$$

Gauss's law enables us to calculate the values of Legendre symbols provided the entries are prime. For example consider $(257/269)$. We have, as $257 \equiv 1 \pmod 4$ and 257 and 269 are prime,

$$\begin{aligned}(257/269)&=(269/257)=(12/257)=(3/257)\\&=(257/3)=(2/3)=-1,\end{aligned}$$

using Theorems 1.3, 1.5, and the reciprocity law.

Finally in this section we give an equivalent formulation of the law. Up to this point we have fixed the modulus q and found those primes p for which

$$x^2 \equiv p \pmod q \tag{6}$$

is soluble. Now consider the reverse problem: fix p and ask for which q the congruence (6) is soluble; Theorem 1.5 deals with the case $p=2$.

THEOREM 2.2 *Let p and q be distinct odd primes and $a \geq 1$. The quadratic reciprocity law is equivalent to:*

$$\textit{if } p \equiv \pm q \pmod{4a} \quad \textit{then} \quad (a/p)=(a/q). \tag{7}$$

Proof. Assume Theorem 2.1. It is sufficient to prove (7) when *a is an odd prime*, for if it holds in this case then the general case follows by unique factorization, Problem 1(i), Chapter 3, Theorem 1.3(i), and Theorem 1.5. If $p \equiv q \pmod{4a}$ then $(p/a)=(q/a)$, and this equation and the law give

$$\begin{aligned}(a/p)&=(-1)^{(p-1)(a-1)/4}(q/a)=(-1)^{(p-1)(a-1)/4}(-1)^{(q-1)(a-1)/4}(a/q)\\&=(-1)^{(a-1)(p+q-2)/4}(a/q).\end{aligned}$$

Now as $p=q+4at$ (for some t), we have $p+q-2=2(q-1+2at)\equiv 0 \pmod 4$ as q is odd. Hence (7) holds in this case. Similarly if $p \equiv -q \pmod{4a}$ we have

$$(a/p)=(-1)^{(a-1)(p+q)/4}(a/q)$$

and $p+q \equiv 0 \pmod 4$.

For the converse assume that (7) holds, and that $p>q$ and $p \equiv q \pmod 4$, so $p=q+4a$ for some $a \geq 1$. We have, using (7) and Theorems 1.3 and 1.5,

$$\begin{aligned}(p/q)&=(q+4a/q)=(a/q)=(a/p)=(4a/p)=(p-q/p)\\&=(-q/p)=(-1)^{(p-1)/2}(q/p).\end{aligned}$$

Hence if $p \equiv 1 \pmod 4$ then $(p-1)/2$ is even and $(p/q)=(q/p)$, and if $p \equiv 3 \pmod 4$ then $q \equiv 3 \pmod 4$ and $(p/q)=-(q/p)$, so the law holds in these cases. Finally suppose $p \equiv -q \pmod 4$, then $p + q = 4a$ for some a and

$$(p/q) = (-q + 4a/q) = (a/q) = (a/p) = (4a/p)$$
$$= (p + q/p) = (q/p).$$

This completes the proof. □

Our two theorems enable us to calculate (a/b) for fixed primes p. For example we have

$$(3/p) = 1 \quad \text{if and only if} \quad p \equiv \pm 1 \pmod{12}.$$

For by Theorem 2.2 $p \equiv \pm q \pmod{12}$ implies $(3/p) = (3/q)$, provided p and q are odd primes. The result follows by noting that $(3/5) = (3/7) = -1$ and $(3/11) = (3/13) = 1$.

4.3 Some further topics

Here we shall consider some extensions of the Legendre symbol and, to begin with, the problem of the distribution of quadratic residues and non-residues. Fix a prime p $(p > 2)$ and consider the set

$$P = \{1, 2, \dots, p-1\}.$$

Let P_r denote the set of quadratic residues $(\text{mod}\, p)$ in P and let $P_n = P \backslash P_r$. Using the results proved so far we have:

(i) $1 \in P_r$ and if a is a square then $a \in P_r$;

(ii) the number of elements in P_r equals the number of elements in P_n (Lemma 1.1);

(iii) if $p \equiv 1 \pmod 4$, $a \in P_r$ if and only if $p - a \in P_r$; and if $p \equiv 3 \pmod 4$, $a \in P_r$ if and only if $p - a \in P_n$, and vice versa [Theorem 1.3(iv)].

Very little is known about the precise distribution of quadratic residues and non-residues in the first half of P. By (i) the maximum number of elements of P_n between successive elements of P_r is $2(\sqrt{p}) + 1$. In 1926 Vinogradov made the

Conjecture Let $\varepsilon > 0$. (i) The number of integers lying between successive elements of P_n is bounded by cp^{ε} for some constant c depending only on ε.

(ii) If $N(p)$ denotes the smallest element of P_n then

$$\lim_{p \to \infty} N(p)/p^{\varepsilon} = 0.$$

Burgess (1957) has shown that, for large p,

$$N(p) < p^{\frac{1}{4}+\varepsilon}$$

for any positive ε, and it is known that $N(p) = c_1 (\ln p)^2$ if the Riemann hypothesis holds. A lower estimate is possible but we do have

$$N(p) > c_2 \ln p$$

for infinitely many p (Salié, 1949). (c_1 and c_2 are constants.) Some further details are given in Gelfond and Linnik (1965).

Using a theorem from Chapter 2, we can prove a weaker version of Burgess's result.

THEOREM 3.1 *$N(p) < \sqrt{p}$ for large p.*

Proof. We have the following:

(i) To each pair $\{x, y\}$ satisfying

$$0 < x,\ y < \sqrt{p} \quad \text{and} \quad (x, y) = 1 \tag{1}$$

there corresponds a unique $z \pmod{p}$ with the property

$$x \equiv yz \pmod{p}. \tag{2}$$

For if $x_1 \equiv y_1 z$ and $x_2 \equiv y_2 z \pmod{p}$ then $x_1 y_2 \equiv x_2 y_1 \pmod{p}$ and (1) implies $x_1 = x_2$ and $y_1 = y_2$.

(ii) The number of pairs satisfying (1) is given by

$$1 + 2 \sum_{n=2}^{\sqrt{p}} \phi(n) = 2 \sum_{n=1}^{\sqrt{p}} \phi(n) - 1,$$

(the pair $\{1, 1\}$ is counted separately).

(iii) We have by Theorem 2.5, Chapter 2,

$$\begin{aligned} 2 \sum_{n=1}^{\sqrt{p}} \phi(n) - 1 &= \frac{6p}{\pi^2} + O(\sqrt{p} \ln p) \\ &> \frac{p}{2} \qquad \text{for large } p. \end{aligned}$$

Now using (i), (ii), and (iii) we have, for large p, the number of different residues classes z satisfying (i) is at least $(p+1)/2$. But as there are only $(p-1)/2$ quadratic residues modulo p it follows that at least one of the residue classes z, z_0 say, must be a quadratic non-residue, and so $(z_0/p) = -1$. Now if $\{x_0, y_0\}$ is the pair corresponding to z_0, we have by (2) either $(x_0/p) = -1$ or $(y_0/p) = -1$ and the result follows. □

Using tables of quadratic residues it can be shown that this result holds for all p except 2, 3, 7, and 23. Further results giving some information about the distribution of quadratic residues can be found in Problems 10 and 18, and Problem 21 of Chapter 10.

The Jacobi symbol

The Legendre symbol is only defined when the second argument is an odd prime, this can be a considerable disadvantage if large numbers are involved. Jacobi introduced an extension of the symbol with odd integer second arguments and Kronecker introduced another extension in which the second argument can be any positive integer but some restriction is required on the first argument. We shall discuss the main properties of both of these symbols. The Jacobi symbol has a number of useful properties whereas Kronecker's symbol is mainly used in quadratic form theory and related questions concerning quadratic fields. Note that the symbols agree with one another on the intersections of their domains.

Definition The Jacobi symbol. Let b be a positive odd integer with prime factorization $\prod_{i=1}^{r} p_i$. For $a \in \mathbb{Z}$ we define $(a/1) = 1$ and

$$(a/b) = (a/p_1) \dots (a/p_r),$$

a product of Legendre symbols.

The Jacobi symbol reduces the amount of calculation in congruence problems. To show that

$$x^2 \equiv a \pmod{b} \qquad (*)$$

is not soluble it is sufficient to prove that $(a/b) = -1$, but if $(a/b) = 1$ we cannot assume that $(*)$ has a solution. Each of the congruences $x^2 \equiv a \pmod{p}$ for $p \mid b$ must be soluble for $(*)$ to have a solution.

The main properties of the Jacobi symbol are given by

Theorem 3.2 *If b, b', b_1, and b_2 are positive and odd then*

(i) $(a_1 a_2/b) = (a_1/b)(a_2/b)$;
(ii) $(a/b_1 b_2) = (a/b_1)(a/b_2)$;
(iii) *if* $a_1 \equiv a_2 \pmod{b}$ *then* $(a_1/b) = (a_2/b)$;
(iv) $(-1/b) = (-1)^{(b-1)/2}$;
(v) $(2/b) = (-1)^{(b^2-1)/8}$;
(vi) $(b/b')(b'/b) = (-1)^{(b-1)(b'-1)/4}$.

Proof. (i), (ii), and (iii) follow from the definition and the corresponding property of the Legendre symbol.

(iv) Note first that if m and n are odd then $(m-1)(n-1) \equiv 0 \pmod{4}$ and so

$$\frac{mn-1}{2} \equiv \frac{m-1}{2} + \frac{n-1}{2} \pmod{2}.$$

Now if $b = \prod_{i=1}^{r} p_i$,

$$(-1/b) = \prod_{i=1}^{r} (-1/p_i) = \prod_{i=1}^{r} (-1)^{(p_i-1)/2}$$
$$= (-1)^{\sum_{i=1}^{r} (p_i-1)/2} = (-1)^{(b-1)/2}$$

using the congruence above $r-1$ times.

(v) This is similar to (iv) for if m and n are odd we have

$$\frac{m^2n^2-1}{8} \equiv \frac{m^2-1}{8} + \frac{n^2-1}{8} \pmod 2.$$

(vi) Let $b = p_1 \ldots p_r$ and $b' = q_1 \ldots q_s$ (p_i and q_j prime) then by the quadratic reciprocity law we have

$$(b/b')(b'/b) = \prod_{i=1}^{r} \prod_{j=1}^{s} (p_i/q_j)(q_j/p_i) = (-1)^{\sum_{i=1}^{r} \sum_{j=1}^{s} (p_i-1)(q_j-1)/4}$$
$$= (-1)^{\sum_{i=1}^{r} (p_i-1)/2 \sum_{j=1}^{s} (q_j-1)/2}$$
$$= (-1)^{(b-1)(b'-1)/4}$$

using the congruence in (iv) again. □

Let us consider two examples using the Jacobi symbol.

(i) Calculate (403/803). Note neither entry is prime.

$$(403/803) = -(803/403) = -(-3/403) = -(-1/403)(3/403)$$
$$= (3/403) = -(403/3) = -(1/3) = -1,$$

using Theorem 3.2.

(ii) A square is a quadratic residue of every prime, we shall show that the converse is also true. This is an instance where the Hasse principle applies, see Section 3 of Chapter 3; the result is false for nth powers and nth power residues when $8|n$ (see p. 83).

THEOREM 3.3 *An integer is a square if and only if it is a quadratic residue of every prime.*

Proof. If $a = b^2$ then $(a/p) = 1$ for all primes p. To prove the converse we construct, for all non-square a, a prime p (depending on a) such that $(a/p) = -1$. Using the Jacobi symbol it is sufficient to find an odd positive integer k such that $(a/k) = -1$. The following three cases cover all possibilities: (i) a is minus a square; (ii) the exponent of 2 in the prime factorization of a is odd; and (iii) the exponent of 2 in the prime factorization of a is even and the exponent of an odd prime factor q of a is odd.

(i) $a = -b^2$. Choose k to satisfy $k > 0$, $k \equiv 3 \pmod 4$ and $(b, k) = 1$. Then

$$(a/k) = (-b^2/k) = (-1/k) = -1.$$

(ii) $a = \pm 2^t b$, where t and b are positive and odd. Using the Chinese remainder theorem (1.8, Chapter 3) choose k so that

$$k > 0,\ k \equiv 5 \pmod 8 \quad \text{and} \quad k \equiv 1 \pmod b.$$

(If $b = 1$, let $k = 5$.) Then $(2^t/k) = (-2^t/k) = -1$ as t is odd, and

$$(b/k) = (k/b) = (1/b) = 1.$$

Hence $(a/k) = -1$ in this case.

(iii) $a = \pm 2^{2u} q^t b$, where t and b are positive and odd, and q is an odd prime such that $(b, q) = 1$. Choose k such that

$$k > 0, \qquad k \equiv 1 \pmod{4b} \quad \text{and} \quad k \equiv c \pmod q,$$

where c is a quadratic non-residue of q. Then $(2^{2u}/k) = (-2^{2u}/k) = 1$, $(b/k) = (k/b) = 1$, and, as t is odd,

$$(q^t/k) = (q/k) = (k/q) = (c/q) = -1,$$

and so $(a/k) = -1$. This completes the proof. □

The Kronecker symbol

Here we give a brief description of this symbol, it will be used in Chapters 9 and 10 when discussing quadratic forms.

Definition Let d satisfy: $d \equiv 0$ or $1 \pmod 4$, and d is not square. The Kronecker symbol (d/n) is defined when $n > 0$ by

(i) $(d/n) = 0$ if $(d, n) > 1$,
(ii) $(d/1) = 1$,
(iii) if d is odd, $(d/2) = (2/|d|)$, a Jacobi symbol, so

$$(d/2) = \begin{cases} 1 & \text{if } d \equiv 1 \pmod 8 \\ -1 & \text{if } d \equiv 5 \pmod 8), \end{cases}$$

(iv) if $n = \prod_{i=1}^{r} p_i$ then $(d/n) = \prod_{i=1}^{r} (d/p_i)$,

a product of Legendre symbols and, if n is even, the symbol $(d/2)$.

An equivalent definition is given by

THEOREM 3.4 *Let $n>0$ and $(d, n)=1$ with d as above.*

(i) *If d is odd we have*

$$(d/n)=(n/|d|),$$

a Jacobi symbol.

(ii) *If $d=2^t b$ where b is odd and $t>0$, we have*

$$(d/n)=(2/n)^t(-1)^{(|b|-1)(n-1)/4}(n/|b|),$$

a product of Jacobi symbols.

Proof. Consider the case $d>0$, $n=2^u c$ where c is odd. We have, as $d\equiv 1 \pmod 4$ in this case,

$$(d/n)=(d/2^u c)=(d/2)^u(d/c)=(2/d)^u(c/d)=(n/|d|),$$

using the definition above and properties of the Jacobi symbol. The remaining cases are similar. □

THEOREM 3.5. *The function $(d/\cdot)$ satisfies*

(i) $(d/1)=1$ *and* $(d/n)=0$ *if* $(d, n)>0$,

(ii) $(d/mn)=(d/m)(d/n)$,

(iii) *if* $m\equiv n \pmod{|d|}$ *then* $(d/m)=(d/n)$.

Proof. (i) and (ii) follow immediately from the definitions. For (iii) we may assume, by (i), that $(d, m)=(d, n)=1$. If d is odd then, by Theorem 3.4 and Theorem 3.2(iii), we have

$$(d/m)=(m/|d|)=(n/|d|)=(d/n).$$

If $d=2^t b$ (b odd) we have

$$(d/m)=(2/m)^t(-1)^{x(m-1)/2}(m/|b|),$$
$$(d/n)=(2/n)^t(-1)^{x(n-1)/2}(n/|b|),$$

where $x=(|b|-1)/2$. Now as above $(m/|b|)=(n/|b|)$. Further as $4|d$ we have $m\equiv n \pmod 4$ and so

$$(-1)^{x(m-1)/2}=(-1)^{x(n-1)/2}.$$

Finally $(2/m)^t=(2/n)^t$ by Theorem 3.2(v); note that $t\geq 2$ and if $t>2$ then $m\equiv n \pmod 8$. Now combine these equations. □

4.4 Problems 4

1. Re-prove Euler's criterion and Lemma 1.1 as follows.

(i) Let $P=\{1, 2, \ldots, p-1\}$ and suppose a is a quadratic non-residue $(\text{mod}\, p)$. Show that P can be written as a disjoint union $(p-1)/2$ pairs $\{x, y\}$

where $xy \equiv a \pmod p$. Deduce

$$(p-1)! \equiv a^{(p-1)/2} \pmod p.$$

(ii) Now suppose a is a quadratic residue $(\bmod\, p)$. Using a similar argument show that in this case

$$(p-1)! \equiv -a^{(p-1)/2} \pmod p$$

and so derive Euler's criterion.

(iii) Finaly use this to re-prove Lemma 1.1.
[Hint. Use the results of Section 2, Chapter 3.]

2. Evaluate $(313/367)$, $(367/401)$, and $(401/313)$.

3. (i) Show that $(a/p) = (b/p)$ if and only if a non-zero integer z can be found satisfying $a \equiv bz^2 \pmod p$.

(ii) Prove that if $p = 4m + 1$ (p a prime) and $d \mid m$ then $(d/p) = 1$.

(iii) Show that the congruences

$$x^6 - 11x^4 + 36x^2 - 36 \equiv 0 \pmod p$$

and

$$x^{2^\alpha} \equiv 2^{2^{\alpha-1}} \pmod p,$$

for $\alpha > 2$, are soluble for every prime p. In the first case how many solutions are there?

4. Let $v(n)$ be the number of solutions of the congruence

$$x^2 \equiv -1 \pmod n.$$

Show that

$$v(n) = \begin{cases} 0 & \text{if } 4 \mid n \text{ or a prime } p \text{ exists satisfying } p \mid n \\ & \text{and } p \equiv 3 \pmod 4, \\ 2^s & \text{if } 4 \nmid n, \text{ all odd prime divisors } p \text{ of } n \text{ satisfy} \\ & p \equiv 1 \pmod 4, \text{ and } s \text{ is the number of distinct} \\ & \text{odd prime factors of } n. \end{cases}$$

5. Prove that there is at least one prime p congruent to $7 \pmod{12}$ dividing $4m^2 + 3$ provided $3 \nmid m$. Deduce that there are infinitely many primes of the form $12n + 7$.

6. Show that all prime factors p of the integer $n^4 - n^2 + 1$ are congruent to $1 \pmod{12}$. [Hint. Factorize $4(n^4 - n^2 + 1)$ in two distinct ways.]

7. (i) Show that $(5/p) = 1$ if and only if $p \equiv \pm 1 \pmod{10}$.

(ii) Using the identity $6119 = 82^2 - 5 \cdot 11^2$ find the prime factors of this number.

(iii) As in (ii) prove that 4751 is prime.

8. Let p be an odd prime.

(i) Using Theorem 2.4, Chapter 3, show that

$$1^2 3^2 \ldots (p-2)^2 \equiv (-1)^{(p+1)/2} \pmod p.$$

(ii) Let R_p be the product of the quadratic residues of p between 1 and p, show that $R_p \equiv -1 \pmod{p}$ if $p \equiv 1 \pmod{4}$ and $R_p \equiv 1 \pmod{p}$ if $p \equiv 3 \pmod{4}$.

9. Given the result: the number of solutions of

$$x^2 - y^2 \equiv a \pmod{p}$$

is $p-1$ if $p \nmid a$ (see Problem 12, Chapter 6) show that

$$\sum_{y=0}^{p-1} (y^2 + a/p) = \begin{cases} -1 & \text{if } p \nmid a \\ p-1 & \text{if } p \mid a. \end{cases}$$

10. Let p be a prime greater than 2.

(i) Show that $\sum_{a=1}^{p-1} (a/p) = 0$.

(ii) Use Theorem 1.4, Chapter 3, to show that

$$\sum_{i=1}^{p-2} (i(i+1)/p) = -1.$$

(iii) Let N be the number of integers n such that $1 \leq n \leq p-2$ and $(n/p) = (n+1/p) = 1$. Prove that

$$N = \frac{1}{4} \sum_{i=1}^{p-2} (1 + (i/p))(1 + (i+1/p)),$$

and so deduce

$$N = \begin{cases} \dfrac{p-3}{4} & \text{if } p \equiv 3 \pmod{4} \\[2ex] \dfrac{p-5}{4} & \text{if } p \equiv 1 \pmod{4}. \end{cases}$$

☆ 11. A counting argument proof of the quadratic reciprocity law.

By counting the lattice points above and below the line $y = qx/p$ inside and on the rectangle $(0, 0)$, $(0, (q-1)/2)$, $((p-1)/2, (q-1)/2)$, $((p-1)/2, 0)$, but excluding the axes, show that

$$\sum_{s=1}^{(p-1)/2} \left[\frac{sq}{p}\right] + \sum_{t=1}^{(q-1)/2} \left[\frac{tp}{q}\right] = (p-1)(q-1)/4.$$

If k satisfies $1 \leq k \leq (p-1)/2$ we can write

$$kq = p\left[\frac{kq}{p}\right] + u_k \quad \text{or} \quad kq = p\left[\frac{kq}{p}\right] + v_k \tag{*}$$

where $1 \leq u_k, p - v_k \leq (p-1)/2$. Using the notation set up in the proof of Gauss's lemma (r_i, s_j, t etc.) prove that

(i) $\sum r_i + \sum s_j = (p^2 - 1)/8$,
(ii) $\sum u_k = \sum r_i$ and $\sum v_k = tp - \sum s_j$,
(iii) summing (*) over k deduce that

$$(p^2 - 1)(q-1)/8 = p \sum_{s=1}^{(p-1)/2} \left[\frac{sq}{p}\right] + 2 \sum v_k - tp,$$

(iv) re-prove Gauss's law. (See Hardy and Wright, 1954, pp.. 76–78.)

☆12. A trigonometrical identity proof of the quadratic reciprocity law using the complex numbers.

(i) Suppose $f(z) = 2i \sin 2\pi z = e^{2\pi i z} - e^{-2\pi i z}$. Show that $f(z+1) = f(z)$, $f(-z) = -f(z)$, and if r is real and $2r \notin \mathbb{Z}$ then $f(r) \neq 0$.

(ii) Using the identity (with $n > 0$)

$$x^n - y^n = \prod_{k=0}^{n-1} (xe^{2\pi i k/n} - ye^{-2\pi i k/n}),$$

show that (i) gives

$$\frac{f(nz)}{f(z)} = \prod_{k=1}^{(n-1)/2} f\left(z + \frac{k}{n}\right) f\left(z - \frac{k}{n}\right).$$

(iii) If p is an odd prime, $a \in \mathbb{Z}$ and $p \nmid a$, show that, using Gauss's lemma and (i),

$$\prod_{j=1}^{(p-1)/2} f\left(\frac{ja}{p}\right) = (a/p) \prod_{j=1}^{(p-1)/2} f\left(\frac{j}{p}\right).$$

(vi) Re-prove Gauss's law in a similar manner to our second proof (see Ireland and Rosen, 1982).

13. Let $t > 2$ and $m \equiv 1 \pmod 8$, show that $x^2 \equiv m \pmod{2^t}$ has four solutions. Hence calculate the number of solutions of $x^2 \equiv n \pmod{p^s}$ for all primes p and positive integers s where $p \nmid n$.

14. Let aRn denote 'there exists x such that $x^2 \equiv a \pmod n$' and let $v(a, n)$ denote the number of solutions of this congruence. Show that if $(m, n) = 1$ then aRm and aRn imply $aRmn$, and $v(a, mn) = v(a, m)v(a, n)$.

15. Let $p^\alpha \parallel n$ stand for $p^\alpha | n$ and $p^{\alpha+1} \nmid n$. Suppose $n > 0$ and $(t, n) = 1$. Using the previous two problems show that $v(t, n) = 0$ if $4 \parallel n$ and $n \not\equiv 1 \pmod 4$, or if $8 | n$ and $n \not\equiv 1 \pmod 8$, or if a prime p exists such that $p > 2$, $p | n$, and $(t/p) = -1$; otherwise if s be the number of distinct odd prime divisors of n then

$$v(t, n) = \begin{cases} 2^s & \text{if } 4 \nmid n \\ 2^{s+1} & \text{if } 4 \parallel n \\ 2^{s+2} & \text{if } 8 | n. \end{cases}$$

(Note, $v(-1, n) = v(n)$ see Problem 4.)

Using the Kronecker symbol deduce that if $d \equiv 0, 1 \pmod 4$ and $(d, k) = 1$ then

$$v(d, 4k) = 2 \sum_{u|k}' (d/u),$$

where the sum is taken over all positive square-free divisors u of k.

16. Justify the following deduction using Theorem 3.2(i)–(iv) and (vi) only.

$$(2/p) = (8 - p/p) = (p/p - 8) = (2/p - 8).$$

This shows that the formula for $(2/p)$ can be derived without using Gauss's lemma. Note that (vi) can also be proved without the use of Gauss's lemma.

17. Show that Theorem 3.3 can be strengthened by replacing the words 'for every prime' by 'for all but finitely many primes'.

18. Let p be a prime congruent to 3 (mod 4). Using the results

(i) $\displaystyle\sum_{n=1}^{\infty} \frac{(2n-1/p)}{2n-1}$ converges and is positive (see Chapter 10),

(ii) $\displaystyle\sum_{n=1}^{p-1} (n/p)\sin(2\pi n/p) = \sqrt{p}$ (see Chapter 6),

(iii) $\displaystyle\sum_{n=1}^{\infty} \frac{\sin(2n-1)x}{2n-1} = \begin{cases} \pi/4 & \text{if } 0<x<\pi \\ -\pi/4 & \text{if } \pi<x<2\pi, \end{cases}$

show that

$$\sum_{n \text{ odd}} \frac{(n/p)}{n} = \frac{1}{\sqrt{p}} \sum_{t=1}^{p-1} (t/p) \sum_{n \text{ odd}} \frac{\sin(2\pi t n/p)}{n} = \frac{\pi}{2\sqrt{p}} \sum_{t=1}^{(p-1)/2} (t/p).$$

Deduce that there are more quadratic residues than non-residues in the interval $[1, (p-1)/2]$.

19. By factorizing $x^4 + 4$, show that $x^4 \equiv -4 \pmod{p}$ is soluble if and only if $p \equiv 1 \pmod 4$.

☆20. Let $p \equiv 1 \pmod 4$ throughout this problem. A famous result of Dirichlet states: $x^4 \equiv 2 \pmod{p}$ is soluble if and only if a and b can be found so that $p = a^2 + 64b^2$. Using

(i) $x^4 \equiv 2 \pmod{p}$ is soluble if and only if $2^{(p-1)/4} \equiv 1 \pmod{p}$, see Theorem 2.8, Chapter 5,

(ii) integers a and b exist such that $p = a^2 + b^2$, see Section 1, Chapter 6,

prove the following in turn.

(iii) If $p = a^2 + b^2$ and $2 \nmid a$ then $(a/p) = 1$.

(iv) Using $2p = (a+b)^2 + (a-b)^2$ show that

$$(a+b/p) = (-1)^{((a+b)^2-1)/8}.$$

(v) If $b \equiv af \pmod{p}$ then $f^2 \equiv -1 \pmod{p}$ and

$$2^{(p-1)/4} \equiv f^{ab/2} \pmod{p}.$$

(vi) Prove Dirichlet's result.

21. For the Kronecker symbol show that, given d, there is an n such that $(d/n) = -1$. Treat the cases (i) d odd, (ii) $\exp(d, 2)$ is odd, and (iii) $\exp(d, 2)$ is even, separately. [$\exp(d, 2)$ is the exponent of 2 in the prime factorization of d.]

22. Using the Kronecker symbol show that $(d/|d|-1) = \operatorname{sgn}(d)$ where $\operatorname{sgn}(d) = 1$ if $d>0$ and $\operatorname{sgn}(d) = -1$ if $d<0$. Deduce that if $m>0$, $n>0$ and $m + n \equiv 0 \pmod{|d|}$ then $(d/n) = (d/m)\operatorname{sgn}(d)$.

5

ALGEBRAIC TOPICS

In this chapter we introduce three topics with an algebraic flavour which will be useful later: algebraic numbers and integers, primitive roots, and characters. Although not closely connected it will be convenient to consider them now.

Algebraic number fields will not be treated in detail; a good introduction can be found in Stewart and Tall (1979). We shall give the basic definitions and some examples of particular number fields. Algebraic number theory can be studied in two ways. First, as a subject in its own right emphasizing the algebraic aspect but it is not our purpose to do this. Secondly, some problems relating to the rational numbers are best treated in this more general context. For example, in Chapter 14 we shall prove that the equation

$$x^3 + y^3 = z^3$$

has no rational solutions by showing that it has no solutions in a field extension of $\mathbb{Q}$ in which $x^3 + y^3$ can be factorized into linear factors.

In Chapter 3 we defined $(\mathbb{Z}/m\mathbb{Z})^*$, the set of reduced residue classes modulo m, and showed that it forms a finite Abelian group under multiplication $(\text{mod } m)$. For some m this group is cyclic; that is a number g, called a primitive root, can be found so that the powers of g generate the reduced system $(\text{mod } m)$. Gauss was the first to show that primitive roots exist for all odd prime power moduli. A weaker result holds for powers of 2 but, except for 2 and 4, primitive roots do not exist in this case. Gauss's result has many applications in number theory. For instance we shall use it to introduce nth power residues. We shall also mention some extensions of this work to general finite fields.

Characters, our third topic, are mappings from systems of residues into the complex numbers: the Legendre symbol is an example. We shall develop their basic properties using primitive roots. Characters have applications in the theory of Gauss sums (Chapter 6) and in the proof of Dirichlet's theorem (Chapter 13) where they are used to codify the congruence relation. They also have many uses in group representation theory.

5.1 Algebraic numbers and integers

We shall make use of the standard algebraic properties of the fields $\mathbb{Q}$ (rational numbers), $\mathbb{R}$ (real numbers) and $\mathbb{C}$ (complex numbers). We shall also use some of the elementary properties of polynomial rings defined over fields; these will be sketched now.

Let K be a field. The set of one variable polynomials defined over K with the usual addition and multiplication operations is a commutative ring with identity denoted by $K[x]$. If $f(x) = a_0 + a_1x + \ldots + a_nx^n \in K[x]$ and $a_n \neq 0$ then n is called the *degree* of f and is denoted by $\deg f$ (note that non-zero constant polynomials have degree 0). As noted on p. 39, $K[x]$ has a division algorithm given by: let $f, g \in K[x]$, where $g \neq 0$, then polynomials r and s can be found in $K[x]$ to satisfy

$$f = gr + s \qquad \text{where } \deg s < \deg g \text{ or } s = 0.$$

If $s = 0$ we write $g \mid f$, that is, g divides f. Further $K[x]$ has a Euclidean algorithm given by

LEMMA 1.1 *Given $f, g \in K[x]$, where $\deg f$ and $\deg g$ are not both zero, polynomials t, u, and d can be found in $K[x]$ to satisfy d is monic, $t \neq 0 \neq u$ and*

$$ft + gu = d.$$

d is the unique monic polynomial of highest degree which divides both f and g.

Either of the derivations of the Euclidean algorithm for $\mathbb{Z}$ given in Chapter 1 can easily be adapted to prove this lemma, see Problem 1. As an example we see that if $f(x) = 2x^2 + 3$ and $g(x) = 5x$ then $d = 1$ and

$$(2x^2 + 3)\left(\frac{1}{3}\right) + (5x)\left(\frac{-2x}{15}\right) = 1.$$

A polynomial $f \in K[x]$ is called *irreducible* over K if no polynomials g and h, both with positive degree, exist in $K[x]$ satisfying $f = gh$. Note that if $f \in \mathbb{C}[x]$ and f is irreducible over $\mathbb{C}$ then the roots of the equation $f(x) = 0$ are distinct, see Problem 2.

We begin the main part of this section with a

DEFINITION If $\alpha \in \mathbb{C}$ and rational numbers $a_1, \ldots, a_n$ exist satisfying

$$\alpha^n + a_1\alpha^{n-1} + \ldots + a_n = 0 \qquad (*)$$

then α is called an *algebraic number*. If $a_i \in \mathbb{Z}$ for all $i = 1, \ldots, n$, then α is called an *algebraic integer*.

When a discussion involves both algebraic integers and elements of $\mathbb{Z}$, we use the term *rational integer* when referring to members of the latter class. Using the definition we note:

(i) If $\alpha \in \mathbb{Q}$ and α is an algebraic integer then $\alpha \in \mathbb{Z}$. For suppose $\alpha = r/q$, where $r, q \in \mathbb{Z}$ and $(r, q) = 1$, and further α satisfies a polynomial equation of the form $(*)$ where $a_1, \ldots, a_n \in \mathbb{Z}$, then we have $r^n + a_1 r^{n-1} q + \ldots + a_n q^n = 0$. This is impossible if $q > 1$ as q does not divide the first term.

(ii) If α is an algebraic number than $c \in \mathbb{Z}$ exists so that $c\alpha$ is an algebraic integer. [c = LCM of denominators of $a_1, \ldots, a_n$ in $(*)$.]

(iii) If α and $a_0\alpha^k + \ldots + a_k$ are algebraic integers, where $a_0, \ldots, a_k \in \mathbb{Q}$, it is not necessary for each a_i to be a rational integer. For example $(1 + \sqrt{5})/2$ is an algebraic integer as it satisfies the equation $x^2 - x - 1 = 0$.

A complex number which is not algebraic is called *transcendental*. Examples are e, π, $\ln 2$ and $2^{\sqrt{2}}$, see Chapter 8.

The set of algebraic numbers forms a field (denoted by $\mathbb{A}$) and the set of algebraic integers forms an integral domain (denoted by $\mathbb{I}$), these facts follow from

LEMMA 1.2 *Let $\beta_1, \ldots, \beta_m$ be algebraic numbers, not all zero, and let V denote the finite dimensional vector space over $\mathbb{Q}$ generated by the set $\{\beta_1, \ldots, \beta_m\}$. Suppose $\alpha \in \mathbb{C}$ and $\alpha\beta \in V$ whenever $\beta \in V$, then α is an algebraic number.*

Proof. We can find rational numbers a_{ij} so that, for $i = 1, \ldots, m$,

$$\alpha\beta_i = \sum_{j=1}^{m} a_{ij}\beta_j.$$

So $\sum_{j=1}^{m} (a_{ij} - \delta_{ij}\alpha)\beta_j = 0$, where $\delta_{ij} = 0$ if $i \neq j$, and $\delta_{ii} = 1$. Hence, as this set of homogeneous linear equations has a solution, $\det(a_{ij} - \delta_{ij}\alpha) = 0$. This is a polynomial equation for α with rational coefficients, and so $\alpha \in \mathbb{A}$. □

LEMMA 1.3 *Let $\gamma_1, \ldots, \gamma_m$ be algebraic integers, not all zero, and let W be the set of sums $\sum_{i=1}^{m} c_i\gamma_i$, where $c_i \in \mathbb{Z}$, with component-wise addition and scalar (from $\mathbb{Z}$) multiplication. W is sometimes called a $\mathbb{Z}$-module. Suppose $\alpha \in \mathbb{C}$ and $\alpha\gamma \in W$ for all $\gamma \in W$, then α is an algebraic integer.*

Proof. We can use the same argument as above. Note that division is not involved in the determinant equation, and the leading coefficient of this equation is ± 1 as it arises from the diagonal entry of the determinant. □

THEOREM 1.4 (i) *The set of algebraic numbers with the operations of complex addition, subtraction, multiplication, and division is a field denoted by $\mathbb{A}$.*

(ii) *The set of algebraic integers with the same operations (except division) is an integral domain denoted by* $\mathbb{I}$.

Proof. (i) Let α_1 and α_2 be algebraic numbers where α_1, α_2 satisfy polynomial equations of degree m and n respectively, and let V denote the vector space generated by the elements $\alpha_1^i\alpha_2^j = \gamma_{ij}$ for $0 \le i < m$ and $0 \le j < n$. Using the polynomial equations for α_1 and α_2 we have $\alpha_k\gamma_{ij} \in V$, for $k = 1, 2$, $(\alpha_1 + \alpha_2)\gamma_{ij} \in V$ and $\alpha_1\alpha_2\gamma_{ij} \in V$ for $0 \le i < m$ and $0 \le j < n$. Hence by Lemma 1.2, $\alpha_1 + \alpha_2$ and $\alpha_1\alpha_2$ are algebraic numbers. Further, if α satisfies

$$\alpha^n + a_1\alpha^{n-1} + \ldots + a_n = 0,$$

where $a_n \neq 0$, then

$$\alpha^{-n} + \frac{a_{n-1}}{a_n}\alpha^{-n+1} + \ldots + \frac{1}{a_n} = 0,$$

and so α^{-1} is also an algebraic number. Finally note that the remaining properties follow because $\mathbb{A}$ is a subset of $\mathbb{C}$.

(ii) This proof is similar to the above using Lemma 1.3. □

The 'degree' of an algebraic number will be defined below using

THEOREM 1.5 *If α is an algebraic number then there is a unique monic irreducible polynomial f with the property $f(\alpha) = 0$.*

Proof. As α is algebraic it satisfies a polynomial equation $f(\alpha) = 0$ where f is monic and has minimum degree. (Note, this implies that f is irreducible.) Suppose α also satisfies the polynomial equation $g(\alpha) = 0$ and $f \nmid g$, then, by Lemma 1.1, polynomials t and u exist such that $ft + gu = 1$. This is impossible as $f(\alpha) = g(\alpha) = 0$, hence $f \mid g$ and the result follows. □

Note that the polynomial f given by the above result has distinct roots in $\mathbb{C}$, see Problem 2.

DEFINITION Let α be an algebraic number and let f be the polynomial for α given by Theorem 1.5.

(i) f is called the *minimum polynomial* for α.

(ii) The *degree* of α is the degree of f.

(iii) A *conjugate* of α is any root of the equation $f(x) = 0$.

Note that, as f is irreducible, all conjugates are distinct.

The algebraic number fields, which are subfields of $\mathbb{A}$, are given by

DEFINITION Let α be an algebraic number of degree n. Let $\mathbb{Q}(\alpha)$ denote the subset of $\mathbb{A}$ consisting of elements of the form

$$a_0 + a_1\alpha + \ldots + a_{n-1}\alpha^{n-1}, \qquad (*)$$

where $a_i \in \mathbb{Q}$ for $i = 0, \ldots, n-1$, with the operations (sum and product etc.) of the complex field $\mathbb{C}$.

THEOREM 1.6 *$\mathbb{Q}(\alpha)$ is a field.*

Proof. Let f be the minimum polynomial for α. It can be used to express α^t in the form $(*)$ for all positive integers t. Further suppose $g(\alpha)$ is an expression of the form $(*)$ where $g(\alpha) \neq 0$. It follows that f and g have no common factors (except 1) and so, by Lemma 1.1, polynomials t and u exist such that $ft + gu = 1$. This gives $1/g(\alpha) = u(\alpha)$, and so $\mathbb{Q}(\alpha)$ is closed under division. The result follows from these two properties. □

The field $\mathbb{Q}(\alpha)$ is called an *algebraic number field* of *degree n* over $\mathbb{Q}$. We can repeat this process and form $\mathbb{Q}(\alpha, \beta) = \mathbb{Q}(\alpha)(\beta)$, the definition is as above except that $\mathbb{Q}$ is replaced by $\mathbb{Q}(\alpha)$. This is unnecessary because it can be shown that a single algebraic number θ suffices. We have, if $\theta = \alpha + c\beta$ for some suitably chosen $c \in \mathbb{Q}$, $\mathbb{Q}(\alpha, \beta) = \mathbb{Q}(\theta)$; see Stewart and Tall (1979, p. 40).

Later we shall use the notion of the 'norm' of an algebraic number, this is given by

DEFINITION (i) Let $\alpha_{(1)}, \ldots, \alpha_{(n)}$ be the conjugates of an nth-degree algebraic number α, where $\alpha_{(1)} = \alpha$. Let $\beta \in \mathbb{Q}(\alpha)$, so $\beta = a_0 + \ldots + a_{n-1}\alpha^{n-1}$ where $a_0, \ldots, a_{n-1} \in \mathbb{Q}$. The *field conjugates* β_i of β, $i = 1, \ldots, n$, are the numbers of the form

$$\beta_i = a_0 + a_1\alpha_{(i)} + \ldots + a_{n-1}\alpha_{(i)}^{n-1}.$$

(ii) If $\beta \in \mathbb{Q}(\alpha)$ then the *norm* of β, $N\beta$, is given by

$$N\beta = \beta_1\beta_2 \ldots \beta_n.$$

Note. The field conjugates need not all be distinct, for instance if β is rational then its field conjugates are n copies of β. Also a field conjugate of $\beta \in \mathbb{Q}(\alpha)$ need not belong to $\mathbb{Q}(\alpha)$. For example if $\alpha = 2^{1/3}$ then its field conjugates in $\mathbb{Q}(\alpha)$ are $2^{1/3}$, $(-1 + \sqrt{-3})/2^{2/3}$ and $(-1 - \sqrt{-3})/2^{2/3}$, clearly the second and third of these are complex and so cannot belong to the real field $\mathbb{Q}(\alpha)$. Note also, by the symmetric function theorem ((ii) on p. viii), $N\alpha \in \mathbb{Q}$ for all α in K. In our example

$$N\alpha = 2^{1/3}(-1 + \sqrt{-3})(-1 - \sqrt{-3})/2^{4/3} = 2 \in \mathbb{Q}.$$

THEOREM 1.7 (i) *The set of field conjugates of $\beta \in \mathbb{Q}(\alpha)$ consists of one or several copies of the set of conjugates of β.*

(ii) *If $\beta \in \mathbb{Q}(\alpha)$ then the degree of β divides the degree of α.*

(iii) *If $\beta \in \mathbb{Q}(\alpha)$ and β is an algebraic integer then $N\beta \in \mathbb{Z}$.*

(iv) *If $\beta, \gamma \in \mathbb{Q}(\alpha)$ then $N\beta\gamma = N\beta N\gamma$.*

We shall not prove this result, proofs are given in the references quoted below and the particular cases we need can be derived easily.

Algebraic integers

Given a field $K = \mathbb{Q}(\alpha)$, let O_K denote the set of algebraic integers in K. This set is closed under sums and products and so is an integral domain which is called the *ring of integers* in K. If α is an algebraic integer, we also define $\mathbb{Z}[\alpha]$ to be the set of polynomial expressions $f(\alpha)$ where each f is a polynomial with rational integer coefficients. Each element of $\mathbb{Z}[\alpha]$ is an algebraic integer, and so $\mathbb{Z}[\alpha] \subseteq O_K$. Note that this inclusion can be strict. For example if $\alpha = \sqrt{5}$ then $\mathbb{Z}[\alpha] = \{a + b\sqrt{5} : a, b \in \mathbb{Z}\}$, but the number $(1 + \sqrt{5})/2$ is an integer in $K = \mathbb{Q}(\alpha)$ and so belongs to O_K but not to $\mathbb{Z}[\alpha]$.

We shall prove a result which characterizes the ring of integers in $K = \mathbb{Q}(\alpha)$. By definition K has a rational basis ($\mathbb{Q}$-basis), that is there exist algebraic numbers $\gamma_1, \ldots, \gamma_n$ such that every element $\beta \in K$ has the representation

$$\beta = a_1\gamma_1 + \ldots + a_n\gamma_n \qquad (*)$$

where $a_i \in \mathbb{Q}$ and K has degree n. (K is a vector space, of dimension n, over $\mathbb{Q}$ with basis $\{\gamma_1, \ldots, \gamma_n\}$.) We show that O_K has a $\mathbb{Z}$-basis, that is every element of O_K can be represented as in $(*)$ except that $a_i \in \mathbb{Z}$ and γ_i is an algebraic integer for $i = 1, \ldots, n$. We begin with a

DEFINITION Let K be an algebraic number field of degree n and let $\gamma_1, \ldots, \gamma_n$ belong to K. We define

$$\Delta(\gamma_1, \ldots, \gamma_n) = [\det(\gamma_{i,j})]^2$$

where $\gamma_{i,j}$ is the jth field conjugate of γ_i.

LEMMA 1.8 *If $\{\gamma_1, \ldots, \gamma_n\}$ is a $\mathbb{Q}$-basis for K then $\Delta(\gamma_1, \ldots, \gamma_n)$ is a non-zero rational number.*

Proof. If $K = \mathbb{Q}(\alpha)$ then $\{1, \alpha, \alpha^2, \ldots, \alpha^{n-1}\}$ is a $\mathbb{Q}$-basis for K, so there exist rational numbers c_{ij} such that, for $1 \le i \le n$,

$$\gamma_i = c_{i1} + c_{i2}\alpha + \ldots + c_{in}\alpha^{n-1}$$

and, taking conjugates (the conjugate of a rational number is itself),

$$\gamma_{i,j} = c_{i1} + c_{i2}\alpha_j + \ldots + c_{in}\alpha_j^{n-1}.$$

Hence

$$\Delta(\gamma_1, \ldots, \gamma_n) = [\det(c_{ij})]^2\Delta(1, \alpha, \ldots, \alpha^{n-1}).$$

Now $\det(c_{ij})$ is rational (as each $c_{ij} \in \mathbb{Q}$) and non-zero (for otherwise $\{\gamma_1, \ldots, \gamma_n\}$ would not be a $\mathbb{Q}$-basis). Further, as $\Delta(1, \alpha, \ldots, \alpha^{n-1})$ is the square of a Vandermonde determinant (see (i) on p. viii), we have

$$\Delta(1, \alpha, \ldots, \alpha^{n-1}) = \left[\prod_{1 \le i < j < n} (\alpha_j - \alpha_i)\right]^2.$$

This product is non-zero (as the conjugates of α are distinct) and is a rational number (because it is a symmetric function in the roots of a polynomial equation with rational coefficients). The lemma follows. □

THEOREM 1.9 *Suppose* $\{\gamma_1, \ldots, \gamma_n\}$ *is a* $\mathbb{Q}$*-basis for* $K = \mathbb{Q}(\alpha)$ *such that each* γ_i $(i = 1, \ldots, n)$ *is an algebraic integer and* $|\Delta(\gamma_1, \ldots, \gamma_n)|$ *is minimal, then* $\{\gamma_1, \ldots, \gamma_n\}$ *is also a* $\mathbb{Z}$*-basis for* O_K.

Proof. Note first that $K = \mathbb{Q}(\alpha)$ has at least one $\mathbb{Q}$-basis by definition. Also, by note (ii) on p. 73, we may assume that the elements of a $\mathbb{Q}$-basis $\{\beta_1, \ldots, \beta_n\}$ are algebraic integers. Hence $|\Delta(\beta_1, \ldots, \beta_n)|$ is a positive rational integer by Lemma 1.8 and note (i) on p. 73. Thus a $\mathbb{Q}$-basis $\{\gamma_1, \ldots, \gamma_n\}$ with minimal integral $|\Delta|$ exists. Suppose it is not a $\mathbb{Z}$-basis, then there is a $\delta \in O_K$ such that

$$\delta = c_1\gamma_1 + \ldots + c_n\gamma_n$$

and at least one c_i, c_1 say, is not integral. Let $c_1 = [c_1] + t$, where $0 < t < 1$. We can choose a new $\mathbb{Q}$-basis $\{\xi_1, \ldots, \xi_n\}$ as follows:

$$\xi_1 = \delta - [c_1]\gamma_1 \quad \text{and} \quad \xi_i = \gamma_i \quad \text{for } 2 \le i \le n.$$

The matrix transforming the first of these bases to the second is

$$T = \begin{pmatrix} t & c_2 & \ldots & c_n \\ 0 & 1 & \ldots & 0 \\ \cdot & \cdot & \ldots & \cdot \\ 0 & 0 & \ldots & 1 \end{pmatrix}$$

and so, as in the lemma above,

$$|\Delta(\xi_1, \ldots, \xi_n)| = (\det T)^2\, |\Delta(\gamma_1, \ldots, \gamma_n)| = t^2\, |\Delta(\gamma_1, \ldots, \gamma_n)|.$$

As $0 < t < 1$, this contradicts the minimality of $|\Delta(\gamma_1, \ldots, \gamma_n)|$ and the theorem is proved. □

The unique number $\Delta(\gamma_1, \ldots, \gamma_n)$ given by this result is called the *discriminant* of the field $\mathbb{Q}(\alpha)$, it is an important invariant for the field.

EXAMPLE Consider $K = \mathbb{Q}(2^{1/3})$. In this case $\{1, 2^{1/3}, 2^{2/3}\}$ is a $\mathbb{Z}$-basis for O_K, and $O_K = \mathbb{Z}[2^{1/3}]$. The discriminant is given by

$$\det\begin{pmatrix} 1 & 2^{1/3} & 2^{2/3} \\ 1 & (-1+\sqrt{-3})/2^{2/3} & (-1+\sqrt{-3})^2/2^{4/3} \\ 1 & (-1-\sqrt{-3})/2^{2/3} & (-1-\sqrt{-3})^2/2^{4/3} \end{pmatrix}^2 = -108.$$

We can develop a 'number theory' for the ring of integers of an algebraic number field similar to that for $\mathbb{Z}$ given in Chapter 1.

(i) By Theorem 1.7, ε is a unit (that is a divisor of 1) if and only if $N\varepsilon = \pm 1$.

(ii) If β, γ, $\varepsilon \in O_K$, ε is a unit, and $\gamma = \varepsilon\beta$ then γ is called an *associate* of β.

The units 1 and -1 belong to all number fields, most fields K have infinitely many units and so an algebraic integer will often have infinitely many associates.

(iii) An element $\xi \in O_K$ is called *irreducible* if $\xi \neq 0$, ξ is not a unit, and whenever $\xi = \beta\gamma$, where β, $\gamma \in O_K$, we have either β or γ as a unit.

This is the extension of the notion of prime number to O_K.

(iv) If β, γ, $\delta \in O_K$ we write

$$\beta \equiv \gamma \pmod{\delta}$$

whenever there exists $\rho \in O_K$ such that $\delta\rho = \beta - \gamma$.

The properties of congruences over $\mathbb{Z}$ given in Chapter 3, especially Lemma 1.1 and Problem 1, are valid in this more general setting.

We noted previously that many domains lack unique factorization; this, and the proliferation of units, can make the interpretation of results difficult. Ideals can be used to overcome this difficulty and to develop the theory further; see the references quoted at the end of this section.

Let d be a square-free rational integer. We shall illustrate the concepts above by considering the ring of integers of the field $L = \mathbb{Q}(\sqrt{d})$. L has the $\mathbb{Q}$-basis $\{1, \sqrt{d}\}$; so if $\alpha \in O_L$ then $\alpha = a + b\sqrt{d}$ where a, $b \in \mathbb{Q}$ and α satisfies a monic quadratic equation with coefficients in $\mathbb{Z}$. This equation has the form

$$(x - a - b\sqrt{d})(x - a + b\sqrt{d}) = x^2 - 2ax + a^2 - b^2d = 0.$$

Hence $2a$ and $a^2 - b^2d = c$ are rational integers. If $a \in \mathbb{Z}$ then $b \in \mathbb{Z}$ as d is square-free. If $a = a_1/2$ where a_1 is an odd integer, and $b = b_1/b_2$ where $(b_1, b_2) = 1$,

$$a_1^2/4 - db_1^2/b_2^2 = c$$

or

$$(a_1b_2)^2 - 4db_1^2 = 4cb_2^2.$$

It follows that, as a_1 is odd, $2|b_2$ but, as d is square-free and $(b_1, b_2) = 1$, $b_2|2$ and so $b_2 = 2$. Finally note that $d \equiv 1 \pmod 4$ as $c \in \mathbb{Z}$. Therefore we have two cases for O_L.

Case 1. $d \not\equiv 1 \pmod 4$. In this case $O_L = \mathbb{Z}[\sqrt{d}]$, that is elements of O_L have the form $a + b\sqrt{d}$, with conjugate $a - b\sqrt{d}$ and norm $a^2 - b^2d$, where a, $b \in \mathbb{Z}$. The discriminant of the field is $4d$.

Case 2. $d \equiv 1 \pmod 4$. In this case $O_L \neq \mathbb{Z}[\sqrt{d}]$. The ring O_L has the $\mathbb{Z}$-basis $\{1, (1 + \sqrt{d})/2\}$, and the field discriminant is d.

In either case, if $d<0$ and not equal to -1 or -3 then the only units are ± 1, and if $d>0$ there are infinitely many units (see Section 3, Chapter 7).

The integral domain $\mathbb{Z}[i]$, known as the *Gaussian integers*, has been extensively studied; we shall describe its basic properties. Suppose α, $\beta \in \mathbb{Z}[i]$ where $\alpha = a + ib$ and $\beta = c + id$, then

$$N\alpha\beta = N\alpha N\beta, \qquad (*)$$

for $N\alpha N\beta = (a^2+b^2)(c^2+d^2) = (ac-bd)^2 + (ad+bc)^2 = N\alpha\beta$. As a unit ε satisfies $N\varepsilon = \pm 1$, the units in this domain are 1, i, -1 and $-i$. $\mathbb{Z}[i]$ has a division algorithm given by

THEOREM 1.10 *If α, $\beta \in \mathbb{Z}[i]$ and $\beta \neq 0$ then γ, δ exist satisfying γ, $\delta \in \mathbb{Z}[i]$, $\alpha = \beta\gamma + \delta$ and $N\delta \le N\beta/2$.*

Proof. Let $\alpha/\beta = r + is$ where r and s are real, and choose rational integers x and y to satisfy $|r-x| \le 1/2$ and $|s-y| \le 1/2$. Let $\gamma = x + iy$ and $\delta = \alpha - \beta\gamma$. Clearly γ and δ belong to $\mathbb{Z}[i]$ and we have

$$N\delta = N\beta N(\alpha/\beta - \gamma) = N\beta((r-x)^2 + (s-y)^2) \le N\beta/2. \qquad \square$$

Using this lemma we can derive a Euclidean algorithm for $\mathbb{Z}[i]$ and so show that $\mathbb{Z}[i]$ has unique factorization as we did for the rational integers in Chapter 1. Further we see, using $(*)$, that the irreducible elements of $\mathbb{Z}[i]$ can be found by factorizing the rational primes in $\mathbb{Z}[i]$. As $Np = p^2$ (p a prime in $\mathbb{Z}$), it follows that in the Gaussian integers p is either (i) irreducible, or (ii) a product of two distinct factors $a+ib$ and $a-ib$, or (iii) a product of identical irreducible factors. Case (i) occurs when $p \equiv 3 \pmod 4$, and case (ii) occurs when $p \equiv 1 \pmod 4$, for we shall show (in Section 1, Chapter 6; see Problem 7) that $p = a^2 + b^2 [= (a+ib)(a-ib)]$ is soluble in $\mathbb{Z}$ if and only if $p \not\equiv 3 \pmod 4$. Case (iii) occurs only when $p=2$, we have $2 = (-i)(1+i)^2$ and so $1+i$ is irreducible as $-i$ is a unit.

The domain $\mathbb{Z}[\omega]$, where $\omega = (-1+\sqrt{-3})/2$, has very similar properties to the Gaussian integers. It is the ring of integers of $\mathbb{Q}(\omega)$, it has unique factorization with a Euclidean algorithm, and it has the units ± 1, $\pm\omega$ and $\pm\omega^2$. We shall use some of these properties in Chapters 6 and 14, they are given in more detail in Problem 6.

Good expositions of algebraic number theory can be found in Stewart and Tall (1979), Ribenboim (1972) and Marcus (1977) amongst others.

5.2 Primitive roots

Returning to our study of $\mathbb{Z}$ we ask the question: can we generate a reduced residue system easily? In particular does a number g exist such that the powers of g form a reduced residue system? We shall show that g exists for some, but not all, moduli m.

DEFINITION (i) Let $(c, m) = 1$. The *order* of c (mod m), denoted by $\operatorname{ord}_m c$, is the least positive integer t such that

$$c^t \equiv 1 \pmod{m}.$$

(ii) For $m > 1$ an integer g is called a *primitive root* (mod m) if $\operatorname{ord}_m g = \phi(m)$.

Using Euler's theorem we have

LEMMA 2.1 (i) $\operatorname{ord}_m c$ *exists, and if* $(c, m) = 1$ *then* $\operatorname{ord}_m c \mid \phi(m)$.

(ii) *If* g *is a primitive root* (mod m) *and* $0 \le u < v \le \phi(m)$ *then* $g^u \not\equiv g^v \pmod{m}$.

Proof. (i) Clearly $\operatorname{ord}_m c$ exists, and $\operatorname{ord}_m c \le \phi(m)$, by Euler's theorem (1.12, Chapter 3). If $(\operatorname{ord}_m c, \phi(m)) = t < \operatorname{ord}_m c$, we can find r and s such that $r \operatorname{ord}_m c + s\phi(m) = t$, and then $a^t = a^{r \operatorname{ord}_m c} a^{s\phi(m)} \equiv 1 \pmod{m}$; this contradicts the definition of $\operatorname{ord}_m c$ and so (i) follows.

(ii) If $g^u \equiv g^v$, then $g^{v-u} \equiv 1 \pmod{m}$ contrary to definition of g. □

This lemma shows that the powers of a primitive root g form a reduced residue system.

EXAMPLE We have $3^2 \equiv 2$, $3^3 \equiv 6$, $3^4 \equiv 4$, $3^5 \equiv 5$, and $3^6 \equiv 1 \pmod{7}$ and so 3 is a primitive root (mod 7). On the other hand $2^3 \equiv 1 \pmod{7}$ and so 2 is not a primitive root (mod 7).

The existence of a primitive root can be interpreted group theoretically. In Section 1 of Chapter 3 we showed that a reduced residue system (mod m) with multiplication (mod m) forms a group $(\mathbb{Z}/m\mathbb{Z})^*$. This group is cyclic if and only if a primitive root (mod m) exists. Also if $(\mathbb{Z}/m\mathbb{Z})^*$ is not cyclic, it can be represented as a direct product of cyclic groups C_i of prime power order. For odd prime powers each C_i is generated by a primitive root and we use Theorem 2.7 in the even case. So for instance $(\mathbb{Z}/7\mathbb{Z})^*$ is cyclic by the example above; but $(\mathbb{Z}/8\mathbb{Z})^*$ is not, as 8 does not possess a primitive root. This latter group is a direct product of two cyclic groups of order 2.

Using the results of Chapter 3 we shall answer the question: which moduli m have primitive roots? These results were conjectured by Euler and first proved by Gauss.

THEOREM 2.2 *A primitive root exists for each prime modulus* p.

Proof. The result is trivial for $p = 2$ so suppose p is an odd prime. If $d \mid p - 1$ let $\psi(d)$ denote the number of elements of a reduced residue system (mod p) which have order d. Suppose $c \mid d$, then a solution of the congruence $x^c \equiv 1 \pmod{p}$ is also a solution of $x^d \equiv 1 \pmod{p}$. By

Theorem 2.3, Chapter 3, this latter congruence has d solutions, hence

$$\sum_{c|d} \psi(c) = d.$$

Now using the Möbius inversion formula (Theorem 1.5, Chapter 2) we obtain

$$\psi(d) = \sum_{c|d} \mu(c)\frac{d}{c}$$
$$= \phi(d)$$

by Theorem 1.8, Chapter 2. It follows that $\psi(p-1) = \phi(p-1) \geq 1$, and so there are $\phi(p-1)$ positive integers less than p and of order $p-1 \pmod p$, that is p has $\phi(p-1)$ primitive roots. □

A table of least positive primitive roots for primes less than 750 is given on p. 337.

Using this important result we can find the other moduli which have primitive roots, we begin with two lemmas.

LEMMA 2.3 *If $\alpha > 1$ and p is an odd prime then*

$$(1 + ap)^{p^{\alpha-2}} \equiv 1 + ap^{\alpha-1} \pmod{p^\alpha}.$$

Proof. By induction. There is nothing to prove if $\alpha = 2$. Suppose the result holds for α. We have

$$b \equiv c \pmod{p^\alpha} \text{ implies } b^p \equiv c^p \pmod{p^{\alpha+1}} \qquad (*)$$

[see (iii) on p. 37]. Applying this to the inductive hypothesis we obtain

$$(1 + ap)^{p^{\alpha-1}} \equiv (1 + ap^{\alpha-1})^p \pmod{p^{\alpha+1}}$$
$$\equiv 1 + \binom{p}{1} ap^{\alpha-1} + T \pmod{p^{\alpha+1}},$$

using the binomial theorem where T is a sum of terms each divisible by $p^{\alpha+1}$ [as $2(\alpha-1)+1 \geq \alpha+1$]. The result follows. □

LEMMA 2.4 *If p is an odd prime, $\alpha > 0$, and $p \nmid a$, then the order of $1 + ap$ modulo p^α is $p^{\alpha-1}$.*

Proof. By the lemma above we have

$$(1 + ap)^{p^{\alpha-1}} \equiv 1 + ap^\alpha \pmod{p^{\alpha+1}}$$
$$\equiv 1 \pmod{p^\alpha},$$

hence the order of $1 + ap \pmod{p^\alpha}$ is a divisor of $p^{\alpha-1}$. But by Lemma 2.3 again

$$(1 + ap)^{p^{\alpha-2}} \equiv 1 + ap^{\alpha-1} \not\equiv 1 \pmod{p^\alpha}$$

as $p \nmid a$. □

THEOREM 2.5 *If p is an odd prime and $\alpha > 0$ then p^α has a primitive root.*

Proof. For some primitive root $(\bmod p)$ g we have $g^{p-1} \not\equiv 1 \pmod{p^2}$. For if g_1 is a primitive root $(\bmod p)$ and $g_1^{p-1} \equiv 1 \pmod{p^2}$, then, as $g_1 + p$ is also a primitive root $(\bmod p)$, we find

$$\begin{aligned}(g_1+p)^{p-1} &\equiv g_1^{p-1} + (p-1)g_1^{p-2}p \pmod{p^2}\\ &\equiv 1 + (p-1)pg_1^{p-2} \not\equiv 1 \pmod{p^2}.\end{aligned}$$

Hence for some primitive root $(\bmod p)$ g, we have $g^{p-1} = 1 + ap$ where $p \nmid a$. This gives, by Lemma 2.4,

$$\operatorname{ord}_{p^\alpha}(g^{p-1}) = p^{\alpha-1}. \qquad (**)$$

We show now that this g is also a primitive root modulo p^α for all $\alpha \geq 1$. Assume that $\alpha > 1$ and $\operatorname{ord}_{p^\alpha} g = n$. As $\phi(p^\alpha) = p^{\alpha-1}(p-1)$, we have, by Lemma 2.1(i), n is a divisor of $\phi(p^\alpha)$ and so, $n = p^\beta v$ where $\beta \leq \alpha - 1$ and $v \mid p-1$. We need to prove that $n = \phi(p^\alpha)$. Suppose first $\beta < \alpha - 1$, then

$$g^{p^\beta(p-1)} \equiv 1 \pmod{p^\alpha}$$

as $v \mid p-1$. This contradicts $(**)$ above, and so $\beta = \alpha - 1$. Secondly suppose $v < p - 1$, then we have

$$1 \equiv g^{p^{\alpha-1}v} \equiv g^{(p^{\alpha-1}-1)v} g^v \equiv g^v \pmod{p},$$

as $p - 1 \mid p^{\alpha-1} - 1$ and $(g, p) = 1$. But this is impossible as g is a primitive root $(\bmod p)$. Hence $v = p - 1$ and g is a primitive root modulo p^α. □

THEOREM 2.6 *The numbers 2, 4, p^α, and $2p^\alpha$, where p is an odd prime and α is a positive integer, are the only integers which possess primitive roots.*

Proof. For an odd number c we have $c^2 \equiv 1 \pmod 8$ and so, by $(*)$ in the proof of Lemma 2.3, $c^{2^{\alpha-2}} \equiv 1 \pmod{2^\alpha}$. Hence 2^α cannot have a primitive root if $\alpha > 2$; note that 3 is the only primitive root $(\bmod 4)$.

Secondly let g be a primitive root $(\bmod p^\alpha)$, $g + p^\alpha$ also has this property and one of these, g say, is odd. Hence, as $\phi(2p^\alpha) = \phi(p^\alpha)$ and $g^t \equiv 1 \pmod 2$ for all t, g is also a primitive root $(\bmod 2p^\alpha)$.

Finally suppose $m = \prod_{i=1}^{k} p_i^{\alpha_i}$ and $\phi(p_i^{\alpha_i}) > 1$ for at least two values of i. Let $z = \operatorname{LCM}\{\phi(p_1^{\alpha_1}), \ldots, \phi(p_k^{\alpha_k})\}$, $z < \phi(m)$ as two or more of the terms in this LCM are even. We have by Euler's theorem and Problem 1 of Chapter 3

$$x^z \equiv 1 \pmod m$$

holds for all x satisfying $(x, m) = 1$, it follows that m cannot have a primitive root. □

Although numbers of the form 2^α where $\alpha > 2$ do not possess primitive roots, we have the following result which shows that $(\mathbb{Z}/2^\alpha\mathbb{Z})^*$ is a direct product of a cyclic group of order $2^{\alpha-2}$ and a group of order 2.

THEOREM 2.7 *The set of numbers* $\{\pm 5^t \mid 0 \le t < 2^{\alpha-2}\}$ *forms a reduced residue system modulo* 2^α, *where the* + *sign* (− *sign*) *applies to a residue congruent to* 1 (−1, *respectively*) *modulo* 4.

We shall leave the proof as an exercise (see Problem 17).

The reader will have noticed that, although we know which numbers possess primitive roots, we have not given a method for finding these roots. Except for trial and error methods, very few general techniques are known (one is given in Problem 13). Artin, in 1927, made the following conjecture: if a is not square and $\neq -1$, 0, or 1, and if $N_a(x)$ is the number of primes less than x for which a is a primitive root, then

$$N_a(x) \sim A \frac{x}{\ln x}$$

where A depends only on a. The numerical evidence for this is strong, and Hooley (1967) has shown that if the extended Riemann hypothesis is true then so is Artin's conjecture, except that the density formula has to be amended slightly when $a \equiv 1 \pmod 4$.

In a remarkable feat of calculation before the age of computers, Jacobi (1956, reprint of the 1839 edition) listed all solutions $\{a, b\}$ of the congruences $g^a \equiv b \pmod p$ where $1 \le a, b < p$, g is the least positive primitive root of p, and $p < 1000$.

Primitive roots have many applications in number theory. Here we use them to introduce nth power residues and the theory of characters. Some other applications are given in the problem section and later in this book. We begin by proving a generalization of Euler's criterion.

DEFINITION Let m and n be positive integers with $n > 1$. Suppose $(a, m) = 1$, then a is an nth *power residue* modulo m if and only if there is an x such that

$$x^n \equiv a \pmod m. \tag{1}$$

THEOREM 2.8 *Suppose m has a primitive root and* $(a, m) = 1$. *Then a is an nth power residue* (mod m) *if and only if*

$$a^{\phi(m)/d} \equiv 1 \pmod m \tag{2}$$

where $d = (n, \phi(m))$. *Further if* (2) *holds then the congruence* (1) *has d solutions.*

COROLLARY 2.9 *If m has a primitive root,* $(a, m) = 1$ *and* $(n, \phi(m)) = 1$, *then a is an nth power residue.*

Proof. The corollary follows immediately from the theorem. Let g be a primitive root modulo m, then we have:

the congruence $x^n \equiv a \pmod{m}$ has d solutions,

if and only if

the congruence $g^{ny} \equiv g^b \pmod{m}$ has d solutions,
where $a \equiv g^b \pmod{m}$,

if and only if

$ny \equiv b \pmod{\phi(m)}$ has d solutions,

if and only if

$d \mid b$

by Theorem 1.4, Chapter 3. Now if $d \mid b$ then $a^{\phi(m)/d} \equiv g^{b\phi(m)/d} \equiv 1 \pmod{m}$. Finally if $a^{\phi(m)/d} \equiv 1 \pmod{m}$ then $g^{b\phi(m)/d} \equiv 1 \pmod{m}$ and so b/d is an integer as g is a primitive root, that is $d \mid b$. □

If $n = 2$ and m is prime, the above result is Euler's criterion (Theorem 1.2, Chapter 4). Further if $n = 3$ and p is a prime congruent to 2 (mod 3) then the congruence $x^3 \equiv a \pmod{p}$ always has a unique solution; and, if p is congruent to 1 (mod 3), a theory of cubic residues (including a cubic reciprocity law) can be developed similar to that in the previous chapter, see for example Ireland and Rosen (1982). The cases $n = 4, 5, \ldots$ can be dealt with in the same manner; see also Problem 19 in this chapter and Problem 20 in Chapter 4.

Finite fields

Many of the results we have discussed for $\mathbb{Z}/p\mathbb{Z}$ hold in all finite fields. Suppose F is a finite field with q elements, then $q = p^n$ for some prime p and F is an n-dimensional vector space over its prime subfield F_0 where F_0 is isomorphic to $\mathbb{Z}/p\mathbb{Z}$.

The multiplicative structure of F follows closely that of $(\mathbb{Z}/p\mathbb{Z})^*$. Let F^* is the multiplicative group of F, then F^* has $q-1$ elements and, if $a \in F^*$, $a^{q-1} = 1_F$ where 1_F is the identity of F. So for all elements c of F we have

$$c^q = c,$$

the analogue of Fermat's theorem. Using this it is a simple matter to derive extensions of Lagrange's theorem and Theorem 2.3, both from Chapter 3, to the field F. As a consequence we have

THEOREM 2.10 *The group F^* is cyclic.*

The proofs of these results are virtually identical to the corresponding ones for $(\mathbb{Z}/p\mathbb{Z})^*$. Note that, as $\mathbb{Z}/p^n\mathbb{Z}$ is not a field (if $n>1$), there is no connection between Theorem 2.10 and Theorem 2.5. Finally we have the analogue of Theorem 2.9: if $a \in F^*$, $x^m = a$ has solutions in F^* if and only if $a^{(q-1)/d} = 1_F$ where $d=(m, q-1)$. For a further discussion and proofs of these results see Ireland and Rosen (1982) or Serre (1973).

5.3 Characters

The third topic in this chapter is the theory of characters, it provides a codification of the congruence relation. We shall use this theory later when discussing Gauss sums and in the proof of Dirichlet's theorem on primes in arithmetic progressions.

DEFINITION Fix a modulus m.

(i) A *multiplicative character* modulo m is a group homomorphism χ from $(\mathbb{Z}/m\mathbb{Z})^*$ into the non-zero complex numbers, that is χ satisfies

$$\chi(ab)=\chi(a)\chi(b).$$

(ii) A *Dirichlet character* modulo m is a function χ mapping the integers into the complex numbers satisfying

(a) $\chi(a) \neq 0$ if and only if $(a, m)=1$,
(b) if $a \equiv b \pmod m$ then $\chi(a)=\chi(b)$,
(c) $\chi(ab)=\chi(a)\chi(b)$.

We can see immediately that if we restrict the domain of definition of a Dirichlet character $(\bmod\, m)$ to a reduced residue system $(\bmod\, m)$, we obtain a multiplicative character. We shall work with Dirichlet characters and use the word 'character' to denote them. Results concerning multiplicative characters can easily be deduced from the corresponding ones for Dirichlet characters.

EXAMPLE The function χ_0 defined by $\chi_0(a)=1$ if $(a, m)=1$ and $\chi_0(a)=0$ if $(a, m)>1$ is a character $(\bmod\, m)$ called the *principal character*. Another example is the Legendre symbol $(\cdot/p)$ which is a character $(\bmod\, p)$.

We shall develop the basic results using the theory of primitive roots introduced in the previous section.

LEMMA 3.1 *If χ is a character* $(\bmod\, m)$ *then*

(i) $\chi(1)=1$,
(ii) *if $(a, m)=1$ then $\chi(a)$ is a $\phi(m)$-root of unity,*
(iii) *the complex conjugate character $\bar{\chi}$ defined by*

$$\bar{\chi}(a)=\overline{\chi(a)}$$

is also a character $(\bmod\, m)$.

Proof. (i) We have $\chi(1)=\chi(1)\chi(1)$ and so $\chi(1)=1$ as $\chi(1)\neq 0$.

(ii) As $(a, m)=1$ we have by Euler's theorem, (b) above and (i)

$$(\chi(a))^{\phi(m)}=\chi(a^{\phi(m)})=\chi(1)=1.$$

(iii) We note that $\bar{\chi}$ satisfies the axioms (a)–(c) above. □

LEMMA 3.2 *If we define the product of two characters* (mod m) χ_1 *and* χ_2 *by*

$$\chi_1\chi_2(a)=\chi_1(a)\chi_2(a),$$

then the set of characters (mod m) *forms a finite Abelian group under this operation. In particular if* χ *runs over the set of characters* (mod m) *then so does* $\chi_1\chi$ *for each fixed character* χ_1.

Proof. It is a simple matter to check that $\chi_1\chi_2$ is a character. Also as the value of a character is 0 or a $\phi(m)$-root of unity there can only be finitely many characters modulo m. They form an Abelian group with identity χ_0 the principal character, and where the inverse of a character is its complex conjugate [see (iii) of Lemma 3.1]. □

Before proceeding we shall show how to construct some characters. If m has a primitive root g and $(a, m)=1$, let $\operatorname{ind}_g a$, the *index* of a to base g, denote the smallest non-negative integer c with the property $a\equiv g^c$ (mod m). We usually write ind a for $\operatorname{ind}_g a$ when the primitive root g has been specified in advance. We have, if $(a, m)=(b, m)=1$,

$$\operatorname{ind} ab\equiv \operatorname{ind} a+\operatorname{ind} b \pmod{\phi(m)}$$

and $\qquad (*)$

$$\operatorname{ind} a\equiv \operatorname{ind} b \pmod{\phi(m)} \quad \text{when} \quad a\equiv b \pmod m,$$

see Problem 16. Also we write $\mathrm{e}(a)$ for $\mathrm{e}^{2\pi i a}$, note that $\mathrm{e}(a)=1$ if and only if $a\in\mathbb{Z}$.

If χ and χ' are characters modulo p^t and p_1^u, respectively, then $\chi\chi'$ is a character modulo $p^tp_1^u$, also χ is a character modulo p^v where $v\geq t$. So many characters can be constructed using Cases 1 and 2 below.

Case 1. $p>2$. Let g be a primitive root (mod p^u) and define χ_1 by

$$\chi_1(a)=\begin{cases}\mathrm{e}(\operatorname{ind} a/\phi(p^u)) & \text{if } (a, p^u)=1\\ 0 & \text{if } (a, p^u)>1.\end{cases}$$

Using $(*)$ it follows immediately that χ_1 is a character modulo p^u.

Case 2. Moduli 2^v, for all characters in this case $\chi(a)=0$ if a is even. If $v=1$, there is only the principal character, and if $v=2$ we also have χ_2 given by $\chi_2(1)=1$ and $\chi_2(3)=-1$. If $v>2$ and, for odd a, t satisfies

$a \equiv \pm 5^t \pmod{2^v}$ (see Theorem 2.7), then

(i) $\chi_3(a) = (-1)^{(a-1)/2}$,

(ii) $\chi_4(a) = e(t/2^{v-2})$

are characters modulo 2^v, the reader should check this.

As a product of characters modulo m is also a character modulo m (Lemma 3.2), we can generate many characters using $\chi_1, \ldots, \chi_4$ and the remark at the beginning of the previous paragraph. In fact all characters can be constructed in this way (see Problem 21), to prove this we need to calculate the number of characters $(\text{mod}\, m)$ as follows.

LEMMA 3.3 *If $d > 0$, $(d, m) = 1$ and $d \not\equiv 1 \pmod{m}$, then there is a character* $(\text{mod}\, m)$ χ *with the property* $\chi(d) \neq 1$.

Proof. This follows from the constructions above. Suppose first m has an odd prime power factor p^u such that $d \not\equiv 1 \pmod{p^u}$, then the character χ_1 has the required property as $\phi(p^u) \nmid \text{ind}\, d$. Similarly if $d \not\equiv 1 \pmod{2^v}$, we can use χ_3 if $d \equiv 3 \pmod 4$, and χ_4 if $d \equiv 1 \pmod 4$. □

THEOREM 3.4 *Let c_m be the order of the group of characters* $(\text{mod}\, m)$ *and let M denote the reduced residue system $\{b : (b, m) = 1,\ 0 < b < m\}$.*

(i) $$\sum_{\chi} \chi(a) = \begin{cases} c_m & \text{if } a \equiv 1 \pmod{m} \\ 0 & \text{if } a \not\equiv 1 \pmod{m}, \end{cases}$$

(ii) $$\sum_{a \in M} \chi(a) = \begin{cases} \phi(m) & \text{if } \chi = \chi_0 \\ 0 & \text{if } \chi \neq \chi_0, \end{cases}$$

(iii) $c_m = \phi(m)$.

(*The first sum is taken over all characters modulo m.*)

Proof. (i) If $a \equiv 1 \pmod{m}$ then $\chi(a) = 1$ and so $\sum_{\chi} \chi(a) = c_m$. If $a \not\equiv 1 \pmod{m}$, use Lemma 3.3 to choose a character χ_1 so that $\chi_1(a) \neq 1$ then

$$\begin{aligned} \sum_{\chi} \chi(a) &= \sum_{\chi} \chi(a)\chi_1(a) && \text{by Lemma 3.2} \\ &= \chi_1(a) \sum_{\chi} \chi(a) \\ &= 0 && \text{as } \chi_1(a) \neq 1. \end{aligned}$$

(ii) If $\chi \neq \chi_0$ we can find $b \in M$ such that $\chi(b) \neq 1$, now proceed as in Case (i).

(iii) We have

$$c_m = \sum_{a \in M} \sum_{\chi} \chi(a) = \sum_{\chi} \sum_{a \in M} \chi(a) = \phi(m).$$ □

COROLLARY 3.5 *If $(t, m)=1$, $t>0$, and $a>0$ then*

$$\frac{1}{\phi(m)}\sum_{\chi}\frac{\chi(a)}{\chi(t)}=\begin{cases}1 & \text{if } a\equiv t \pmod m\\ 0 & \text{if } a\not\equiv t \pmod m,\end{cases}$$

where the sum is taken over all characters modulo m.

Proof. Choose u so that $tu\equiv 1 \pmod m$, then $\chi(t)\chi(u)=1$, and so

$$\sum_{\chi}\frac{\chi(a)}{\chi(t)}=\sum_{\chi}\chi(au)=\begin{cases}\phi(m) & \text{if } au\equiv 1, \text{ that is } a\equiv t \pmod m,\\ 0 & \text{if } au\not\equiv 1, \text{ that is } a\not\equiv t \pmod m.\end{cases}\qquad\square$$

This corollary provides a codification of the congruence relation.

The set of characters modulo m can be split into three classes: (i) the principal character χ_0; (ii) characters χ taking only real values, they satisfy $\chi^2=\chi_0$; and (iii) complex valued characters, these occur in conjugate pairs χ and $\bar{\chi}$. The real characters are determined by the following result; note that, by Theorem 3.5, Chapter 4, the Kronecker symbol $(d/\cdot)$ is a character modulo $|d|$.

THEOREM 3.6 *Dirichlet's lemma.*
If χ is a real character $\pmod m$ *then a number h can be found so that*

$$\chi(a)=(h/a) \text{ and } h\equiv 0 \text{ or } 1 \pmod 4,$$

where $(h/\cdot)$ is a Kronecker symbol.

Proof. We shall give the proof when m is an odd prime p, the remaining cases are similar, see Problem 22. First choose a primitive root g modulo p. Then we have $(a/p)=(-1)^{\operatorname{ind} a}$, if $(a, p)=1$. For as $g^{\operatorname{ind} a}\equiv a \pmod p$,

$$\begin{aligned}(a/p)=1 &\text{ if and only if } g^{2x}\equiv g^{\operatorname{ind} a} \pmod p \text{ is soluble}\\ &\text{ if and only if } 2x\equiv \operatorname{ind} a \pmod{p-1} \text{ is soluble}\\ &\text{ if and only if } 2=(2,p-1)\mid \operatorname{ind} a\\ &\text{ if and only if } (-1)^{\operatorname{ind} a}=1.\end{aligned}$$

Let χ_1 be the character $\pmod p$ given by Case 1 on p. 86 (with $u=1$). By Lemma 3.2, χ_1^v is also a character $\pmod p$ for $v=1,\ldots,p-1$, and each of these is distinct. Further, by Theorem 3.4 there are only $p-1$ characters modulo p, hence a general character $\pmod p$ has the form

$$\chi(a)=\begin{cases}\mathrm{e}\left(\dfrac{v \operatorname{ind} a}{p-1}\right) & \text{if } (a,p)=1\\ 0 & \text{if } (a,p)>1\end{cases}$$

for $v\in\{1, 2, \ldots, p-1\}$. Now if χ is real then $p-1 \mid 2v$, hence $v=(p-1)/2$ or $p-1$. In the latter case $\chi=\chi_0$ and in the former we have

$$\chi(a)=\mathrm{e}^{\pi i \operatorname{ind} a}=(-1)^{\operatorname{ind} a}=(a/p)$$

if $(a, p)=1$; and if $(a, p)>1$, $\chi(a)=0=(a/p)$. Finally let $p^*=p$ if $p\equiv 1 \pmod 4$, and $p^*=-p$ if $p\equiv 3 \pmod 4$, then by the quadratic reciprocity law (Chapter 4) we find that

$$\chi(a)=(p^*/a)$$

and $p^*\equiv 1 \pmod 4$. □

An important consequence of this result is: if p is an odd prime then there are only two real characters modulo p, the principal character and the Legendre symbol (but see Problem 24).

5.4 Problems 5

1. Prove Lemma 1.1 and find the polynomials t, u, and d when (i) $f(x)=x^5-1$, $g(x)=x^2-1$, and (ii) $f(x)=x^5-1$, $g(x)=x^2+1$.

2. (i) Let Df denote the derivative of the polynomial f. Suppose K is a subfield of $\mathbb{C}$, $f\in K[x]$, and $f\neq 0$, show that f is divisible by the square of a polynomial with positive degree if and only if f and Df have a common factor g which satisfies $\deg g>0$.
(ii) Deduce that an irreducible polynomial defined over K has no repeated roots in the complex field.

3. (i) Which of the following numbers are algebraic integers and what are their degrees?

$$\text{(a) } \sqrt{3}+\sqrt{7}, \quad \text{(b) } e^{\pi i/17}, \quad \text{(c) } \sqrt{(1+\sqrt{2})}+\sqrt{(1-\sqrt{2})}.$$

(ii) Let α satisfy the polynomial equation $x^2+\beta x+\gamma=0$ where β and γ are algebraic integers of degree 2. Show that α is an algebraic integer. What is its degree?

4. Let $K=\mathbb{Q}(\alpha)$ and suppose $\{\gamma_1, \ldots, \gamma_n\}$ is a $\mathbb{Z}$-basis for O_K. Use Cramer's rule to show that a positive constant c exists so that, if $\beta=a_1\gamma_1+\ldots+a_n\gamma_n\in O_K$, then, for $1\leq k\leq n$,

$$|a_k|\leq c\max_{1\leq j\leq n}|\beta_j|$$

where β_j denotes the jth field conjugate of β.

5. (i) Let $K=\mathbb{Q}(\alpha)$ where α has minimum polynomial f and degree n. Show that the discriminant of the $\mathbb{Q}$-basis $\{1, \alpha, \ldots, \alpha^{n-1}\}$ satisfies

$$\Delta(1, \alpha, \ldots, \alpha^{n-1})=(-1)^{n(n-1)/2}N(Df(\alpha))$$

where Df denotes the derivative of f.
(ii) Now let $\alpha=e^{2\pi i/p}$ where p is an odd prime. Assuming that the ring of integers of $\mathbb{Q}(\alpha)$ equals $\mathbb{Z}[\alpha]$, show that the discriminant of $\mathbb{Q}(\alpha)$ is

$$(-1)^{(p-1)/2}p^{p-2}.$$

6. Prove the following properties of the integral domain $\mathbb{Z}[\omega]$ where $\omega = (-1 + \sqrt{-3})/2$.

(i) $\omega^2 + \omega + 1 = 0$, and so $\omega^3 = 1$.

(ii) If $\alpha \in \mathbb{Z}[\omega]$, it can be written in the form $a + \omega b$ where $a, b \in \mathbb{Z}$, the conjugate of α is $a + \omega^2 b = a - b - \omega b$, and $N\alpha = a^2 - ab + b^2$ where $N\alpha$ is the norm of α.

(iii) Using the method of proof of Theorem 1.10, show that $\mathbb{Z}[\omega]$ has a division algorithm. Hence show that $\mathbb{Z}[\omega]$ is a unique factorization domain. Note that the GCD and congruences can be defined exactly as for $\mathbb{Z}$ providing associates are taken into account.

(iv) Let $\lambda = 1 - \omega$. Show that λ^2 is an associate of 3. Deduce that if $\alpha \in \mathbb{Z}[\omega]$, then $\alpha \equiv -1$, 0, or 1 (mod λ).

(v) Use Theorem 1.3, Chapter 6, to show that the irreducible elements of $\mathbb{Z}[\omega]$ are (a) λ, (b) the rational primes p if $p \equiv 2 \pmod 3$, and (c) $r + \omega s$ and $r + \omega^2 s$ if $p \equiv 1 \pmod 3$ and $p = r^2 - rs + s^2$.

7. Use the fact that $\mathbb{Z}[i]$ has unique factorization to show that, if $p \equiv 1 \pmod 4$, rational integers a and b exist satisfying $p = a^2 + b^2$.
[Hint. Begin with Theorem 1.3(iv), Chapter 4.] Further proofs of this result will be given in the next chapter.

8. (i) Find the orders (mod 23) of 2, 3, and 5.

(ii) Find all the primitive roots of 7, 14, and 49.

9. If $p = 2^m + 1$, where p is a prime, and $(a/p) = -1$, show that a is a primitive root modulo p.

10. Show that

$$1^k + 2^k + \ldots + (p-1)^k \equiv \begin{cases} 0 \pmod p & \text{if } p - 1 \nmid k \\ -1 \pmod p & \text{if } p - 1 \mid k. \end{cases}$$

11. Prove that, if m has at least one primitive root, then it has $\phi(\phi(m))$ of them modulo m, and their product is congruent to 1 modulo m, provided $m > 6$.

12. Let g be a primitive root modulo p, prove that no k exists satisfying $g^{k+2} \equiv g^{k+1} + 1 \equiv g^k + 2 \pmod p$.

13. Show that 2 is a primitive root of the primes p and p' where $p = 4q + 1$ for some prime q, and $p' = 2r + 1$ for some prime r congruent to 1 (mod 4). Note: although we know that there are infinitely many primes of the form $4k + 1$, we do not know if infinitely many of the corresponding k are prime. Hence we cannot use this result for Artin's primitive root conjecture, see p. 83.

14. In the statement of Artin's conjecture why did we leave out the cases $a = -1$, and a is a square?

15. Suppose 10 is a primitive root modulo p. Let $a_1, a_2, \ldots$ and $b_0, b_1, \ldots$ be given by: $b_0 = 1$ and, if $k > 0$,

$$10b_{k-1} = pa_k + b_k; \qquad 0 \le b_k < p.$$

Show that $1/p = 0 \cdot a_1 a_2 \ldots$ (decimal expansion) and $b_k \equiv 10^k \pmod p$. Show further that this decimal expansion is periodic with period $p - 1$. Gauss's work on primitive roots began with this problem, he wanted to find infinitely many primes with this property. This is of course related to Artin's conjecture.

16. Prove that, if $(a, m) = (b, m) = 1$ and g is a primitive root modulo m,
 (i) $\operatorname{ind}_g a \equiv \operatorname{ind}_g b \pmod{\phi(m)}$, if $a \equiv b \pmod m$,
 (ii) $\operatorname{ind}_g ab \equiv \operatorname{ind}_g a + \operatorname{ind}_g b \pmod{\phi(m)}$.
Use these results to solve the congruence $7x \equiv 11 \pmod{18}$.

17. Prove Theorem 2.7 by showing that 5 has order 2^{t-2} modulo 2^t.

18. Find all the solutions of $x^5 \equiv 1 \pmod{41}$ and hence solve the congruence

$$(x^2+1)(x+1)x \equiv -1 \pmod{41}.$$

19. (i) Show that if $p \equiv 3 \pmod 4$ then $x^4 \equiv a \pmod p$ is soluble if and only if $(a/p) \geq 0$, and find the solutions of

$$x^4 \equiv 3 \pmod{11}.$$

(ii) Prove that the congruence $x^4 \equiv -1 \pmod p$ is soluble if and only if $p \equiv 1 \pmod 8$.

20. Write out the tables of values for all characters modulo 7 and 8.

21. Given a modulus m show that all characters modulo m can be constructed using powers and products of the characters $\chi_1, \ldots, \chi_4$ given on p. 86.

22. Find the numbers h so that $\chi(a) = (h/a)$ for a real character χ modulo $2^{\alpha_0} p^{\alpha_1} q^{\alpha_2}$ where $\alpha_0 > 2$, $\alpha_1, \alpha_2 > 0$ and p and q are distinct primes.

23. Is the group $(\mathbb{Z}/m\mathbb{Z})^*$ isomorphic to the group of characters modulo m?

24. Let $m = 2^{\alpha_0} p_1^{\alpha_1} \ldots p_k^{\alpha_k}$. Show that all characters with moduli m are real if and only if each of $\phi(2^{\alpha_0 - 1}), \phi(p_1^{\alpha_1}), \ldots, \phi(p_k^{\alpha_k})$ equals 1 or 2. Deduce that this property holds if and only if $m = 1, 2, 3, 4, 6, 8, 12$, or 24.

25. Let N denote the number of solutions of the congruence $x^n \equiv a \pmod p$. Show that, if $n \mid p-1$,

$$N = \begin{cases} 1 & \text{if } a \equiv 0 \pmod p \\ \sum_\chi^n \chi(a) & \text{if } a \not\equiv 0 \pmod p, \end{cases}$$

where $\sum_\chi^n \chi(a)$ denotes the sum of all $\chi(a)$ such that $\chi^n = \chi_0$.

26. Let $\{f_n\}$ be a non-increasing sequence of positive real numbers and let χ be a character modulo k such that $\chi \neq \chi_0$. Show that

$$\left|\sum_{n=r}^{s} \chi(n) f_n\right| < 2\phi(k) f_r$$

using the following method. Define $Q(x) = \sum_{n \leq x} \chi(n)$ [so $\chi(n) = Q(n) - Q(n-1)$], and use partial summation and Theorem 3.4.

6

SUMS OF SQUARES AND GAUSS SUMS

The concepts introduced so far form the basis for the major topics to be discussed in this chapter, they are Gauss sums and, to begin with, the representation of integers as sums of squares. Consider the proposition: if p is a prime and $p \equiv 1 \pmod 4$, then the Diophantine equation

$$x^2 + y^2 = p$$

has an integer solution. This result, which is generally attributed to Fermat, is one of the seminal theorems of number theory. Both the ideas associated with its many proofs and the study of its extensions and generalizations have had a major influence on the development of the subject. We shall discuss the infinite descent method and use it to prove Fermat's result and some related results including Lagrange's four square theorem. In Chapters 9 and 10 we shall treat the general arithmetic theory of quadratic forms.

The interplay between congruence theory, the primes, and Diophantine problems which is central to the material above is also the starting point for Gauss sums, our second topic. These are finite sums of roots of unity with some nice formal properties; they are an essential tool in several branches of number theory. Gauss introduced them to give a new derivation of his quadratic reciprocity law, one which can be generalized to higher degree congruences, and to answer some questions concerning the ruler and compass construction of regular polygons (Problem 16). They also have applications in the analytic theory of quadratic forms and, via Jacobi sums, to the estimation of the number of solutions of a wide class of congruences. We shall derive the elementary properties of these sums and discuss the applications mentioned above. (For applications to quadratic form theory see Chapter 10.) The tricky question of the sign of the quadratic Gauss sum will be considered in the final section.

6.1 Sums of squares

Let f be a quadratic form. Consider the equation

$$f(x_1, \ldots, x_n) = m. \tag{1}$$

The *infinite descent method,* which is a type of mathematical induction, gives solutions to this equation in a few important cases. For the initial step we look for a positive integer k such that

$$f(x_1, \ldots, x_n) = km \tag{2}$$

is soluble in integers $x_1, \ldots, x_n$; this is often achieved using congruence results. Secondly we show that, given a solution of equation (2) for some $k > 1$, we can always find another solution of this equation with k' replacing k and where $0 < k' < k$; this is called the reduction step. Hence a solution of equation (1) can be found using the initial step and a finite number of reduction steps.

We begin with Fermat's result discussed in the introduction. The first extant proofs are due to Euler.

THEOREM 1.1 *Let p be a prime. The Diophantine equation*

$$x^2 + y^2 = p$$

is soluble in integers x and y if and only if $p = 2$ or $p \equiv 1 \pmod 4$.

Proof. Note first that $2 = 1^2 + 1^2$. Secondly if $p \equiv 3 \pmod 4$, there is no solution because $z^2 \equiv 0$ or $1 \pmod 4$ for all integers z. Finally suppose $p \equiv 1 \pmod 4$, we use the infinite descent method in this case. By Theorem 1.3, Chapter 4, there is an integer x satisfying $0 < x < p/2$ and $x^2 + 1 \equiv 0 \pmod p$, hence

$$x^2 + y^2 = kp \tag{3}$$

has a solution $\{x_0, y_0\}$ for some positive $k < p$.

Now we make use of the identity

$$(z_1^2 + z_2^2)(z_3^2 + z_4^2) = (z_1 z_3 + z_2 z_4)^2 + (z_1 z_4 - z_2 z_3)^2. \tag{4}$$

Let x_1 and y_1 be the absolute least residues modulo k of x_0 and y_0, so we have for some c and d

$$\begin{aligned} x_1^2 + y_1^2 &= (x_0 - ck)^2 + (y_0 - dk)^2 \\ &\equiv x_0^2 + y_0^2 \pmod k \\ &\equiv 0 \pmod k \qquad \text{by (3).} \end{aligned}$$

Hence $x_1^2 + y_1^2 = k'k$ for some $k' < k$ as $x_1^2 + y_1^2 \leq 2(k/2)^2$. Applying (4) to this equation and (3), we have

$$(x_0^2 + y_0^2)(x_1^2 + y_1^2) = (x_0 x_1 + y_0 y_1)^2 + (x_0 y_1 - x_1 y_0)^2 = k'k^2 p.$$

Now

$$x_0 x_1 + y_0 y_1 = x_0(x_0 - ck) + y_0(y_0 - dk) \equiv x_0^2 + y_0^2 \equiv 0 \pmod k$$

and

$$x_0y_1 - x_1y_0 = x_0(y_0 - dk) - y_0(x_0 - ck) \equiv 0 \pmod k.$$

Thus if $x_2 = (x_0x_1 + y_0y_1)/k$ and $y_2 = (x_0y_1 - x_1y_0)/k$, x_2 and y_2 are integers and

$$x_2^2 + y_2^2 = k'p.$$

The result follows by the infinite descent method with (3) as the initial step and with the argument just described as the reduction step. □

In the case $p \equiv 1 \pmod 4$ the solution is unique if we assume that x and y are positive and x is even, see Problem 1, Problem 2, or Corollary 3.8, Chapter 9. Further proofs of this result are given by (i) on p. 103 and by Problem 7, Chapter 5.

THEOREM 1.2 *The equation*

$$x^2 + y^2 = m$$

has an integer solution if and only if each prime factor of m congruent to 3 *modulo* 4 *occurs to an even power in the prime factorization of m.*

Proof. If the condition on the prime factorization of m holds then use Theorem 1.1 and (4) above with the identity $n^2 = n^2 + 0^2$.

For the converse suppose $p \equiv 3 \pmod 4$, $x^2 + y^2 = p^{2k+1}n$, $(p, n) = 1$, $k \geq 0$, $p \nmid x$ and $p \nmid y$. (If $p \mid x$ then $p \mid y$ and $p^2 \mid x^2 + y^2$, hence we can divide the equation by p^2.) Therefore we have $x^2 \equiv -y^2 \pmod p$ and so $(-1/p) = 1$. This is impossible by Theorem 1.3, Chapter 4, and so the proof is complete. □

An equally striking but much less well known result is

THEOREM 1.3 *Let p be a prime. The equations*

(i) $x^2 + 3y^2 = p$,

(ii) $x^2 - xy + y^2 = p$

are both soluble in integers if and only if $p \equiv 1 \pmod 3$ *or* $p = 3$.

Proof. Assume $p > 3$ as the cases $p = 2, 3$ are trivial. We use the infinite descent method for equation (i) and solutions of (ii) will follow by substitution.

(i) This is very similar to the proof of Theorem 1.1 as we have the identity

$$(z_1^2 + 3z_2^2)(z_3^2 + 3z_4^2) = (z_1z_3 + 3z_2z_4)^2 + 3(z_1z_4 - z_2z_3)^2. \tag{5}$$

The congruence

$$x^2 \equiv -3 \pmod p$$

is soluble if and only if $(-3/p) = 1$. We have $(-3/p) = (-1/p)(3/p) = 1$ if and only if $p \equiv 1$ or 7 (mod 12) (see p. 61); but as p is odd this is equivalent to $p \equiv 1 \pmod 3$. Hence, as in Theorem 1.1, if $p \not\equiv 1 \pmod 3$, equation (i) has no solution, and if $p \equiv 1 \pmod 3$, integers x_0, y_0, and k can be found to satisfy $0 < k < p$ and

$$x_0^2 + 3y_0^2 = kp. \tag{6}$$

Let x_1 and y_1 be the absolute least residues modulo k of x_0 and y_0, respectively, so

$$|x_1| \leq k/2 \quad \text{and} \quad |y_1| \leq k/2. \tag{7}$$

Suppose we have equality in both cases of (7). Then k is even and integers c and d exist satisfying $2x_0 = k(1+2c)$ and $2y_0 = k(1+2d)$. Using (6) these give, dividing by k,

$$k[(1+2c)^2 + 3(1+2d)^2] = 4p, \tag{8}$$

and so $k = 2$ or 4 because $k < p$. If $k = 2$ then, as p is odd, $kp \equiv 2 \pmod 4$, but this is impossible as $x^2 + 3y^2 \not\equiv 2 \pmod 4$ for all x and y. If $k = 4$, (8) gives a solution to the equation (i). Hence we may assume that at least one of the inequalities in (7) is strict. Thus

$$x_1^2 + 3y_1^2 < k^2.$$

Now proceed as in the proof of Theorem 1.1.

(ii) If $p \equiv 2 \pmod 3$ then the equation (ii) has no solutions because $x^2 - xy + y^2 \not\equiv 2 \pmod 3$ for all x and y [use $z^2 \not\equiv 2 \pmod 3$ for all z]. If $p \equiv 1 \pmod 3$ we can find, by (i), x and y so that $x^2 + 3y^2 = p$. Let t and u be given by $t + u = 2x$ and $t - u = 2y$. Then

$$4p = (t+u)^2 + 3(t-u)^2 = 4(t^2 - tu + u^2). \qquad \square$$

Example As 61 is congruent to 1 modulo 3 and 4, it has the representations

$$61 = 6^2 + 5^2 = 7^2 + 3 \cdot 2^2 = 9^2 - 9 \cdot 5 + 5^2.$$

Some further results are given in the problem section. We can use the infinite descent method for the equations $x^2 + ny^2 = m$ when $n = 1$, 2, 3, or 4, but a different approach is needed for the general case and this will be considered in Chapters 9 and 10.

Finally we come to the famous 'four squares theorem' first proved by Lagrange in 1770. It was conjectured by Bachet in 1621, and possibly much earlier, see p. 263.

Theorem 1.4 *Every non-negative integer can be expressed as a sum of four squares.*

Proof. As above we have an identity which reduces the result to the prime case: namely,

$$(x_1^2+x_2^2+x_3^2+x_4^2)(y_1^2+y_2^2+y_3^2+y_4^2)=(x_1y_1+x_2y_2+x_3y_3+x_4y_4)^2$$
$$+(x_1y_2-x_2y_1+x_3y_4-x_4y_3)^2+(x_1y_3-x_3y_1+x_4y_2-x_2y_4)^2$$
$$+(x_1y_4-x_4y_1+x_2y_3-x_3y_2)^2. \tag{9}$$

Now as $1=1^2+0^2+0^2+0^2$ and $2=1^2+1^2+0^2+0^2$ the result will follow if we can show that every odd prime is a sum of four squares. We do this as follows.

No two distinct elements of the set $S=\{x^2:x=0, 1, \ldots, (p-1)/2\}$ are congruent $(\operatorname{mod} p)$, this also holds for the set $T=\{-1-y^2:y=0, 1, \ldots, (p-1)/2\}$. $S\cup T$ has $p+1$ elements, so integers $x\in S$ and $y\in T$ can be found to satisfy

$$x^2\equiv -1-y^2 \pmod p$$

with $x, y<p/2$. This gives $x^2+y^2+1^2+0^2\equiv 0 \pmod p$ and so the equation

$$x_1^2+x_2^2+x_3^2+x_4^2=kp \tag{10}$$

is soluble with $k<p$ and $x_i<p/2$ for $i=1, 2, 3$, and 4. For the reduction step there are two cases to consider.

Case 1. k is even. Here, by (10), we have x_1, x_2, x_3, and x_4 are either all even, all odd, or two are even and two are odd. Hence, relabelling if necessary, we have

$$\frac{x_1+x_2}{2}, \quad \frac{x_1-x_2}{2}, \quad \frac{x_3+x_4}{2}, \quad \text{and} \quad \frac{x_3-x_4}{2}$$

are all integers and

$$\left(\frac{x_1+x_2}{2}\right)^2+\left(\frac{x_1-x_2}{2}\right)^2+\left(\frac{x_3+x_4}{2}\right)^2+\left(\frac{x_3-x_4}{2}\right)^2=\frac{k}{2}p.$$

Case 2. k is odd. Let y_i be the absolute least residue modulo k of x_i, $i=1, 2, 3$, and 4, so

$$y_1^2+y_2^2+y_3^2+y_4^2\equiv x_1^2+x_2^2+x_3^2+x_4^2 \pmod k$$
$$\equiv 0 \pmod k \text{ by (10)}.$$

Thus $y_1^2+y_2^2+y_3^2+y_4^2=k'k$ where $k'<k$ as k is odd. Now if we let z_i equal the term in the ith bracket on the right-hand side of (9) we have, using (10),

$$z_1^2+z_2^2+z_3^2+z_4^2=k^2k'p.$$

Further

$$z_1 = x_1y_1 + x_2y_2 + x_3y_3 + x_4y_4 \equiv x_1^2 + x_2^2 + x_3^2 + x_4^2 \pmod{k}$$
$$\equiv 0 \pmod{k},$$

and

$$z_2 = x_1y_2 - x_2y_1 + x_3y_4 - x_4y_3 \equiv 0 \pmod{k},$$

with similar identities for z_3 and z_4. Hence z_i/k is an integer for each i and

$$(z_1/k)^2 + (z_2/k)^2 + (z_3/k)^2 + (z_4/k)^2 = k'p.$$

The result follows by applying the infinite descent method to Cases 1 and 2 above, with (10) as the initial step. □

This important result has applications both inside and outside number theory. For example an application in mathematical logic shows that the notion of a natural number can be defined in the first-order theory of the addition and multiplication of integers; that is, n is a natural number if and only if the equation $n = x^2 + y^2 + z^2 + t^2$ is soluble in $\mathbb{Z}$ (see Robinson, 1949).

The infinite descent method has provided solutions to the two- and four-square problems; using more precise arguments involving the σ and τ functions, it is possible to give formulas for the number of representations, see Problems 2 and 4. So far we have not considered the three square problem. This is more difficult because there is no product formula, similar to (4) or (9) above, in this case. For example $3 = 1^2 + 1^2 + 1^2$ and $13 = 3^2 + 2^2 + 0^2$, but 39 cannot be represented as a sum of three squares. It is a simple matter to check that if $n = 4^t(8k + 7)$, where t and k are non-negative integers, then n cannot be represented as a sum of three squares. [Use the fact that, for all x, $x^2 \equiv 0$, 1, or 4 (mod 8).] If n is not of this form then it can be represented as a sum of three squares, we shall prove this in Chapter 9.

More generally we can ask whether all integers can be expressed as a sum of kth powers with g summands where g depends only on k. This is usually known as Waring's problem. In 1770, Waring postulated that $g(2) = 4$, $g(3) = 9$, $g(4) = 19$, and so on. In 1909, Hilbert was the first to show that g exists for all k, an elementary proof of this result is given in Ellison (1971). Earlier Wieferich had shown that $g(3) = 9$ using the three-square theorem mentioned above, see Bachmann (1966). More recently Chen proved that $g(5) = 37$.

We shall not consider Waring's problem in detail, a good exposition is given in Hardy and Wright (1954, Chapter 21). There is one aspect which distinguishes the cases $k = 2$ and $k > 2$; in the latter case there are a few exceptional integers which require more summands than usual. For instance, if $k = 3$ only 23 and 239 need nine cubes and if $n > 8042$ then, in all probability, seven cubes are sufficient. Further if $k = 4$ it is most likely that 79, 159, 239, 319, 399, 479, and 559 are the only integers requiring nineteen fourth powers, note $79 = 4 \cdot 2^4 + 15 \cdot 1^4$, etc. If we let $G(k)$ denote the least number of summands needed to express all sufficiently large positive integers as a sum of kth powers then clearly

$$G(k) \leq g(k) \quad \text{and} \quad G(2) = g(2) = 4.$$

For most k the precise value of $G(k)$ is not known. Upper and lower bounds exist for all k, $G(3)$ lies between 4 and 7 with the most likely value 4 or 5, and $G(4) = 16$ (see Problem 9).

6.2 Gauss and Jacobi sums

Throughout this section let p denote an odd prime and ζ a pth root of unity with $\zeta \neq 1$. We have immediately, factorizing $\zeta^p - 1$,

LEMMA 2.1

$$\sum_{t=0}^{p-1} \zeta^t = 0.$$

Other sums of powers of ζ have been considered, for example

$$\sum_{x=0}^{p-1} \zeta^{x^2}.$$

Gauss was one of the first to study this sum, he derived its main properties and gave some remarkable applications as we shall see below.

Note first that, as the congruence $x^2 \equiv t \pmod{p}$ has $1 + (t/p)$ solutions and $\zeta^t = \zeta^u$ if $t \equiv u \pmod{p}$,

$$\sum_{x=0}^{p-1} \zeta^{x^2} = \sum_{t=0}^{p-1} (1 + (t/p))\zeta^t = \sum_{t=0}^{p-1} (t/p)\zeta^t \qquad (*)$$

using Lemma 2.1. As the Legendre symbol $(\cdot/p)$ is a character modulo p, this suggests the following

DEFINITION Suppose χ is a fixed non-principal character and χ_0 is the principal character, both modulo p. If $a \in \mathbb{Z}$ let g_a be given by

$$g_a(\chi) = \sum_{t=0}^{p-1} \chi(t)\zeta^{at},$$

and

$$g_a(\chi_0) = \sum_{t=0}^{p-1} \zeta^{at}.$$

$g_a(\chi)$ is called a *Gauss sum*. We write $g(\chi)$ for $g_1(\chi)$.

Three comments on this definition:

(i) The sum given in (*) above is $g(\chi_1)$ where χ_1 is the Legendre symbol, that is the real non-principal character (mod p), see Theorem 3.6 of Chapter 5.

(ii) By Lemma 2.1 we have

$$g_a(\chi_0) = \sum_{t=0}^{p-1} \chi_0(t)\zeta^{at} + 1 = \begin{cases} p & \text{if } a \equiv 0 \pmod p \\ 0 & \text{if } a \not\equiv 0 \pmod p. \end{cases}$$

(iii) Using the periodic structure of χ and ζ we see that if $a \equiv b \pmod p$ then $g_a(\chi) = g_b(\chi)$.

The basic properties of the Gauss sum are given by the next two results.

THEOREM 2.2 *Let $a \in \mathbb{Z}$, χ be a non-principal character* (mod p), *and a^* satisfy $aa^* \equiv 1 \pmod p$ if $a \not\equiv 0 \pmod p$. Then*

$$g_a(\chi) = \begin{cases} \chi(a^*)g(\chi) & \textit{if } a \not\equiv 0 \pmod p \\ 0 & \textit{if } a \equiv 0 \pmod p. \end{cases}$$

COROLLARY

$$\sum_{t=0}^{p-1} (t/p)\zeta^{at} = (a/p) \sum_{t=0}^{p-1} (t/p)\zeta^{t}.$$

Proof. If $a \not\equiv 0 \pmod p$ we have

$$\chi(a)g_a(\chi) = \sum_{t=0}^{p-1} \chi(at)\zeta^{at} = g(\chi)$$

for as t ranges over $\mathbb{Z}/p\mathbb{Z}$ so does at. Now note that the multiplicative inverse of $\chi(a)$ is $\chi(a^*)$. Using comment (ii) above, $a \equiv 0 \pmod p$ implies

$$g_a(\chi) = g_0(\chi) = \sum_{t=0}^{p-1} \chi(t) = 0 \qquad (*)$$

by Theorem 3.4 of Chapter 5 as χ is not principal. The Corollary is a restatement of the theorem with χ as the Legendre symbol. □

THEOREM 2.3 *If χ is a non-principal character modulo p then $|g(\chi)| = \sqrt{p}$ where as usual $|z|$ denotes the modulus of the complex number z. Further if χ is real, and $g^k(\chi) = (g(\chi))^k$, then*

$$g^2(\chi) = (-1)^{(p-1)/2}p.$$

Proof. This result is derived by evaluating the sum T below in two ways. If $a \not\equiv 0 \pmod p$ and $aa^* \equiv 1 \pmod p$ we have, by Lemma 3.1, Chapter 5, $\chi(a^*)\chi(a) = 1$ and so, as $|\chi(a)| = 1$, $\overline{\chi(a^*)} = \chi(a)$. Taking the conjugate of Theorem 2.2 we obtain

$$\overline{g_a(\chi)} = \overline{\chi(a^*)g(\chi)} = \chi(a)\overline{g(\chi)},$$

and so

$$g_a(\chi)\overline{g_a(\chi)} = \chi(a^*)g(\chi)\chi(a)\overline{g(\chi)} = |g(\chi)|^2.$$

Hence, as $g_0(\chi) = 0$ (if $\chi \neq \chi_0$) by (*) above,

$$T = \sum_{a=0}^{p-1} g_a(\chi)\overline{g_a(\chi)} = (p-1)\,|g(\chi)|^2. \qquad (**)$$

Further we have, for $0 \leq x, y \leq p-1$,

$$\sum_{a=0}^{p-1} \chi(x)\overline{\chi(y)}\zeta^{a(x-y)} = \begin{cases} 0 & \text{if } x \neq y \text{ or if } x = y = 0 \\ p & \text{if } x = y \neq 0. \end{cases}$$

[Use Lemma 2.1 and $\chi(0) = 0$ for the upper part, and $\chi(x)\overline{\chi(x)}\zeta^0 = 1$ for the lower part.] This gives

$$\begin{aligned} T &= \sum_{a=0}^{p-1}\left(\sum_{x=0}^{p-1} \chi(x)\zeta^{ax}\right)\left(\sum_{y=0}^{p-1} \overline{\chi(y)}\zeta^{-ay}\right) \\ &= \sum_{x=0}^{p-1}\sum_{y=0}^{p-1}\sum_{a=0}^{p-1} \chi(x)\overline{\chi(y)}\zeta^{a(x-y)} \\ &= p(p-1). \end{aligned}$$

The first part follows using (**).

For the second part we have $\chi(a) = (a/p)$ by Theorem 3.6, Chapter 5. Hence, as $\overline{\zeta^t} = \zeta^{-t}$,

$$\begin{aligned} \overline{g(\chi)} &= \sum_{t=0}^{p-1} (t/p)\zeta^{-t} = (-1/p)\sum_{t=0}^{p-1} (-t/p)\zeta^{-t} \\ &= (-1/p)g(\chi) \end{aligned}$$

and so by the first part

$$p = g(\chi)\overline{g(\chi)} = (-1/p)g^2(\chi).$$

This completes the proof as $(-1/p) = (-1)^{(p-1)/2}$. □

We shall consider the exact value of $g(\chi)$ in the next section.

Example Let $p = 3$ and so $\zeta^3 = 1$. The only non-principal character is given by $\chi(0) = 0$, $\chi(1) = 1$ and $\chi(2) = -1$, hence

$$g^2(\chi) = \left(\sum_{t=0}^{2} \chi(t)\zeta^t\right)^2 = (\zeta - \zeta^2)^2 = -3.$$

Using Gauss sums we give now a new proof of the quadratic reciprocity law and introduce Jacobi sums, which in turn will enable us to estimate the number of solutions that some congruences possess.

Quadratic reciprocity

Our final proof of this result is based on Gauss's sixth proof; unlike those given in Chapter 4 it does not rely on Gauss's lemma and has the advantage that it suggest an approach to more general reciprocity laws for the cubic, quartic, . . . cases.

Let p and q be distinct odd primes and let $p^* = (-1)^{(p-1)/2}p$. Gauss's quadratic reciprocity law can be stated in the form

$$(q/p) = (p^*/q),$$

for

$$(p^*/q) = (-1/q)^{(p-1)/2}(p/q) = (-1)^{(p-1)(q-1)/4}(p/q)$$

using Theorem 1.3 of Chapter 4.

We shall work with congruences modulo q in the ring of algebraic integers (see p. 78). By Theorem 2.3 we have, writing g for $g(\chi)$,

$$g^{q-1} = g^{2(q-1)/2} = (p^*)^{(q-1)/2} \equiv (p^*/q) \pmod q$$

using Euler's criterion (Theorem 1.2, Chapter 4), hence

$$g^q \equiv (p^*/q)g \pmod q.$$

But by the definition of g, and as the binomial coefficients are either equal to 1 or divisible by q,

$$g^q = \left[\sum_{t=0}^{p-1} (t/p)\zeta^t\right]^q \equiv \sum_{t=0}^{p-1} (t/p)^q \zeta^{qt} = g_q \pmod q.$$

Hence these two congruences give

$$g_q \equiv (p^*/q)g \pmod q.$$

Now applying the corollary of Theorem 2.2 to g_q we have

$$(q/p)g \equiv (p^*/q)g \pmod q,$$

and, multiplying by g, we obtain

$$(q/p)p^* \equiv (p^*/q)p^* \pmod q$$

by Theorem 2.3. Finally as p and q are coprime and $q > 2$ we have

$$(q/p) = (p^*/q). \qquad \square$$

Jacobi sums

For our second application of Gauss sums we shall consider the problem of counting the number of solutions of the congruence

$$x^n + y^n \equiv 1 \pmod{p},$$

an example of a general result of Weil (see below). The proofs use Jacobi sums which are closely related to Gauss sums and whose properties can be derived easily from the theorems above.

DEFINITION Let χ and ψ be non-principal characters modulo p. Define the *Jacobi sum J* by

$$J(\chi, \psi) = \sum_{s+t=1} \chi(s)\psi(t)$$

where the sum is taken over all pairs $\{s, t\}$ such that $0 \le s,t \le p-1$ and $s+t \equiv 1 \pmod{p}$. For the principal character χ_0 set

$$J(\chi_0, \chi_0) = p \quad \text{and} \quad J(\chi_0, \chi) = J(\chi, \chi_0) = 0.$$

For example if χ is the Legendre symbol modulo 3 then

$$J(\chi, \chi) = (0/3)(1/3) + (1/3)(0/3) + (2/3)(2/3) = 1.$$

The main properties of this sum are given by

THEOREM 2.4 *Let χ and ψ be non-principal characters* $(\operatorname{mod} p)$.

(i) $J(\chi, \bar{\chi}) = -\chi(-1)$.

(ii) *If $\chi\psi \neq \chi_0$ then*

$$J(\chi, \psi) = \frac{g(\chi)g(\psi)}{g(\chi\psi)} \quad \textit{and} \quad |J(\chi, \psi)| = \sqrt{p}.$$

(iii) *If $p \equiv 1 \pmod{n}$, $n > 2$ and χ is a character of order n (that is $\chi^n = \chi_0$) then*

$$g^n(\chi) = p\chi(-1)J(\chi, \chi)J(\chi, \chi^2) \ldots J(\chi, \chi^{n-2}).$$

Proof. (i) Let t^* satisfy $tt^* \equiv 1 \pmod{p}$ so $\bar{\chi}(t) = \chi(t^*)$. Note that

$$\text{as } t \text{ ranges between 2 and } p-1,\ (p+1-t)t^* \text{ ranges between 1 and } p-2. \tag{**}$$

(**) holds for if $(p+1-t)t^* \equiv (p+1-u)u^* \pmod{p}$ then $(p+1)t^* \equiv (p+1)u^*$ and so $t^* \equiv u^* \pmod{p}$. This is only possible if $t \equiv u \pmod{p}$ by Theorem 1.6 of Chapter 3. Now

$$\begin{aligned} J(\chi, \bar{\chi}) &= \chi(1)\bar{\chi}(0) + \chi(0)\bar{\chi}(1) + \sum_{\substack{s+t=1 \\ s,t \not\equiv 0 \, (\operatorname{mod} p)}} \chi(s)\bar{\chi}(t) \\ &= \sum_{t=2}^{p-1} \chi((p+1-t)t^*) \\ &= \sum_{u=1}^{p-2} \chi(u) = -\chi(-1) \end{aligned}$$

by (**) and Theorem 3.4 of Chapter 5.

(ii) We have, using the periodic properties of χ, ψ and ζ,

$$g(\chi)g(\psi) = \left[\sum_{s=0}^{p-1} \chi(s)\zeta^s\right]\left[\sum_{t=0}^{p-1} \psi(t)\zeta^t\right]$$

$$= \sum_{t=0}^{p-1}\left[\sum_{x+y=t} \chi(x)\psi(y)\right]\zeta^t.$$

For $t=0$ the inner sum is

$$\sum_{x=0}^{p-1} \chi(x)\psi(-x) = \chi(-1)\sum_{x=0}^{p-1} \chi\psi(x) = 0$$

as $\chi\psi \neq \chi_0$. For $t \neq 0$ we can find u and v so that $x \equiv tu$ and $y \equiv tv \pmod p$ and then $x+y \equiv t \pmod p$ if and only if $u+v \equiv 1 \pmod p$. Hence, if $t \neq 0$,

$$\sum_{x+y=t} \chi(x)\psi(y) = \sum_{u+v=1} \chi(tu)\psi(tv) = \chi\psi(t)J(\chi, \psi)$$

and the first part of (ii) follows. For the second part use Theorem 2.3.

(iii) By (ii) we have

$$g^2(\chi) = J(\chi, \chi)g(\chi^2)$$

and, multiplying by $g(\chi)$ repeatedly,

$$g^{n-1}(\chi) = J(\chi, \chi)J(\chi, \chi^2)\ldots J(\chi, \chi^{n-2})g(\chi^{n-1}).$$

But $\chi^{n-1} = \bar{\chi}$ in this case, so

$$g(\chi)g(\chi^{n-1}) = g(\chi)g(\bar{\chi}) = p\chi(-1)$$

by Theorem 2.3 as $\overline{g(\chi)} = \chi(-1)g(\bar{\chi})$. Hence the result follows if we multiply the equation for $g^{n-1}(\chi)$ by $g(\chi)$. □

We use the Jacobi sums first to re-prove two results from Section 1 and give solutions of a related Diophantine equation.

(i) The equation $p = a^2 + b^2$ is soluble if $p \equiv 1 \pmod 4$. There is a character χ modulo p of order 4. For if ψ has order $p-1$ let $\chi = \psi^{(p-1)/4}$, and the values of χ belong to the set $\{1, i, -1, -i\}$. Hence the Jacobi sum $J(\chi, \chi)$ is a Gaussian integer, that is $J(\chi, \chi) = a + ib$ where $a, b \in \mathbb{Z}$. Now by Theorem 2.4

$$p = |J(\chi, \chi)|^2 = a^2 + b^2.$$

(ii) The equation $p = c^2 - cd + d^2$ is soluble if $p \equiv 1 \pmod 3$. We proceed as above, let χ be a character of order 3, so $J(\chi, \chi)$ has the form $c + \omega d$ where $\omega = e^{2\pi i/3}$ and $c, d \in \mathbb{Z}$. Again using Theorem 2.4 we have

$$p = |J(\chi, \chi)|^2 = (c + \omega d)(c + \bar{\omega} d) = c^2 - cd + d^2.$$

THEOREM 2.5 *If $p \equiv 1 \pmod 3$ then integers u and v can be found to satisfy*

$$4p = u^2 + 27v^2 \quad \textit{and} \quad u \equiv 1 \pmod 3.$$

Proof. By (ii) above we have

$$\begin{aligned} 4p &= 4(c^2 - cd + d^2) \\ &= (c+d)^2 + 3(c-d)^2 = (2c-d)^2 + 3d^2 = (2d-c)^2 + 3c^2. \end{aligned}$$

One of c, d, or $c - d$ is divisible by 3. For suppose $3 \nmid c$ and $3 \nmid d$ then if either $c \equiv 1$ and $d \equiv 2 \pmod 3$, or $c \equiv 2$ and $d \equiv 1 \pmod 3$, we have $3 | c^2 - cd + d^2$. This is impossible as $p \equiv 1 \pmod 3$. Hence the equation $4p = u^2 + 27v^2$ has a solution, and if $u \equiv 2 \pmod 3$ we can replace it with $-u$. □

It can be shown that the solution $\{u, v\}$ above is unique using the fact that $\mathbb{Z}[\omega]$ has unique factorization, see Problem 3. This equation has connections with two problems involving cubic congruences. The first is given in Theorem 2.7 below and the second states that, if $p \equiv 1 \pmod 3$, a solution exists to the equation

$$p = x^2 + 27y^2$$

if and only if

$$z^3 \equiv 2 \pmod p$$

is soluble. A proof can be found in Ireland and Rosen (1982), note that a similar result is given in Problem 20 of Chapter 4.

We come now to the question: how many solutions exist for the congruence

$$x^n + y^n \equiv 1 \pmod p?$$

We shall answer this exactly when $n = 2$ or 3, and give an estimate when $n > 3$ (Problem 14). Weil (1949) has shown that for a very general class of polynomials f, if N denotes the number of solutions of the congruence $f \equiv 0 \pmod p$ then

$$|N - (p+1)| \leq 2g\sqrt{p}$$

where g is a constant called the genus of f (see Chapter 15); so provided that g is small compared with p, this congruence will always have many solutions. Returning to the congruence above we need a lemma for the case $n = 3$.

LEMMA 2.6 *If* $p \equiv 1 \pmod 3$, χ *is a character of order* 3 *and* $J(\chi, \chi) = a + \omega b$, *where* $\omega = e^{2\pi i/3}$ *and* $a, b \in \mathbb{Z}$, *then*

(i) $a \equiv -1 \pmod 3$;
(ii) $3|b$;
(iii) $4p = (2a - 3b_1)^2 + 27b_1^2$ *where* $3b_1 = b$;
(iv) *the real part of* $J(\chi, \chi)$ *is* $\frac{1}{2}(2a - b)$.

Proof. We have

$$g^3(\chi) \equiv \sum_{t=0}^{p-1} \chi(t)^3 \zeta^{3t} = \sum_{t=1}^{p-1} \zeta^{3t} = -1 \pmod 3)$$

by Lemma 2.1. But by Theorem 2.4, as $\chi(-1) = \chi((-1)^3) = 1$,

$$g^3(\chi) = pJ(\chi, \chi) \equiv a + \omega b \pmod 3$$

as $p \equiv 1 \pmod 3$. Combining these we obtain

$$a + \omega b \equiv -1 \pmod 3$$

and, taking conjugates,

$$a + \bar{\omega} b \equiv -1 \pmod 3.$$

Hence $a \equiv -1 \pmod 3$ and $3|b(\omega - \bar{\omega})$, that is $3^2|-3b^2$ and so $3|b$. For (iii) note that

$$4p = 4|J(\chi, \chi)|^2 = (2a - b)^2 + 3b^2,$$

and (iv) follows similarly. □

Let $N(f(x, y) \equiv 0(p))$ denote the number of pairs of integers $\{x, y\}$ satisfying $0 \le x, y \le p - 1$ and $f(x, y) \equiv 0 \pmod p$.

THEOREM 2.7 (i) $N(x^2 + y^2 \equiv 1 (p)) = p - (-1/p)$.

(ii) *If* $p \equiv 1 \pmod 3$ *then* $N(x^3 + y^3 \equiv 1 (p)) = p - 2 + u$ *where* u *is the solution to the equation* $4p = u^2 + 27v^2$ *given in Theorem* 2.5.

Proof. (i) We have

$$N_1 = N(x^2 + y^2 \equiv 1 (p)) = \sum_{a+b=1} N(x^2 \equiv a (p))N(x^2 \equiv b (p))$$

where this sum has the same form as that in the definition of the Jacobi sum on p. 102. But $N(x^2 \equiv a (p)) = 1 + \psi(a)$ where ψ is the Legendre symbol $(\cdot/p)$ and so, multiplying out, we obtain

$$N_1 = p + \sum_{a=0}^{p-1} \psi(a) + \sum_{b=0}^{p-1} \psi(b) + J(\psi, \psi)$$
$$= p - \psi(-1)$$

by Lemma 2.1 and Theorem 2.4(i).

(ii) There are two characters of order exactly 3, χ and χ^2 where $\chi^3 = \chi_0$, so using Problem 25 of Chapter 5

$$N(x^3 \equiv a\ (p)) = 1 + \chi(a) + \chi^2(a).$$

Hence

$$\begin{aligned}
N(x^3 + y^3 \equiv 1\ (p)) &= \sum_{a+b=1} N(x^3 \equiv a\ (p))N(x^3 \equiv b\ (p)) \\
&= \sum_{a+b=1} (1 + \chi(a) + \chi^2(a))(1 + \chi(b) + \chi^2(b)) \\
&= p + J(\chi, \chi) + J(\chi, \chi^2) + J(\chi^2, \chi) + J(\chi^2, \chi^2) \\
&\qquad \text{by Lemma 2.1 and the definition of } J, \\
&= p - \chi(-1) - \chi^2(-1) + J(\chi, \chi) + J(\chi^2, \chi^2) \\
&\qquad \text{by Theorem 2.4(i) as } \chi^2 = \bar{\chi}, \\
&= p - 2 + 2[\text{real part}(J(\chi, \chi))] \\
&\qquad \text{as } \chi(-1) = 1 \text{ and } \chi^2 = \bar{\chi}, \\
&= p - 2 + u
\end{aligned}$$

by Lemma 2.6. □

Note that in part (i) above each solution $\{x, y\}$ is associated with seven (three if either x or y is zero or $x = y$) other solutions formed by either interchanging x and y or by replacing x or y with $p - x$ or $p - y$, respectively. In part (ii) the solutions are in pairs $\{x, y\}$ and $\{y, x\}$, unless $x = y$. As an example consider the case $p = 13$.

(i) We have $(-1/13) = 1$ and so $N(x^2 + y^2 \equiv 1\ (13)) = 12$, and the solutions are generated by $\{0, 1\}$ and $\{2, 6\}$.

(ii) We have $52 = (-5)^2 + 27(1)^2$ and $-5 \equiv 1 \pmod 3$, therefore $N(x^3 + y^3 \equiv 1\ (13)) = 6$. The solutions are $\{0, 1\}$, $\{1, 0\}$, $\{0, 3\}$, $\{3, 0\}$, $\{0, 9\}$, and $\{9, 0\}$.

6.3 The sign of the quadratic Gauss sum

Theorem 2.3 states that if ψ is the real non-trivial character modulo p, that is the Legendre symbol $(\cdot/p)$, then

$$g(\psi) = \begin{cases} \pm\sqrt{p} & \text{if } p \equiv 1 \pmod 4 \\ \pm i\sqrt{p} & \text{if } p \equiv -1 \pmod 4. \end{cases} \tag{1}$$

The + sign is the correct one in both cases but there is no simple or direct proof of this. Gauss records in his diary that this result was one of the most challenging he faced during his whole career, he worked on it for four years finally proving it in 1805. A number of proofs now exist using a variety of methods. The one given here is due to Schur, it uses some elementary facts about matrices and their determinants.

We begin by giving a sketch of the proof. A matrix S is defined below so that

$$g(\psi) = \text{trace}(S) = \lambda_1 + \lambda_2 + \ldots + \lambda_p,$$

where λ_j $(j = 1, \ldots, p)$ are the eigenvalues of S, and we show that each λ_j equals $\sqrt{p}$, $-\sqrt{p}$, $i\sqrt{p}$, or $-i\sqrt{p}$. Suppose these values occur with multiplicities r, s, t, and u, respectively, so $r + s + t + u = p$. We derive three further relations for the integers r, s, t, and u from which we can deduce the value of $g(\psi)$.

THEOREM 3.1 *If ψ is the real non-principal character modulo p then the value of the quadratic Gauss sum is given by*

$$g(\psi) = \begin{cases} \sqrt{p} & \text{if } p \equiv 1 \pmod 4 \\ i\sqrt{p} & \text{if } p \equiv -1 \pmod 4. \end{cases}$$

Let S be the $p \times p$ matrix whose (j, k)th element is ζ^{jk} where $\zeta = e^{2\pi i/p}$, that is

$$S = \begin{pmatrix} 1 & 1 & \ldots & 1 \\ 1 & \zeta & \ldots & \zeta^{p-1} \\ \cdot & \cdot & \cdot\ \cdot & \cdot \\ 1 & \zeta^{p-1} & \ldots & \zeta^{(p-1)^2} \end{pmatrix}.$$

Now as

$$\sum_{t=0}^{p-1} \zeta^{jt}\zeta^{kt} = \sum_{t=0}^{p-1} \zeta^{(j+k)t} = \begin{cases} p & \text{if } j + k \equiv 0 \pmod p \\ 0 & \text{if } j + k \not\equiv 0 \pmod p), \end{cases}$$

we have immediately

$$S^2 = \begin{pmatrix} p & 0 & \ldots & 0 \\ 0 & 0 & \ldots & p \\ \cdot & \cdot & \cdot\ \cdot & \cdot \\ 0 & p & \ldots & 0 \end{pmatrix}.$$

LEMMA 3.2 $\det(S) = i^{p(p-1)/2} p^{p/2}$.

Proof. As the interchange of rows in a determinant changes the sign, we have

$$\det(S^2) = p^p(-1)^{(p-1)/2} = (-1)^{p(p-1)/2} p^p,$$

[as $-1 = (-1)^p$], hence

$$\det(S) = \pm i^{p(p-1)/2} p^{p/2}. \tag{2}$$

To determine the sign we proceed as follows. Note first that

$$U = \sum_{0\le j<k\le p-1} j+k = \sum_{k=1}^{p-1}\sum_{j=0}^{k-1} j+k = \sum_{k=1}^{p-1} k^2 + k(k-1)/2$$
$$= 2p((p-1)/2)^2, \tag{3}$$

that is U is divisible by $2p$. Now let $\eta = e^{i\pi/p}$ (so $\eta^U = 1$) and we have

$$\zeta^k - \zeta^j = \eta^{j+k}(\eta^{k-j} - \eta^{-(k-j)}) = \eta^{j+k} 2i \sin((k-j)\pi/p).$$

So, as $\det(S)$ is a Vandermonde determinant (see (i) on p. viii),

$$\det(S) = \prod_{0\le j<k\le p-1} (\zeta^k - \zeta^j)$$
$$= \eta^U \prod_{0\le j<k\le p-1} 2i \sin((k-j)\pi/p)$$
$$= i^{p(p-1)/2} \prod_{0\le j<k\le p-1} 2\sin((k-j)\pi/p)$$

by (3). Now $2\sin((k-j)\pi/p) > 0$ for all j, k such that $0 \le j < k \le p-1$, hence the + sign is correct in (2) and the lemma follows. □

Proof of Theorem 3.1. By (*) on p. 98 and the definition of g we have

$$g(\psi) = \text{trace}(S) = \lambda_1 + \ldots + \lambda_p$$

where λ_j $(j = 1, \ldots, p)$ are the eigenvalues of the matrix S. Now λ_j^2 $(j = 1, \ldots, p)$ are the eigenvalues of S^2 [as $S^2 - \lambda^2 I = (S - \lambda I)(S + \lambda I)$] and the characteristic polynomial of S^2 is

$$\det\begin{pmatrix} p-x & 0 & \ldots & 0 \\ 0 & -x & \ldots & p \\ \cdot & \cdot & \cdot & \cdot \\ 0 & p & \ldots & -x \end{pmatrix} = -(x-p)^{(p+1)/2}(x+p)^{(p-1)/2}.$$

So the set of eigenvalues $\{\lambda_1^2, \ldots, \lambda_p^2\}$ consists of $(p+1)/2$ copies of p and $(p-1)/2$ copies of $-p$. Therefore

$$\lambda_j = \pm\sqrt{p} \quad \text{or} \quad \pm i\sqrt{p} \qquad (j = 1, \ldots, p)$$

Suppose $\sqrt{p}$ occurs r times, $-\sqrt{p}$ occurs s times, $i\sqrt{p}$ occurs t times and $-i\sqrt{p}$ occurs u times, then we have

$$r + s = \frac{p+1}{2} \quad \text{and} \quad t + u = \frac{p-1}{2}. \tag{4}$$

Secondly as $g(\psi) = \text{trace}(S)$ we have

$$g(\psi) = (r - s + (t-u)i)\sqrt{p}, \tag{5}$$

therefore by (1)

$$\begin{aligned} r-s=\pm 1 \quad &\text{and} \quad t=u \qquad \text{if } p\equiv 1 \pmod 4\\ r=s \quad &\text{and} \quad t-u=\pm 1 \qquad \text{if } p\equiv -1 \pmod 4. \end{aligned} \tag{6}$$

Finally we have

$$\det(S)=\prod_{j=1}^{p}\lambda_j=(-1)^s i^t(-i)^u p^{p/2}=i^{2s+t-u}p^{p/2},$$

so by Lemma 3.2

$$2s+t-u\equiv p(p-1)/2 \pmod 4. \tag{7}$$

We can now derive the result using (4)–(7). Suppose first $p\equiv 1 \pmod 4$, so by (6) and (7) $2s\equiv(p-1)/2 \pmod 4$ and hence, by (4),

$$r-s=\frac{p+1}{2}-2s\equiv\frac{p+1}{2}-\frac{p-1}{2}=1 \pmod 4.$$

By (6) this gives $r-s=1$. The first half of the result now follows by (5). Secondly suppose $p\equiv -1 \pmod 4$. By (7) we have

$$t-u\equiv -\frac{p-1}{2}+2s=-\frac{p-1}{2}+\frac{p+1}{2}=1 \pmod 4)$$

by (4) and (6) as $r=s$ in this case. Hence by (6), $t-u=1$ and the theorem is proved. □

This result can be extended to all composite moduli: namely, if $\zeta=e^{2\pi i/n}$

$$\sum_{j=0}^{n-1}\zeta^{j^2}=\frac{(1+i)(1+i^{-n})}{2}\sqrt{n}=\begin{cases}(1+i)\sqrt{n} & \text{if } 4|n\\ \sqrt{n} & \text{if } n\equiv 1 \pmod 4\\ 0 & \text{if } n\equiv 2 \pmod 4\\ i\sqrt{n} & \text{if } n\equiv 3 \pmod 4.\end{cases}$$

We shall not require this more general form, see for example Landau (1958) where four proofs of Gauss's result are given. Case 3 is treated in Problem 17.

6.4 Problems 6

1. Assume the result: if α and β are real numbers and $\beta\geq 1$ then integers u and y exist satisfying

$$(u,y)=1, \qquad 0<y\leq\beta \quad \text{and} \quad \left|\alpha-\frac{u}{y}\right|<\frac{1}{y\beta},$$

(see Chapter 7). Using this prove a strengthened version of Theorem 1.2: if $n>1$

and $z^2 \equiv -1 \pmod{n}$ is soluble then unique integers x and y exist satisfying

$$(x, y) = 1, \quad x > 0, \quad y > 0, \quad x \equiv zy \pmod{n} \quad \text{and} \quad n = x^2 + y^2.$$

☆2. The number of representations of a positive integer as a sum of two squares.

(i) Show that the number of solutions of

$$n = x^2 + y^2, \qquad (x, y) = 1$$

is $4v(n)$ where $v(n)$ is the number of solutions of $x^2 \equiv -1 \pmod{n}$; $v(n) = v(-1, n)$, see Problems 4 and 15, Chapter 4.

(ii) Using this show that the number of solutions of the equation $n = x^2 + y^2$ is $u(n)$ where

$$u(n) = 4 \sum_{d^2 | n} v\left(\frac{n}{d^2}\right).$$

(iii) Deduce $u(n) = 4 \sum_{d|n} \chi_2(d)$ where χ_2 is the character defined on p. 86.

(iv) Show further that

$$u(n) = \begin{cases} 0 & \text{if } p \equiv 3 \pmod 4 \text{ and } p \text{ divides } n \text{ with odd multiplicity,} \\ 4\tau(\hat{n}) & \text{otherwise,} \end{cases}$$

where $\hat{n}$ is the product of the prime powers p^{α} in the prime factorization of n for those p satisfying $p \equiv 1 \pmod 4$.

(v) Finally deduce that if $p \equiv 1 \pmod 4$ then $p = x^2 + y^2$ has 'essentially' only one solution, i.e. one where $x > 0$, $y > 0$ and $2|x$.

3. Let $p \equiv 1 \pmod 3$. Show that the equation $4p = x^2 + 27y^2$ has a unique solution provided that we assume $x \equiv 1 \pmod 3$. (You will need to consider the integral domain $\mathbb{Z}[\omega]$ where $\omega = (-1 + \sqrt{-3})/2$, assume it has unique factorization, see Problem 6, Chapter 5.)

☆☆4. In this problem we shall find the number of ways $Q(n)$ of expressing a positive integer n as a sum of four squares counting separately changes of order and sign.

(i) Using (ii) and (iii) of Problem 2 show that if n is odd and $a(n)$ is the number of solutions of

$$4n = u_1^2 + u_2^2 + u_3^2 + u_4^2$$

where each u_i is odd, then $a(n) = \sigma(n)$.

(ii) Using substitutions show that, if n is odd,

$$Q(2n) = 3Q(n), \quad Q(2n) = Q(4n), \quad \text{and} \quad Q(4n) = 16\sigma(n) + Q(n).$$

(iii) Finally show that, if n is odd and $s > 0$,

$$Q(n) = 8\sigma(n), \qquad Q(2^s n) = 24\sigma(n).$$

(iv) Check this result when $n = 105$ and when $n = 210$.

5. For $a > 0$, find directly all solutions of

$$2^a = x_1^2 + x_2^2 + x_3^2 + x_4^2.$$

Deduce that there are infinitely many n such that

$$n = x_1^2 + x_2^2 + x_3^2 + x_4^2$$

has no solution if (i) $(x_1, x_2, x_3, x_4) = 1$, or if (ii) $x_1^2 > x_2^2 > x_3^2 > x_4^2$.

6. Show that for all large n

$$n = x_1^2 + x_2^2 + x_3^2 + x_4^2 + x_5^2, \quad x_1^2 > x_2^2 > x_3^2 > x_4^2 > x_5^2$$

are jointly soluble. Use an estimation argument and Problems 2 and 4.

7. Show that if $n > 169$ then n can be written as a sum of five non-zero squares. By direct computation find all integers n which cannot be written as a sum of four non-zero squares.

8. Consider the equation

$$n = \pm x_1^2 \pm x_2^2 \pm \ldots \pm x_k^2,$$

show that this is soluble for all n with suitable choices of the signs when $k = 3$ but not always when $k = 2$.

9. Using the identity

$$6(x_1^2 + x_2^2 + x_3^2 + x_4^2)^2 = \sum_{i,j} \{x_i + x_j)^4 + (x_i - x_j)^4\},$$

where the sum on the right-hand side is taken over all pairs $\{i, j\}$ such that $1 \le i < j \le 4$, show that $g(4) \le 53$—every positive integer can be expressed as a sum of 53 fourth powers. Can the number 53 be replaced by 50?

10. Let χ be a non-principal character and ψ be the Legendre symbol both modulo p. Using Lemma 2.1 and the fact that $t^2 \equiv u(p)$ has $1 + \psi(u)$ solutions, prove that

$$\sum_{t=0}^{p-1} \chi(1 - t^2) = J(\chi, \psi).$$

By making a suitable substitution deduce that, if $k \not\equiv 0$ and $2 \cdot 2^* \equiv 1 \pmod{p}$,

$$\sum_{v=0}^{p-1} \chi(v(k - v)) = \chi((2^*)^2 k^2) J(\chi, \psi).$$

If $\chi^2 \ne \chi_0$, deduce

$$J(\chi, \chi) = \chi((2^*)^2) J(\chi, \psi).$$

[Hint. Expand $g^2(\chi)$ explicitly.]

11. Find all the solutions of $x^2 + y^2 \equiv 1 \pmod{19}$ and $x^3 + y^3 \equiv 1 \pmod{19}$.

12. Show that the number of solutions of the congruence

$$x^2 - y^2 \equiv a \pmod{p}$$

is $p - 1$ if $p \nmid a$, and $2p - 1$ if $p \mid a$.

13. Show that if $p \equiv 2 \pmod 3$ then $x^3 + y^3 \equiv 1 \pmod{p}$ has exactly p solutions. Find them when $p = 5$ and 17.

14. Let $p \equiv 1 \pmod{n}$, let χ be a character of order n, and let

$$I_n = \sum_{j=0}^{n-1} \chi^j(-1).$$

(i) Show that

$$I_n = \begin{cases} n & \text{if } x^n \equiv -1 \pmod{p} \text{ is soluble} \\ 0 & \text{otherwise.} \end{cases}$$

(ii) Deduce $|N(x^n + y^n \equiv 1(p)) + I_n - (p+1)| \leq (n-1)(n-2)\sqrt{p}$.

☆15. Suppose $p \equiv 1 \pmod{4}$, ψ be the Legendre character $(\bmod\, p)$, χ be a character $(\bmod\, p)$ of order 4 (so $\chi^2 = \psi$), and $z = J(\psi, \chi)$. Prove in turn:

(i) z is a Gaussian integer $a + ib$, and

$$J(\psi, \chi) + J(\psi, \chi^3) = z + \bar{z} = 2a$$

where $p = a^2 + b^2$.

(ii) By considering the sum $\sum_{s+t=1} (\psi(s) - 1)(\chi(t) - 1)$ show that if $s \not\equiv 0$ $(\bmod\, p)$ then $\chi(s) - 1 \equiv 0 \pmod{1+i}$ and $z \equiv 1 \pmod{2+2i}$.

(iii) Deduce that a is odd and b is even, $4 | b$ implies $a \equiv 1 \pmod{4}$, and $4 \nmid b$ implies $a \equiv -1 \pmod{4}$.

(iv) $N(x^2 + y^4 \equiv 1\ (p)) = p - 1 - 2a$.

(v) Finally using the transformation $(x, y) \to ((1 + x^2)y, x)$, show that $N(x^2 + y^2 + x^2y^2 \equiv 1\ (p)) = p - 3 - 2a$.

A version of this result was conjectured by Gauss, it was the last entry in his mathematical diary.

(vi) Find solutions for the congruences in (iv) and (v) when $p = 17$.

16. A complex number z is called *constructible* if there exist subfields K_i of the complex field $\mathbb{C}$ such that $\mathbb{Q} = K_0 \subset K_1 \subset \ldots \subset K_m$, $z \in K_m$ and $K_i = K_{i-1}(\sqrt{z_i})$ for some $z_i \in K_{i-1}$, $i = 1, \ldots, m$. Note that z is constructible if and only if both its real and imaginary parts are constructible.

(i) Show that ζ_{2^n} is constructible where $\zeta_t = e^{2\pi i/t}$.

Now let p be a Fermat prime, that is a prime of the form $2^n + 1$ (see p. 10).

(ii) If χ be a character $(\bmod\, p)$, show that $g(\chi)$ is constructible using Theorem 2.4(iii).

(iii) Deduce that ζ_p is constructible.

As all rational operations and the formation of square roots can be constructed geometrically using only a ruler and compass, it follows that if p is a Fermat prime then a regular polygon with p sides can be constructed in this manner.

17. Show that if $n \equiv 2 \pmod{4}$ and $\zeta = e^{2\pi i/n}$ then $\sum_{j=0}^{n-1} \zeta^{j^2} = 0$.

18. In this problem let d, d_1, and d_2 be fundamental discriminants (see Problem 20, Chapter 1), let $\sqrt{k} = \sqrt[+]{k}$ if $k \geq 0$, and $i\sqrt[+]{-k}$ if $k < 0$, and let $e_t(a) = e^{2\pi i a/t}$. Show that, if $n > 0$,

$$\sum_{j=1}^{|d|} (d/j) e_{|d|}(nj) = (d/n)\sqrt{d} \qquad (*)$$

where $(d/\cdot)$ denotes the Kronecker symbol, by proving the following in turn:

(i) $$(d_1/|d_2|)(d_2/|d_1|) = \begin{cases} -1 & \text{if } d_1 < 0 \text{ and } d_2 < 0 \\ 1 & \text{otherwise,} \end{cases}$$

and so deduce

$$(d_1/|d_2|)(d_2/|d_1|)(\sqrt{d_1})(\sqrt{d_2}) = \sqrt{(d_1 d_2)}.$$

(ii) If (*) holds when $d = d_1$, and when $d = d_2$ then it also holds when $d = d_1 d_2$, provided $(d_1, d_2) = 1$. Consider a complete residue system modulo $|d_1 d_2|$ of the form $j_1 |d_2| + j_2 |d_1|$ and use (i).

(iii) It is sufficient to prove (*) when $n = 1$.

(iv) Proposition (*) is true when $d = -4$, ± 8, and $(-1)^{(p-1)/2} p$.

7

CONTINUED FRACTIONS

Some number theoretic properties of the real numbers will be considered in this chapter and its sequel. We begin by asking how close a rational number a/b can approximate to a real number α assuming some restriction on the size of b; that is, given α, can integers a and b be found to satisfy

$$\left|\alpha - \frac{a}{b}\right| < \frac{1}{f(b)}$$

for a suitably chosen function f? This is the basic question in Diophantine approximation theory. One of the main results is Hurwitz's theorem which states that, for all irrational α, the inequality above has infinitely many solutions a and b provided $f(b) \leq b^2\sqrt{5}$, and this is best possible. These considerations have led to applications in many fields, in particular to Diophantine equations and to the theory of transcendental numbers; see Section 3 and Chapter 8.

We shall start by discussing continued fractions, they were introduced in Chapter 1; all good rational approximations to a real number α are continued fraction convergents to α. In Section 1 we shall derive the basic properties including the result relating periodic continued fractions with quadratic numbers. The remaining sections give applications to (i) best approximations, (ii) the Diophantine problem known as Pell's equation, and (iii) sets of reals modulo 1. Some further applications will be given in the next chapter. We use Greek letters to denote real numbers and Roman letters to denote integers throughout this chapter. Some of our proofs rely on the following simple but useful result: if b, d and y are positive,

$$\frac{a}{b} < \frac{x}{y} < \frac{c}{d},$$

and $bc - ad = 1$, then $y > d$. This follows because $ay < bx$ and $dx < cy$ give $y = (bc - ad)y > (bx - ay)d \geq d$.

7.1 Basic properties

As many of the proofs in this chapter are purely manipulatory, we shall leave some of the details to the reader. We begin by restating the basic

DEFINITION A *continued fraction* is an expression of the form

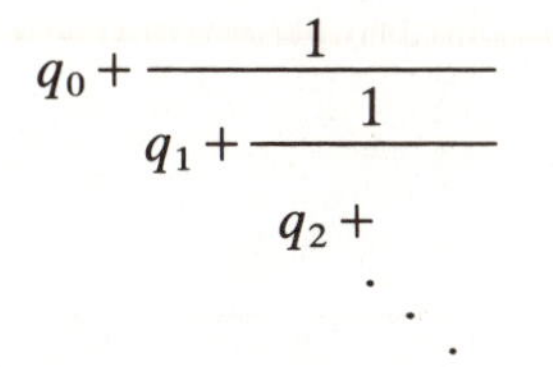

with either a finite or infinite number of entries q_i. We shall use the notation $[q_0, q_1, q_2, \ldots]$ for this expression. If there are infinitely many entries q_i then, for all $i \geq 0$, $q_i \in \mathbb{Z}$ and $q_{i+1} \geq 1$. This also applies if there are only finitely many entries except that the last entry may be any real number greater than or equal to 1. Continued fractions of either of these types are called *simple*.

An infinite continued fraction $[q_0, q_1, q_2, \ldots]$ is said to *converge* when the sequence of finite continued fractions $[q_0]$, $[q_0, q_1]$, $[q_0, q_1, q_2], \ldots$ converges. In Lemma 1.2 we show that all simple continued fractions converge. (In general, a continued fraction $[q_0, q_1, \ldots]$ converges if and only if $\sum_{i=0}^{\infty} q_i$ diverges, see Khinchin (1964).) Note that for finite or convergent infinite continued fractions and for all k,

$$[q_0, q_1, \ldots, q_k, \ldots] = [q_0, q_1, \ldots, [q_k, \ldots]].$$

The whole theory depends upon the following basic result.

THEOREM 1.1 *Let $q_0, q_1, \ldots$ be a finite ($k+1$ element) or infinite sequence of positive integers, with the exception that q_0 can be zero, and let a_n and b_n be given by*

$$\begin{aligned} a_0 &= q_0, \quad a_1 = q_0 q_1 + 1, \quad a_{n+2} = a_{n+1} q_{n+2} + a_n, \\ b_0 &= 1, \quad b_1 = q_1, \quad b_{n+2} = b_{n+1} q_{n+2} + b_n, \end{aligned} \tag{*}$$

where $n \leq k$ in the finite case. If α is a real number greater than 1, then

(i) $[q_0, \ldots, q_n, \alpha] = \dfrac{\alpha a_n + a_{n-1}}{\alpha b_n + b_{n-1}}$ *provided $n > 0$,*

(ii) $[q_0, \ldots, q_n] = \dfrac{a_n}{b_n}$.

Proof. By induction. Equation (i) clearly holds when $n = 1$. For $n > 1$ we have

$$[q_0, \ldots, q_{n+1}, \alpha] = \left[q_0, \ldots, q_n, q_{n+1} + \frac{1}{\alpha}\right]$$

$$= \frac{\left(q_{n+1} + \dfrac{1}{\alpha}\right)a_n + a_{n-1}}{\left(q_{n+1} + \dfrac{1}{\alpha}\right)b_n + b_{n-1}} \quad \text{by the inductive hypothesis,}$$

$$= \frac{\alpha a_{n+1} + a_n}{\alpha b_{n+1} + b_n} \quad \text{by } (*),$$

and (i) follows. For (ii) let $\alpha = q_{n+1}$ in (i) and use (*). □

LEMMA 1.2 *Using the notation above we have, for* $n \geq 0$,

(i) $a_n b_{n+1} - a_{n+1} b_n = (-1)^{n+1}$, *and so* $\dfrac{a_n}{b_n} - \dfrac{a_{n+1}}{b_{n+1}} = \dfrac{(-1)^{n+1}}{b_n b_{n+1}}$,

(ii) $(a_n, b_n) = 1$,

(iii) *if* $n > 0$ *then* $b_{n+1} > b_n$, *and so* $b_n \geq n$,

(iv) $\dfrac{a_0}{b_0} < \dfrac{a_2}{b_2} < \ldots < \dfrac{a_{2n}}{b_{2n}} < \ldots < \dfrac{a_{2n+1}}{b_{2n+1}} < \ldots < \dfrac{a_1}{b_1}$,

(v) *all infinite simple continued fractions converge.*

Proof. (i) Using the equations (*) we have $a_0 b_1 - a_1 b_0 = -1$ and $a_n b_{n+1} - a_{n+1} b_n = -(a_{n-1} b_n - a_n b_{n-1})$, now use induction.

(ii) This follows immediately from (i).

(iii) Use (*) and induction, noting that $q_n \geq 1$ for $n > 0$.

(iv) Substituting q_{n+1} for α in Theorem 1.1 we see that, as $q_{n+1} \geq 1$, a_{n+2}/b_{n+2} lies between a_n/b_n and a_{n+1}/b_{n+1}. But $a_0/b_0 < a_1/b_1$, so $a_0/b_0 < a_2/b_2 < a_1/b_1$. The result follows by induction.

(v) By (i) and (iv), $\{a_n/b_n\}$ is a Cauchy sequence and so converges. □

As in Chapter 2, let $[\alpha]$ denote the integer part of α. If α is a non-integral positive real number then

$$\alpha = [\alpha] + \frac{1}{\alpha_1} = [[\alpha], \alpha_1]$$

for some $\alpha_1 \in \mathbb{R}$ where $\alpha_1 > 1$. Similarly, if $\alpha_1 \notin \mathbb{Z}$, we have

$$\alpha_1 = [\alpha_1] + \frac{1}{\alpha_2} = [[\alpha_1], \alpha_2]$$

for some $\alpha_2 \in \mathbb{R}$ where $\alpha_2 > 1$, and $\alpha = [[\alpha], [\alpha_1], \alpha_2]$. Continuing this process we obtain the sequence $\alpha, \alpha_1, \alpha_2, \alpha_3, \ldots$ which will terminate at α_k if α_k is the first integer in this sequence, and it will continue indefinitely if no term in this sequence is an integer. From now on we shall write q_i for $[\alpha_i]$ and we have

LEMMA 1.3 *Using the above notation, if none of* $\alpha, \alpha_1, \ldots, \alpha_{n-1}$ *are integers, then*

$$\alpha = [q_0, \ldots, q_{n-1}, \alpha_n].$$

Proof. This follows easily by induction. □

DEFINITION Let α be a positive real number. Using the notation above and the equations (*) in Theorem 1.1, the rational number a_n/b_n is called the nth *convergent* to α provided none of $\alpha, \alpha_1, \ldots, \alpha_{n-1}$ belong to $\mathbb{Z}$.

We shall see below that a_n/b_n is a 'good' approximation to α but we treat the finite case first.

LEMMA 1.4 *α is a rational number if and only if it has a finite continued fraction representation.*

Proof. A finite continued fraction all of whose entries are integers clearly represents a rational number. Conversely if $\alpha = a/b$ we have, using the equations of the Euclidean algorithm, see (2) on p. 3.

$$\frac{a}{b} = q_1 + \frac{1}{\frac{b}{r_1}} \quad \text{if} \quad r_1 > 0, \qquad \frac{b}{r_1} = q_2 + \frac{1}{\frac{r_1}{r_2}} \quad \text{if} \quad r_2 > 0, \ldots.$$

Here $q_1 = \left[\frac{a}{b}\right]$, $q_2 = \left[\frac{b}{r_1}\right], \ldots$, and so $\frac{a}{b} = [q_1, \ldots, q_k]$ where k is such that r_{k-1} is the last non-zero remainder in the Euclidean algorithm. □

This lemma shows that, if α is irrational, then α_k is irrational for all k.

THEOREM 1.5 *Suppose α is an irrational number. Using the notation above we have*

(i) $\displaystyle\lim_{n\to\infty} \frac{a_n}{b_n} = \alpha$,

(ii) $\displaystyle\left|\alpha - \frac{a_n}{b_n}\right| < \frac{1}{b_n b_{n+1}} < \frac{1}{b_n^2}.$

Proof. By Lemma 1.2(v) $\lim_{n\to\infty} \{a_n/b_n\}$ exists. Now by Theorem 1.1 (with α_{n+1} for α) and Lemma 1.3, we see that α lies between a_{n-1}/b_{n-1} and a_n/b_n. Both parts of the theorem follow by Lemma 1.2(iii) and (iv). □

We come now to the question of the uniqueness of representation. Clearly $[q_0, \ldots, q_n, 1] = [q_0, \ldots, q_{n-1}, q_n + 1]$ so each rational number has at least two representations. Apart from this, representations are unique for we have

THEOREM 1.6 (i) *If* $[q_0, \ldots, q_n] = [q_0', \ldots, q_m']$, $q_n > 1$ *and* $q_m' > 1$ *then* $m = n$ *and* $q_i = q_i'$ *for all* $i \leq n$.

(ii) *If* $[q_0, \ldots, q_n, \ldots] = [q_0', \ldots, q_n', \ldots]$ *then* $q_n = q_n'$ *for all* n.

Proof. (i) We have

$$[q_0, \ldots, q_n] = q_0 + \frac{1}{[q_1, \ldots, q_n]} = q_0' + \frac{1}{[q_1', \ldots, q_m']}.$$

Now as q_n and q_m' are greater than 1 each fraction is proper, so $q_0 = q_0'$, and then $[q_1, \ldots, q_n] = [q_1', \ldots, q_m']$. Continue this process.

(ii) This is similar to the above. □

EXAMPLE Find the first few terms in the continued fraction representation of (i) 41/9, (ii) $\sqrt{41}$, and (iii) e.

For rational numbers we can use the Euclidean algorithm as in the proof of Lemma 1.4.

(i) We have $41 = 9 \cdot 4 + 5$, $9 = 5 \cdot 1 + 4$, $5 = 4 \cdot 1 + 1$, $4 = 1 \cdot 4 + 0$, and so $41/9 = [4, 1, 1, 4]$.

Given the decimal expansion of a number we can find its continued fraction representation by alternatively taking fractional parts and reciprocals.

(ii) We have $\sqrt{41} = 6.403124\ldots$, $\sqrt{41} - 6 = 0.403124\ldots$, $\alpha_1 = 1/((\sqrt{41}) - 6) = 2.480625\ldots$, $\alpha_2 = 1/(\alpha_1 - 2) = 2.080625\ldots$, $\alpha_3 = 1/(\alpha_2 - 2) = 12.403124\ldots$, etc. Hence $\sqrt{41} = [6, 2, 2, 12, \ldots]$. Note that $\alpha_3 - 6$ is approximately equal to $\sqrt{41}$, and so the entries in the representation are cyclic, that is $\sqrt{41} = [6, 2, 2, 12, 2, 2, 12, 2, \ldots]$. We shall prove this below.

(iii) $\mathrm{e} = 2.7182818284\ldots$ and so $\mathrm{e} = [2, 1, 2, 1, 1, 4, 1, 1, 6, 1, \ldots]$, see Problem 16.

Finally we consider the following general question. Are there any non-trivial sets of real numbers A all of whose elements have continued fraction representations obeying some simple property? Most of these questions involve measure theoretic considerations (see Khinchin, 1964). For example it can be shown that if all q_n in the representation of each α in A are bounded by a fixed constant, then A may be countable or uncountable but has Lebesgue measure zero. One result which can be proved using elementary methods is: α is quadratic if and only if it has an eventually periodic continued fraction representation, see Example (ii) above. We also noted above [in (iii)] that e has a continued fraction representation with a regular pattern; but π has a very irregular representation (see Problem 1), so it seems likely that no generalization of this result is possible.

For $k \geq 0$ and $n > 0$ we write $[q_0, \ldots, q_{k-1}, \overset{*}{q}_k, q_{k+1}, \ldots, \overset{*}{q}_{k+n}]$ for the periodic continued fraction, with infinitely many entries, $[q_0, \ldots, q_{k-1}, q_k, q_{k+1}, \ldots, q_{k+n}, q_k, q_{k+1}, \ldots]$. Further we write $[q_0, \ldots, q_{k-1}, \overset{*}{q}_k]$ for $[q_0, \ldots, q_{k-1}, q_k, q_k, \ldots]$.

THEOREM 1.7 *The continued fraction representation of α is eventually periodic if and only if α is a quadratic number, that is α satisfies a quadratic polynomial equation with rational coefficients.*

Proof. Suppose $\alpha = [q_0, \ldots, q_{k-1}, \overset{*}{q}_k, \ldots, \overset{*}{q}_{k+n-1}]$ then we have, by Lemma 1.3 and Theorem 1.1,

$$\alpha = [q_0, \ldots, q_{k-1}, \alpha_k] = \frac{\alpha_k a_{k-1} + a_{k-2}}{\alpha_k b_{k-1} + b_{k-2}}$$

and

$$[q_k, \ldots, q_{k+n-1}, \alpha_k] = \alpha_k = \frac{\alpha_k a_{k+n-2} + a_{k+n-3}}{\alpha_k b_{k+n-2} + b_{k+n-3}}.$$

Combining these we obtain a quadratic equation for α with rational coefficients, that is α is quadratic.

Conversely suppose α is positive and $\alpha = \alpha_0 = (c_0 + \sqrt{d})/e_0$, where c_0, d, and e_0 are integers, d is not square, $e_0 \neq 0$, and $e_0 | d - c_0^2$. [For this last condition rewrite α_0 as $(c_0 e_0 + \sqrt{(d e_0^2)})/e_0^2$.] Define c_{i+1}, e_{i+1}, α_i, and q_i by

$$\begin{aligned} c_{i+1} &= q_i e_i - c_i & e_{i+1} &= (d - c_{i+1}^2)/e_i \\ \alpha_i &= (c_i + \sqrt{d})/e_i & q_i &= [\alpha_i]. \end{aligned} \qquad (**)$$

It is a simple matter to check (by induction) that c_i and e_i are integers, $e_i \neq 0$ and $e_i | d - c_i^2$ using the equations

$$e_{i+1} = \frac{d - c_{i+1}^2}{e_i} = \frac{d - c_i^2}{e_i} + 2q_i c_i - q_i^2 e_i.$$

Also $\alpha_i - q_i = 1/\alpha_{i+1}$ by definition. Hence $\alpha_{i+1} > 0$ and $\alpha = [q_0, q_1, \ldots]$.

Now let $\alpha_i' = (c_i - \sqrt{d})/e_i$. By Theorem 1.1 and Lemma 1.3 we have, taking conjugates and rewriting,

$$\alpha_i' = -\frac{b_{i-2}}{b_{i-1}} \left(\frac{\alpha_0' - a_{i-2}/b_{i-2}}{\alpha_0' - a_{i-1}/b_{i-1}} \right).$$

By Theorem 1.5 the term in brackets tends to 1 as i tends to infinity, and so there is an n_0 such that $\alpha_n' < 0$ if $n > n_0$. Hence $\alpha_n - \alpha_n' = \dfrac{2\sqrt{d}}{e_n} > 0$, and thus $e_n > 0$, if $n > n_0$. Also using (**) we have

$$e_n \le e_n e_{n+1} = d - c_{n+1}^2 < d$$

and

$$c_{n+1}^2 < c_{n+1}^2 + e_n e_{n+1} = d.$$

Hence if $n > n_0$ there can only be finitely many distinct pairs $\{c_n, e_n\}$, and so there is a $k > 0$ such that, if $n > n_0$, $q_{n+k} = q_n$. As this implies $q_{n+t} = q_{n+k+t}$ where $t \ge 0$, the result follows. □

This proof provides a bound for b_α, the period of the continued fraction representation of α. For example if $\alpha = \sqrt{d}$ then $b_\alpha < 2d^{3/2}$. In practice this is usually a very crude estimate as can be seen from the tables of continued fraction representations given on pp. 338–9.

A quadratic number α is called *purely periodic* if its first period begins with q_0, see Problem 9. As an example consider $\alpha = [\overset{*}{q}_0, \overset{*}{q}_1]$ where $q_1 | q_0$. We have, by Theorem 1.1,

$$\alpha = \frac{(q_0 q_1 + 1)\alpha + q_0}{q_1 \alpha + 1}$$

and so $2\alpha = q_0 + \sqrt{(q_0^2 + 4t)}$, where $tq_1 = q_0$. This gives

$$[1, 1, 1, \ldots] = (1 + \sqrt{5})/2,$$

$$[2, 2, 2, \ldots] = 1 + \sqrt{2} \quad \text{and so} \quad [1, 2, 2, \ldots] = \sqrt{2},$$

$$[2, 1, 2, 1, \ldots] = 1 + \sqrt{3} \quad \text{and so} \quad [1, 1, 2, 1, 2, \ldots] = \sqrt{3}.$$

7.2 Best approximation

For our first application we consider the problem of finding good rational approximations to real numbers.

Definition The *best approximation* to a real number α relative to n is the rational number a/b closest to α satisfying $0 < b \le n$.

Continued fraction convergents are best approximations relative to their denominators. For example 22/7 is a convergent to π; in fact 22/7 is the best approximation to π relative to 56; see comments below the proof of Theorem 2.2. We prove first

Theorem 2.1 *Suppose α is an irrational number and a_n/b_n is a convergent to α with $n > 1$. If $0 < b \le b_n$ and $a/b \ne a_n/b_n$ then*

$$|a_n - b_n\alpha| < |a - b\alpha|.$$

COROLLARY *Under the same conditions we have*

$$\left|\alpha - \frac{a_n}{b_n}\right| < \left|\alpha - \frac{a}{b}\right|,$$

that is the convergent a_n/b_n is the best approximation to α relative to b_n.

Proof. The corollary is a weakened version of the theorem. Suppose $\alpha = [q_0, q_1, \ldots]$. Note first that $|a_{n-1} - b_{n-1}\alpha| > |a_n - b_n\alpha|$.

$$|a_{n-1} - b_{n-1}\alpha| = \left|\frac{a_{n-1}b_n - a_nb_{n-1}}{b_n} + b_{n-1}\left(\frac{a_n}{b_n} - \alpha\right)\right|.$$

By Lemma 1.2, if $a_n/b_n > \alpha$ then n is odd and $a_{n-1}b_n - a_nb_{n-1} = -1$. Hence

$$\begin{aligned} |a_{n-1} - b_{n-1}\alpha| &= \left|\frac{-1}{b_n} + b_{n-1}\left(\frac{a_n}{b_n} - \alpha\right)\right| \\ &\geq \left|\frac{-1}{b_n} + \frac{b_{n-1}}{b_nb_{n+1}}\right| \qquad \text{by assumption and Lemma 1.5} \\ &= \frac{q_{n+1}}{b_{n+1}} > |a_n - b_n\alpha| \qquad (*) \end{aligned}$$

as $q_{n+1} \geq 1$. We can use a similar argument if $a_n/b_n < \alpha$, hence the theorem will follow by induction if we prove it when b satisfies $b_{n-1} < b \leq b_n$. We may assume that $(a, b) = 1$.

Case 1. $b = b_n$. Here, as $a \neq a_n$, we have $|a_n - a| \geq 1$ and, as $n > 1$, $|a_n - \alpha b_n| < 1/b_{n+1} < 1/2$. Combining these two inequalities we obtain $|a_n - \alpha b_n| < |a - \alpha b|$.

Case 2. $b_{n-1} < b < b_n$. We have in this case $a_{n-1}/b_{n-1} \neq a/b \neq a_n/b_n$, so we can choose non-zero rational numbers γ and δ to satisfy

$$\gamma a_n + \delta a_{n-1} = a, \qquad \gamma b_n + \delta b_{n-1} = b.$$

By Lemma 1.2 these equations give $\gamma = \pm(ab_{n-1} - ba_{n-1})$ and $\delta = \pm(ab_n - ba_n)$, that is γ and δ are integers. Further as $b < b_n$, γ and δ have opposite signs but, by Lemma 1.2 again, $a_n - b_n\alpha$ and $a_{n-1} - b_{n-1}\alpha$ also have opposite signs, hence $\gamma(a_n - b_n\alpha)$ and $\delta(a_{n-1} - b_{n-1}\alpha)$ have the same sign. Thus finally, as

$$a - b\alpha = \gamma(a_n - b_n\alpha) + \delta(a_{n-1} - b_{n-1}\alpha),$$

we have

$$\begin{aligned} |a - b\alpha| &> |a_{n-1} - b_{n-1}\alpha| \qquad \text{as } \delta \text{ is integral,} \\ &> |a_n - b_n\alpha| \qquad \text{by } (*). \end{aligned}$$

□

The corollary to the above theorem can be improved to

THEOREM 2.2 *Suppose α is an irrational number. Then the continued fraction convergent a_n/b_n for α is the best approximation to α relative to any n satisfying*

(i) $n < b_{n+1}$ *if* $q_{n+1} = 1$,

(ii) $n < b_{n-1} + q_{n+1}b_n/2$ *if* $q_{n+1} > 1$.

Proof. We shall consider case (ii) when n is even (the remaining cases follow in a similar manner). In this case, if $\beta = 2\alpha - a_n/b_n$ (and so $\alpha - a_n/b_n = \beta - \alpha$), we have

$$\frac{a_n}{b_n} < \alpha < \frac{a_{n+1}}{b_{n+1}} < \beta < \frac{a_{n-1}}{b_{n-1}}.$$

We shall prove the theorem by showing that no rational number in an interval I, which includes the interval $(a_n/b_n, \beta)$, has a denominator less than $b_{n-1} + q_{n+1}b_n/2$. The interval I is $(a_n/b_n, \delta)$ where δ lies midway between a_{n+1}/b_{n+1} and a_{n-1}/b_{n-1}. Hence our first objective is to prove that $\delta > \beta$.

By Lemma 1.2 a rational number lying strictly between a_n/b_n and a_{n-1}/b_{n-1} has the form

$$T(s, t) = \frac{sa_{n-1} + ta_n}{sb_{n-1} + tb_n}$$

where s and t are positive integers (see Case 2 in the proof above). Note $\delta = T(2, q_{n+1}) = T(1, q_{n+1}/2)$. The proposition

$$\beta < T(1, \theta), \tag{*}$$

is equivalent to

$$2\alpha - \frac{a_n}{b_n} < T(1, \theta) = \frac{a_{n-1}}{b_{n-1}} - \frac{\theta}{b_{n-1}(b_{n-1} + \theta b_n)}.$$

But as (Lemma 1.2 and Theorem 1.5)

$$\frac{a_{n-1}}{b_{n-1}} = \frac{a_n}{b_n} + \frac{1}{b_n b_{n-1}} \quad \text{and} \quad \alpha - \frac{a_n}{b_n} < \frac{1}{b_n b_{n+1}},$$

(*) will hold if

$$\frac{2}{b_n b_{n+1}} < \frac{1}{b_n b_{n-1}} - \frac{\theta}{b_{n-1}(b_{n-1} + \theta b_n)} = \frac{1}{b_n(b_{n-1} + \theta b_n)}.$$

Using the definition of b_{n+1}, this is equivalent to

$$b_{n-1} + 2\theta b_n < q_{n+1} b_n$$

or

$$\frac{b_{n-1}}{2b_n} + \theta < \frac{q_{n+1}}{2}.$$

Hence, as $b_{n-1} < b_n$, (*) holds if $\theta \le [q_{n+1}/2]$; that is $\delta > \beta$.

Returning to the main proof suppose u/v is a rational number in I. We have, as the length of this interval is greater than $u/v - a_n/b_n$,

$$0 < \frac{ub_n - va_n}{b_n v} < \frac{1}{b_n(b_{n-1} + q_{n+1}b_n/2)}.$$

The numerators and denominators of these fractions are integers, hence by the note on p. 114, $v > b_{n-1} + q_{n+1}b_n/2$. □

Returning to our example above, the first three convergents to π are 3, 22/7, and 333/106 as $\pi = [3, 7, 15, \ldots]$. Our theorem shows that 22/7 is the best approximation to π relative to any integer less than $1 + 15 \cdot 7/2 = 53\frac{1}{2}$. In fact a better approximation is 179/57, see Problem 10.

Our next few results will show exactly how accurate (relative to the size of the denominator) rational approximations can be to a real number; the results will have useful applications later.

THEOREM 2.3 (Hurwitz) *There are infinitely many convergents a_n/b_n to an irrational number α which satisfy*

$$\left|\alpha - \frac{a_n}{b_n}\right| < \frac{1}{b_n^2\sqrt{5}}.$$

Proof. We derive this result by showing that at least one of any three consecutive continued fraction convergents to α satisfies the inequality. We have by Theorem 1.1 and Lemma 1.2

$$\left|\alpha - \frac{a_n}{b_n}\right| = \frac{1}{b_n(\alpha_{n+1}b_n + b_{n-1})} = \frac{1}{b_n^2(\alpha_{n+1} + \rho_{n+1})}$$

where $\rho_{n+1} = b_{n-1}/b_n$. We shall show that

$$\alpha_i + \rho_i \le \sqrt{5} \qquad (**)$$

cannot hold for three consecutive values of i: this gives the result. Suppose (**) holds for $i = n-1$ and $i = n$. As $\alpha_{n-1} = q_{n-1} + 1/\alpha_n$ and (if $n > 2$)

$$\frac{1}{\rho_n} = \frac{b_{n-1}}{b_{n-2}} = \frac{q_{n-1}b_{n-2} + b_{n-3}}{b_{n-2}} = q_{n-1} + \rho_{n-1},$$

we have

$$\frac{1}{\alpha_n}+\frac{1}{\rho_n}=\alpha_{n-1}+\rho_{n-1}\leq\sqrt{5}.$$

So

$$1=\frac{1}{\alpha_n}\alpha_n\leq\left(\frac{-1}{\rho_n}+\sqrt{5}\right)(-\rho_n+\sqrt{5})$$

as (**) holds for $i=n$. Thus as ρ_n is rational we have

$$\rho_n+1/\rho_n<\sqrt{5},\quad\text{that is}\quad \rho_n>(-1+\sqrt{5})/2.$$

If we now assume that (**) holds for $i=n$ and $i=n+1$ these inequalities will hold when ρ_n is replaced by ρ_{n+1}. Using these four inequalities we have

$$q_n=\frac{b_n-b_{n-2}}{b_{n-1}}=\frac{1}{\rho_{n+1}}-\rho_n<(-\rho_{n+1}+\sqrt{5})-\rho_n<\sqrt{5}\ -\ (-1+\sqrt{5})=1.$$

This is a contradiction as $q_n\geq 1$ if $n>0$, so the result follows. □

This result cannot be improved for we have

THEOREM 2.4 *A real number γ exists with the property: if $\theta>\sqrt{5}$ then the inequality*

$$\left|\gamma-\frac{a}{b}\right|<\frac{1}{b^2\theta} \qquad (*)$$

has only finitely many rational solutions a/b.

Proof. Let $\gamma=(-1+\sqrt{5})/2$ and suppose solutions to (*) exist with arbitrarily large b. Rewriting (*) we have

$$\frac{-1+\sqrt{5}}{2}=\frac{a}{b}+\frac{\delta}{b^2}\quad\text{where}\quad|\delta|<1/\theta<1/\sqrt{5},$$

that is

$$-\frac{\delta}{b}+\frac{\sqrt{5}}{2}b=\frac{b}{2}+a.$$

Squaring this gives

$$\frac{\delta^2}{b^2}-\delta\sqrt{5}=a^2+ab-b^2.$$

Now, as $-1<-\delta\sqrt{5}<1$, we have, for sufficiently large b,

$$-1<\delta^2/b^2-\delta\sqrt{5}<1$$

and hence $a^2+ab-b^2=0$. This gives $(2a+b)^2=5b^2$ which is impossible. □

We showed at the end of the last section that the number $(-1+\sqrt{5})/2$ used in this proof has the continued fraction representation $[0, 1, 1, \ldots]$. In fact, any real number whose continued fraction representation is eventually all one's can be used for γ in Theorem 2.4. Further it can be shown that, for all numbers not of this type, Theorem 2.3 holds with $\sqrt{8}$ replacing $\sqrt{5}$. Also $\sqrt{8}$ is best possible (in the sense of Theorem 2.4 with $\sqrt{8}$ replacing $\sqrt{5}$) when $\gamma = \sqrt{2}$ or any number whose continued fraction representation is eventually all two's. This process continues; finally it can be shown that there are uncountably many real numbers β such that the inequality

$$\left|\beta - \frac{a}{b}\right| < \frac{1}{3b^2}$$

has infinitely many rational solutions a/b, but only finitely many solutions if the number 3 is replaced by any larger number. (See for example LeVeque, 1977.)

7.3 Pell's equation

The Diophantine equation $x^2 - dy^2 = 1$, usually known as Pell's equation, is of considerable importance in number theory. It is an accident that Pell's name has been attached to this equation; its properties have been studied by many mathematicians and the first published proof establishing its solubility is due to Lagrange. The main applications are in quadratic form theory (see Chapters 9 and 10) and to the unit problem for real quadratic fields—if $d \not\equiv 1 \pmod 4$ and $\gamma = x + y\sqrt{d}$, then γ is a unit in $\mathbb{Z}[\sqrt{d}]$ (that is $N\gamma = \pm 1$) if and only if $x^2 - dy^2 = \pm 1$; there is a similar result if $d \equiv 1 \pmod 4$ (see Chapter 5).

Throughout this section we shall assume that d is positive and not square (Pell's equation has only trivial solutions otherwise).

THEOREM 3.1 *The equation*

$$x^2 - dy^2 = 1 \qquad (*)$$

is soluble in integers x and y.

Proof. By Theorems 1.1 and 1.7 with $\alpha = \sqrt{d}$ we have $\alpha_n = (c_n + \sqrt{d})/e_n$ [see (**) on p. 119] and

$$\sqrt{d} = \frac{\alpha_n a_{n-1} + a_{n-2}}{\alpha_n b_{n-1} + b_{n-2}} = \frac{(c_n + \sqrt{d})a_{n-1} + e_n a_{n-2}}{(c_n + \sqrt{d})b_{n-1} + e_n b_{n-2}}.$$

Equating irrational and rational parts we obtain

$$b_{n-1}c_n + b_{n-2}e_n = a_{n-1}, \qquad a_{n-1}c_n + a_{n-2}e_n = db_{n-1}$$

and, multiplying by a_{n-1} and b_{n-1}, respectively,

$$a_{n-1}^2 - db_{n-1}^2 = (-1)^n e_n$$

by Lemma 1.2. Again using Theorem 1.7 we have $(c_n + \sqrt{d})/e_n = (c_{n+kt} + \sqrt{d})/e_{n+kt}$ where k is the period of the continued fraction representation of $\sqrt{d}$ and t is a positive integer. Hence, as $e_n \leq d$ (see the proof of Theorem 1.7), there is a non-zero integer e such that

$$u^2 - dv^2 = e$$

has infinitely many solutions $\{u, v\}$. We may assume that u and v are positive. We partition this set of solutions into classes by: $\{u, v\}$ is in thc same class as $\{u', v'\}$ if and only if $u \equiv u'$ and $v \equiv v' \pmod{|e|}$. At least one of these classes has two (in fact, infinitely many) distinct members $\{u_1, v_1\}$ and $\{u_2, v_2\}$. Now let x and y be given by

$$x = \frac{u_1 u_2 - dv_1 v_2}{e}, \qquad y = \frac{u_1 v_2 - u_2 v_1}{e},$$

we claim that x and y are integers and they solve (*). This follows because

$$u_1 u_2 - dv_1 v_2 \equiv u_1^2 - dv_1^2 \equiv 0 \pmod{|e|},$$
$$0 \neq u_1 v_2 - u_2 v_1 \equiv 0 \pmod{|e|},$$

and

$$e^2(x^2 - dy^2) = (u_1^2 - dv_1^2)(u_2^2 - dv_2^2) = e^2.$$

Hence $\{x, y\}$ is a non-trivial solution of Pell's equation. □

Pell's equation has infinitely many solutions generated by a *fundamental solution* $\{x_0, y_0\}$ given by

THEOREM 3.2 *Suppose $\{x_0, y_0\}$ is the solution of Pell's equation (*) with x_0 and y_0 positive and $D = x_0 + y_0\sqrt{d}$ minimal, then $\{x, y\}$ is a solution of (*) if and only if x and y satisfy*

$$x + y\sqrt{d} = \pm(x_0 + y_0\sqrt{d})^n$$

for some integer n positive, negative, or zero.

Proof. Suppose first $\{x_1, y_1\}$ and $\{x_2, y_2\}$ are solutions of (*) and x_3 and y_3 are given by

$$x_3 + y_3\sqrt{d} = (x_1 + y_1\sqrt{d})(x_2 + y_2\sqrt{d}),$$

then $\{x_3, y_3\}$ is a solution of (*) for

$$x_3^2 - dy_3^2 = (x_1 x_2 + dy_1 y_2)^2 - d(x_1 y_2 + x_2 y_1)^2 = (x_1^2 - dy_1^2)(x_2^2 - dy_2^2) = 1.$$

Secondly note that if $n = 0$ we have $x = 1$, $y = 0$ (the trivial solution).

Also

$$(x_1 + y_1\sqrt{d})^{-1} = \frac{x_1 - y_1\sqrt{d}}{x_1^2 - dy_1^2} = x_1 + (-y_1)\sqrt{d}.$$

It follows, by induction, that if $\{x_1, y_1\}$ is a solution of (*) and x and y are given by $x + y\sqrt{d} = \pm(x_1 + y_1\sqrt{d})^n$ then $\{x, y\}$ is a solution of (*) for every integer n.

Now suppose $\{t, u\}$ is a solution of (*) with t and u positive. Choose m to satisfy

$$D^m \le t + u\sqrt{d} < D^{m+1},$$

that is

$$1 \le (t + u\sqrt{d})D^{-m} < D.$$

By the argument above if $(t + u\sqrt{d})D^{-m} = X + Y\sqrt{d}$ then $\{X, Y\}$ is a solution of (*) with X and Y non-negative (for we have $1 \le X + Y\sqrt{d} < D$ and $0 < X - Y\sqrt{d} \le 1$, hence $2X \ge 1$ and $2Y\sqrt{d} \ge 0$). It follows by the minimality of D that $X = 1$ and $Y = 0$. Hence $t + u\sqrt{d} = D^m$ and the proof is complete. □

Suppose we have a solution $\{x, y\}$ of Pell's equation with x and y positive, then

$$\left|\frac{x}{y} - \sqrt{d}\right| = \frac{1}{y(x + y\sqrt{d})} < \frac{1}{2y^2}$$

if $d > 4$; hence, by Problem 13, x/y is a (continued fraction) convergent to $\sqrt{d}$; the cases $d = 2, 3$ can be checked directly. If k is the period of the continued fraction representation of $\sqrt{d}$ and k is even then $\{a_{k-1}, b_{k-1}\}$ is a solution of Pell's equation (*), see the proof of Theorem 3.1; and, using a similar argument, it follows that the equation $x^2 - dy^2 = -1$ is insoluble. If k is odd the situation is slightly different: $\{a_{k-1}, b_{k-1}\}$ is a solution of the equation $x^2 - dy^2 = -1$, and $\{a_{2k-1}, b_{2k-1}\}$ is a solution of (*), see Problems 20 and 21. Two tables of continued fraction representations of $\sqrt{d}$ for $d < 200$ are given on pp. 338–9.

As an example consider the cases $d = 13$ and $d = 14$. In the first case we have $\sqrt{13} = [3, \overset{*}{1}, 1, 1, 1, \overset{*}{6}]$ with period 5, and $[3, 1, 1, 1, 1] = 18/5$. This gives $\{18, 5\}$ is a solution of the equation $x^2 - 13y^2 = -1$ and, as $649 + 180\sqrt{13} = (18 + 5\sqrt{13})^2$, $\{649, 180\}$ is a solution (in fact the fundamental solution) of Pell's equation with $d = 13$. In the second case $\sqrt{14} = [3, \overset{*}{1}, 2, 1, \overset{*}{6}]$ with period 4, and $[3, 1, 2, 1] = 15/4$. Hence $\{15, 4\}$ is the fundamental solution of Pell's equation with $d = 14$. The equation $x^2 - 14y^2 = -1$ is not soluble as can be checked by reducing modulo 7.

Lastly we shall briefly consider the equation $x^2 - dy^2 = m$ and give a criterion for its solubility. Note that a general indefinite binary quadratic form equation can be reduced to an equation of this type by rational transformations (see Chapter 9).

THEOREM 3.3 *If the equation*

$$u^2 - dv^2 = m \tag{**}$$

has a solution, $\{u, v\}$*, then it has infinitely many solutions at least one of which satisfies*

$$0 < u < \sqrt{((x_0 + 1)\,|m|/2)}$$

where $\{x_0, y_0\}$ *is the fundamental solution of Pell's equation* (*).

Proof. The first part follows in a similar manner to the first part of Theorem 3.2. We consider the case $m > 0$; a virtually identical argument can be used if $m < 0$. Let us make the assumption: the equation (**) is soluble *and* all solutions $\{u, v\}$ of (**) are such that $u \geq \sqrt{((x_0 + 1)m/2)}$. Choose u and v so that $\{u, v\}$ is a solution of (**), both u and v are positive and u is minimal. Define u_1 and v_1 by

$$u_1 + v_1\sqrt{d} = (u + v\sqrt{d})(x_0 - y_0\sqrt{d}).$$

As above $\{u_1, v_1\}$ is also a solution to (**), and

$$u_1 = ux_0 - dvy_0 = u\left(x_0 - y_0\sqrt{d}\,\frac{v\sqrt{d}}{u}\right) = u[D^{-1} + y_0(1 - \sqrt{(1 - m/u^2)})\sqrt{d}]$$

where $D = x_0 + y_0\sqrt{d}$. As $0 < m/u^2 < 1$, it follows that

$$0 < 1 - \sqrt{(1 - m/u^2)} < m/(2u^2 - m)$$

and so, as $y_0\sqrt{d} < x_0$,

$$0 < u_1 < u(D^{-1} + x_0 m/(2u^2 - m)).$$

Hence, as u is minimal, $D^{-1} + x_0 m/(2u^2 - m) > 1$, and this implies $\sqrt{((x_0 + 1)m/2)} > u$ (as $0 < D^{-1} < 1$) which contradicts our assumption. Thus *either* the equation (**) is not soluble *or* there is a solution $\{u, v\}$ with $u < \sqrt{((x_0 + 1)m/2)}$. This completes the proof. □

In the proof of Theorem 3.1 we showed that the equation $u^2 - dv^2 = (-1)^n e_n$ is soluble for all n. Using the methods of this chapter it can be shown that if no n exists such that $m = (-1)^n e_n$, and $|m| < \sqrt{d}$, then the equation

$$u^2 - dv^2 = m \tag{*}$$

has no solution in integers. On the other hand, Theorem 1.8 of Chapter 10 gives a method solving (*) which is applicable in a number of important cases.

7.4 A set of real numbers modulo 1

Let $((\alpha))$ denote the fractional part of α, that is $((\alpha)) = \alpha - [\alpha]$. Chebyshev was the first mathematician to consider the question: suppose

α is a fixed irrational number, what form does the set $\{((n\alpha)): n = 1, 2, \ldots\}$ take? We shall show that it is dense in the interval $[0, 1]$ and 'uniformly distributed'. Many extensions and generalizations of this problem have been studied; a good introduction is given in Hardy and Wright (1954).

THEOREM 4.1 *If α is an irrational number and β is a real number, then there are infinitely many pairs of integers x and y such that*

$$|x\alpha - y - \beta| < 3/x.$$

Proof. By Theorem 2.3 there are infinitely many integers a and b satisfying $(a, b) = 1$ and

$$\alpha = \frac{a}{b} + \frac{\gamma}{b^2}$$

for some γ satisfying $|\gamma| < 1$. Also let c be the integer closest to $b\beta$, so

$$\beta = \frac{c}{b} + \frac{\delta}{2b}$$

where $|\delta| \le 1$. Further as $(a, b) = 1$, integers x and y can be found to satisfy $ax - by = c$ and $b \le 2x < 3b$ (see Theorem 1.5, Chapter 1). Using these equations we have

$$\begin{aligned} |x\alpha - y - \beta| &= \left|\frac{xa}{b} + \frac{x\gamma}{b^2} - y - \frac{c}{b} - \frac{\delta}{2b}\right| = \left|\frac{x\gamma}{b^2} - \frac{\delta}{2b}\right| \\ &< \frac{x}{b^2} + \frac{1}{2b} \\ &< \frac{9}{4x} + \frac{3}{4x} = \frac{3}{x} \end{aligned}$$

as $b > 2x/3$. The result follows because $x \ge b/2$ and b can be arbitrarily large. □

This theorem gives the Chebyshev result mentioned above for if β is a real number between 0 and 1, and x and y satisfy the inequality, then $y = [x\alpha]$. A novel application is given in Problem 25. This result can be strengthened in two ways. First the constant 3 can be replaced by $(1 + \varepsilon)/\sqrt{5}$ for $\varepsilon > 0$. Secondly, not only is the set $S = \{((n\alpha)): n = 1, 2, \ldots\}$ dense in the unit interval, but its members are distributed 'evenly' over this interval in the following sense.

DEFINITION Let $R = \{\alpha_0, \alpha_1, \ldots\}$ be a set of real numbers in the unit interval and let $0 \le \beta_1 < \beta_2 \le 1$. Define T by: $T(n, \beta_1, \beta_2)$ equals the number of elements α_i of R satisfying $i \le n$ and $\beta_1 \le \alpha_i \le \beta_2$. R is said to be *uniformly distributed* in the unit interval if and only if

$$\lim_{n\to\infty}\frac{T(n, \beta_1, \beta_2)}{n}=\beta_2-\beta_1$$

holds for all β_1 and β_2.

We shall show that the Chebyshev set has this strong property, some other examples are given in Problem 24.

THEOREM 4.2 *If α is irrational, then the set $S=\{((n\alpha)):n=1, 2, \ldots\}$ is uniformly distributed in the unit interval.*

Proof. As in the proof above we begin by noting that there are infinitely many pairs of integers a and b, with $b>0$, such that

$$\alpha=\frac{a}{b}+\frac{\delta}{b^2}, \qquad |\delta|<1, \qquad (a, b)=1.$$

For b to be chosen, let J_j be the set $\{jb+k:k=0, 1, \ldots, b-1\}$ and consider the set $K_j=\{((i\alpha)):i\in J_j\}$. We have

$$\alpha(jb+k)=\left(\frac{a}{b}+\frac{\delta}{b^2}\right)(jb+k)=ja+\frac{ka+j\delta}{b}+\frac{k\delta}{b^2},$$

and so (as $k<b$)

$$((\alpha(jb+k)))=\left(\left(\frac{ka+[j\delta]}{b}+\frac{\gamma}{b}\right)\right)$$

where $|\gamma|<2$. $[j\delta]$ is independent of k and so, as k ranges over a complete system of residues modulo b, $ka+[j\delta]$ also ranges over this system. It follows that

$$K_j=\left\{\left(\left(\frac{k+\gamma}{b}\right)\right):k=0, 1, \ldots, b-1\right\}. \qquad (*)$$

Let $[\beta_1, \beta_2]$ be a subinterval of the unit interval and let b be sufficiently large so that we can choose integers v and w to satisfy

$$v-1<b\beta_1\le v<w\le b\beta_2<w+1.$$

Let the number of elements of K_j in the interval $[\beta_1, \beta_2]$ be denoted by W_j. Then W_j is bounded below by the number of elements of K_j in the interval $(v/b, w/b)$; that is, using (*), W_j is bounded below by the number of integers k which satisfy

$$v\le k+\gamma\le w.$$

It follows that $W_j\ge w-v-4$ as $|\gamma|<2$. Similarly $W_j\le w-v+6$; note that these bounds are independent of j.

Let $n = rb + s$ where $0 \le s < b$, and let $T(n, \beta_1, \beta_2)$ be the number of reals $((x\alpha))$ in the interval $[\beta_1, \beta_2]$ for $x \le n$. We have by the inequalities above

$$r(w - v - 4) \le T(n, \beta_1, \beta_2) \le (r + 1)(w - v + 6).$$

But

$$r(w - v - 4) = \frac{n - s}{b}(w - v - 4) \ge n(\beta_2 - \beta_1) - \frac{6n}{b} - (w - v - 4),$$

and

$$(r + 1)(w - v + 6) \le \left(\frac{n}{b} + 1\right)(w - v + 6) \le n(\beta_2 - \beta_1) + \frac{6n}{b} + (w - v + 6).$$

Given $\varepsilon > 0$ choose b so that $b > 1/\varepsilon$ and choose n so that $n > b/\varepsilon$. This gives $(w - v)/n < \varepsilon$ and $1/n < \varepsilon$. Combining the inequalities above we obtain

$$n(\beta_2 - \beta_1) - 7n\varepsilon \le T(n, \beta_1, \beta_2) \le n(\beta_2 - \beta_1) + 13n\varepsilon.$$

Hence

$$\lim_{n\to\infty} \frac{1}{n} T(n, \beta_1, \beta_2) = \beta_2 - \beta_1. \qquad \square$$

7.5 Problems

1. Find the first ten terms in the continued fraction representations of e and π. Deduce

$$\left|\pi - \frac{355}{113}\right| < \frac{1}{2 \cdot 10^6}.$$

This approximation to π was first discovered by the Chinese mathematician Chao Jung-Tze about AD 500.

2. Let a_n/b_n be a convergent to α where $\alpha = [q_0, q_1, \ldots]$.
   (i) Show that

$$a_n = \det\begin{pmatrix} q_0 & -1 & 0 & \ldots & 0 & 0 \\ 1 & q_1 & -1 & \ldots & 0 & 0 \\ \cdot & \cdot & \cdot & \cdots & \cdot & \cdot \\ 0 & 0 & 0 & \cdots & 1 & q_n \end{pmatrix}.$$

Is there a similar expression for b_n?
   (ii) If

$$\alpha = \frac{r\alpha_k + s}{t\alpha_k + u},$$

prove that

$$\begin{pmatrix} r & s \\ t & u \end{pmatrix} = \begin{pmatrix} q_0 & 1 \\ 1 & 0 \end{pmatrix}\begin{pmatrix} q_1 & 1 \\ 1 & 0 \end{pmatrix} \cdots \begin{pmatrix} q_{k-1} & 1 \\ 1 & 0 \end{pmatrix}.$$

3. If β is a positive real number, $q_0, q_1, \ldots$ are positive integers and n is odd show that

$$[q_0, \ldots, q_n] > [q_0, \ldots, q_{n-1}, q_n + \beta].$$

Is this true if n is even?

4. Let u_n be the nth term in the Fibonacci sequence (see Problem 19, Chapter 1). Show that

(i) u_{n+2}/u_{n+1} is the nth convergent to $(1+\sqrt{5})/2$,

(ii) $p \mid u_{p-1}$ if $p \equiv \pm 1 \pmod 5$, and $p \mid u_{p+1}$ if $p \equiv \pm 2 \pmod 5$, hence show that every prime divides infinitely many Fibonacci numbers.

5. Suppose α be a real number in the interval $(0, 1)$, and suppose $\alpha = [0, q_1, q_2, \ldots]$.

(i) What is the probability that (a) $q_1 \geq m$, (b) $q_1 = m$?

(ii) Show that the probability that $q_2 \geq m$ is

$$\sum_{q_1=1}^{\infty} \frac{1}{q_1(mq_1+1)}.$$

(iii) Using (ii) find the probability that $q_2 = 1$.

[In this question probability is defined in terms of the total lengths of the associated intervals.]

6. Use Lemma 1.2 to find a solution of the equation $167x + 43y = 2$.

7. Using the notation of this chapter show that

(i) $a_{n+2}b_n - a_n b_{n+2} = (-1)^n q_{n+2}$,

(ii) $a_{n+1}/a_n = [q_{n+1}, q_n, \ldots, q_0]$,

(iii) $b_{n+1}/b_n = [q_{n+1}, q_n, \ldots, q_1]$.

8. The solar year is $365\frac{20929}{86400}$ days. At the present time the 365 day year is corrected by adding 97 days (during leap years) in four centuries. Find a better approximation.

9. Assume that α is a quadratic irrational number which satisfies $\alpha > 1$ and $-1 < \alpha' < 0$ where α' is the conjugate of α, that is $\alpha' = (a - b\sqrt{d})/c$, if $\alpha = (a + b\sqrt{d})/c$. Deduce in turn

(i) $-1 < \alpha_n' < 0$, where α_n satisfies $\alpha = [q_0, \ldots, q_{n-1}, \alpha_n]$,

(ii) $q_n = [-1/\alpha_{n+1}']$,

(iii) α is purely periodic, that is the first period of the continued fraction representation of α begins with q_0.

Now assume that α is purely periodic; note, this implies that $q_0 > 0$.

(iv) Show that α satisfies the polynomial equation $f(\alpha) = 0$ where $f(x) = x^2 b_{n-1} + x(b_{n-2} - a_{n-1}) - a_{n-2}$.

(v) Hence show that the second root α' of this equation satisfies $-1 < \alpha' < 0$. It follows that the continued fraction representation of a quadratic irrational number α is purely periodic if and only if $\alpha > 1$ and $-1 < \alpha' < 0$.

10. Suppose n is even and a_n/b_n is a convergent to α. Show that if $q_{n+2}>1$ then

$$\frac{a_n}{b_n}<\frac{a_n+a_{n+1}t}{b_n+b_{n+1}t}<\frac{a_{n+2}}{b_{n+2}}$$

for $0<t<q_{n+2}$. Hence show how to find the best rational lower estimate for α whose denominator is less than T. An exactly similar method can be used for upper estimates. Find the best upper and lower rational estimates for $x=779/207$ whose denominators are less than 60. Which is closer to x?

11. Find the rational approximation to π with smallest denominator which is closer to π than 355/113.

12. Use the results on Farey series given in Problem 21, Chapter 1, to re-prove Theorem 1.5.

13. (i) Show that for any pair of consecutive convergents to α at least one member of the pair satisfies

$$\left|\alpha-\frac{a}{b}\right|<\frac{1}{2b^2}.$$

(Hint. α lies between consecutive convergents.)

(ii) Let α be an irrational number and suppose the rational number a/b satisfies: $(a, b)=1$ and

$$\left|\alpha-\frac{a}{b}\right|<\frac{1}{2b^2}.$$

If n is given by $b_n\le b<b_{n+1}$, show that

$$\left|\alpha-\frac{a_{n+1}}{b_{n+1}}\right|<\frac{1}{2bb_{n+1}}.$$

Hence show that a/b is a convergent to α.

14. By writing x^3-dy^3 as a product of two factors, show that if $\{x_0, y_0\}$ is a solution of the equation $x^3-dy^3=n$ then x_0/y_0 is a convergent to $d^{1/3}$ provided $y_0>8\,|n|/3d^{2/3}$ and d is positive and not a cube.

☆15. Suppose δ is a real number which is neither zero nor a negative integer and let

$$f(\delta, x)=\sum_{n=0}^{\infty}\frac{x^n}{\delta(\delta+1)\ldots(\delta+n-1)n!}.$$

Find a recursive relation for f and use it to show that, for $y\in\mathbb{R}$,

$$\frac{y}{\delta}\frac{f(\delta+1, y^2)}{f(\delta, y^2)}=\left[0, \frac{\delta}{y}, \frac{\delta+1}{y}, \ldots, \frac{\delta+n}{y}, \alpha_{n+2}\right]$$

where $\alpha_{n+2}=\dfrac{\delta+n+1}{y}\dfrac{f(\delta+n+1, y^2)}{f(\delta+n+2, y^2)}$. By putting $\delta=1/2$ and $y=1/2t$, and expressing $e^z \pm e^{-z}$ in terms of f, deduce that

$$\frac{e^{2/t}-1}{e^{2/t}+1}=[0, t, 3t, \ldots, (2n+1)t, \ldots].$$

(Note that the continued fractions in this question are not simple.)

☆ 16. Using the previous problem we have

$$\alpha=\frac{e+1}{e-1}=[2, 6, 10, \ldots, 2+4n, \ldots]$$

and $e=(\alpha+1)/(\alpha-1)$. Let $\gamma=[2, 1, 2, 1, 1, 4, 1, \ldots, 1, 2n, 1, \ldots]$, let q_n/r_n be the nth convergent for γ, and let s_n/t_n be the nth convergent for α. Show that

$$q_{3n+1}=s_n+t_n \quad \text{and} \quad r_{3n+1}=s_n-t_n$$

by considering the recursive relations for s_n and t_n and the values of q_{3n+m} (and r_{3n+m}) for $m=-3, -2, -1, 0$, and 1. Finally deduce that $e=\gamma$. (This result is due to Euler, similar results are known for e^x, $e^{1/x}$, etc., see Perron, 1950.)

17. Find, where possible, the fundamental solutions of the equations

$$x^2-31y^2=1, \quad x^2-30y^2=-1, \quad x^2-29y^2=4.$$

18. Suppose d is positive and not square. Show that the equation

$$v^2-dw^2=4$$

is always soluble and that there is a fundamental solution $\{v_0, w_0\}$ from which all other solutions are generated by $(n \in \mathbb{Z})$

$$\frac{v+w\sqrt{d}}{2}=\pm\left(\frac{v_0+w_0\sqrt{d}}{2}\right)^n.$$

19. Let p be a prime number. By considering $x'+1$ and $x'-1$, where $\{x', y'\}$ is a solution of Pell's equation, show that the equation

$$z^2-pt^2=-1$$

is soluble if and only if $p=2$ or $p\equiv 1 \pmod 4$.

20. In this, and the next, problem let d be a positive non-square integer and let k be the period of the continued fraction representation of $\sqrt{d}$. Suppose $\alpha=[\sqrt{d}]+\sqrt{d}$, $c_0=[\sqrt{d}]$, $e_0=1$ and c_{i+1}, e_{i+1}, α_i, and q_i are as defined in (**) on p. 119. Show that

(i) $e_i=1$ if and only if $i|k$,
(ii) $e_j \neq -1$ for all j,
(iii) $q_t<\sqrt{d}$ for $t=1, 2, \ldots, k-1$.

21. (Continuation.) Show that the continued fraction representation of $\sqrt{d}$ has the form $[q_0, \overset{*}{q}_1, \ldots, q_{k-1}, 2\overset{*}{q}_0]$ where $q_0=[\sqrt{d}]$, and $q_t=q_{k-t}$ for $t=1, 2, \ldots, k-1$. (For the last part use Problem 7.)

22. What form does d take when the continued fraction representation of $\sqrt{d}$ has a period of length (i) one, (ii) three, and (iii) two? Give some examples.

23. Show that there are infinitely many integers n such that the sum $1+2+\ldots+n$ is a square.

24. Which of the following sequences are uniformly distributed in the unit interval?

(i) $\frac{0}{1}, \frac{0}{2}, \frac{1}{2}, \frac{0}{3}, \frac{1}{3}, \frac{2}{3}, \ldots, \frac{0}{t}, \frac{1}{t}, \ldots, \frac{t-1}{t}, \ldots,$

(ii) $((1!\,\mathrm{e})), ((2!\,\mathrm{e})), \ldots, ((n!\,\mathrm{e})), \ldots,$

(iii) $((\theta)), ((2\theta)), \ldots, ((n\theta)), \ldots,$

where $\theta = 0.123456789101112\ldots$, that is the decimal formed by writing the sequence of natural numbers in increasing order.

25. Let S be a square in the plane and suppose the sides of S act as a mirror to a ray of light which is emitted from a point P inside the square with angle α to one of the sides, and which is reflected repeatedly by the mirrors. If the path passes through a corner assume that it is reflected directly back along the same path. Show that the path either passes arbitrarily close to every point in the square or is periodic in which case $\tan \alpha$ is rational.

8

TRANSCENDENTAL NUMBERS

A real or complex number which satisfies no polynomial equation with algebraic coefficients is called *transcendental* (see Section 1, Chapter 5). Liouville, in 1844, was the first to show that transcendental numbers exist, although we now know that almost all real or complex numbers have this property. This does not mean that every problem has been solved for there are many specific numbers whose transcendental status is unknown, examples are Euler's constant γ, $\mathrm{e}+\pi$ and $\pi^{\sqrt{2}}$. In the last 25 years an important new theory has been developed, by Baker and others, from ideas associated with the Gelfond–Schneider theorem (see below). Using this theory we can show, for example, that

$$\mathrm{e}^{\alpha}\alpha_1^{\beta_1}\ldots\alpha_k^{\beta_k}$$

is transcendental if α, $\alpha_1, \ldots, \alpha_k$, $\beta_1, \ldots, \beta_k$ are algebraic. This work is beyond the scope of an undergraduate text; the reader will find a good account in Baker (1975).

We shall discuss the Liouville method in Section 1 and give an application to some Diophantine problems. This method has the disadvantage that no commonly used number, like e or π, has been shown to be transcendental using it. A different method was introduced by Hermite in 1873 and it forms the basis for many modern transcendence proofs. In the second section we give two examples of results proved using this method. From these we can deduce the transcendence of e, π, e^{α}, $\sin\alpha$, $\ln\alpha, \ldots$ for most algebraic α. The Gelfond–Schneider theorem states that α^{β} is transcendental if α and β are algebraic, $\alpha\neq 0$ or 1, and β is irrational. A proof will be given in the final section; it is difficult and can be left out on first reading as no further results depend upon it.

8.1 Liouville's theorem and applications

Let us reconsider the Diophantine approximation inequality

$$\left|\alpha-\frac{x}{y}\right|<\frac{\beta}{y^k}$$

where x and y are integers. By Hurwitz's theorem (2.3, Chapter 7) this has infinitely many solutions if α is irrational, $k=2$ and $\beta=1/\sqrt{5}$, but the situation is different if $k>2$. Our first result was discovered by Liouville, using it he gave the original proof of the existence of transcendental numbers.

THEOREM 1.1 *Suppose α is an algebraic number of degree $n>1$. Then for all $\beta>0$ and $\delta>0$ the inequality*

$$\left|\alpha-\frac{x}{y}\right|<\frac{\beta}{y^{n+\delta}} \qquad (*)$$

has only finitely many integer solutions x and y.

Proof. Let α satisfy the polynomial equation, with rational integer coefficients, $f(z)=a_0z^n+a_1z^{n-1}+\ldots+a_n=0$. We have

$$\left|f\left(\frac{x}{y}\right)\right|=\frac{1}{y^n}\left|a_0x^n+a_1x^{n-1}y+\ldots+a_ny^n\right|\geq\frac{1}{y^n}$$

as the term between the modulus signs is a non-zero integer. By the mean value theorem

$$f(\alpha)-f\left(\frac{x}{y}\right)=\left(\alpha-\frac{x}{y}\right)f'(\eta)$$

for some η lying between α and x/y. Hence, as $f'(\eta)\neq 0$ and we may assume that x/y is bounded, these equations give

$$\left|\alpha-\frac{x}{y}\right|=\left|f\left(\frac{x}{y}\right)\right|\Big/|f'(\eta)|>\frac{1}{my^n}$$

for some constant m. Now if $(*)$ has infinitely many solutions, some solutions will involve arbitrarily large y. But if $y^\delta>m\beta$, $(*)$ contradicts the last inequality. □

COROLLARY 1.2 *The number $\xi=\sum_{n=1}^{\infty}\frac{1}{2^{n!}}$ is transcendental.*

Proof. Suppose on the contrary ξ is algebraic of degree n. Let $\frac{x}{y}=\sum_{i=1}^{n+t-1}\frac{1}{2^{i!}}$ so $y=2^{(n+t-1)!}$. We have

$$\left|\alpha-\frac{x}{y}\right|=\frac{1}{2^{(n+t)!}}\left(1+\frac{1}{2^{n+t+1}}+\ldots\right)<\frac{2}{2^{(n+t)!}}<\frac{2}{y^{n+t}}.$$

This holds for all positive integers t contradicting Theorem 1.1. □

Real numbers α which possess rational approximations x_n/y_n such that $|\alpha-x_n/y_n|<1/y_n^{\delta_n}$, where $\limsup_{n\to\infty}\delta_n=\infty$, are called *Liouville numbers*.

The number ξ given above is Liouville, some further examples are considered in the problem section.

This result establishes the existence of transcendental numbers, but it was Cantor, about half a century after Liouville, who first pointed out that almost all real (and complex) numbers are transcendental. His famous diagonal argument shows that the set of real numbers is uncountable whilst it is well known that the set of algebraic numbers is countable as there are only countably many polynomials with integer coefficients (see Problem 3).

Mahler has classified all complex numbers into four classes A, S, T, and U by the scheme below.

DEFINITION (i) Given a polynomial f with integer coefficients, not all zero, let h be the maximum of the absolute values of the coefficients of f, h is called the *height* of f.

(ii) Given γ, n, and $h>1$, let f be the polynomial with integer coefficients, degree at most n, and height at most h, such that $|f(\gamma)|$ takes the least non-zero value.

(iii) Using f as defined in (ii), let $\omega(n, h)$, ω_n, and ω be given by

$$|f(\gamma)| = h^{-n\omega(n,h)},$$

$$\omega_n = \limsup_{h\to\infty} \omega(n, h),$$

$$\omega = \limsup_{n\to\infty} \omega_n.$$

(iv) Finally let ν be the least positive integer n such that $\omega_n = \infty$, and let $\nu = \infty$ if $\omega_n < \infty$ for all n.

There are four possibilities for the values of ω and ν (note ω and ν cannot both be finite). Mahler's scheme defines

γ to be an A-number if and only if $\omega = 0$ and $\nu = \infty$,

γ to be an S-number if and only if $0 < \omega < \infty$ and $\nu = \infty$,

γ to be an T-number if and only if ω and ν are infinite,

γ to be an U-number if and only if $\omega = \infty$ and ν is finite.

So if ω is finite then $\omega(n, h)$ is uniformly bounded for all n and h; this is true for most real numbers. But if γ is a Liouville number then $\omega_1 = \infty$. It can be shown that

(i) if two numbers α and β are algebraicly dependent [that is, there is a polynomial g with integer coefficients such that $g(\alpha, \beta) = 0$] then they belong to the same class,

(ii) almost all real and complex numbers are S-numbers,

(iii) α is an A-number if and only if α is algebraic (see Problem 4),

(iv) the Liouville numbers are U-numbers,

(v) T-numbers exist.

This last result was proved by Schmidt about 20 years ago. Many problems remain, for example it is known that π is an S- or T-number, but which? For further details see Baker (1975), Le Veque (1955), or Schmidt (1980).

Thue–Siegel–Roth theorem

Before considering other methods for establishing transcendence we shall describe briefly another topic, initiated by Thue, which again takes our standard Diophantine inequality as its starting point. Suppose for the moment Liouville's theorem (1.1) can be improved to

CONJECTURE Let α be an algebraic number of degree $n > 1$ then a number c exists such that

$$\left|\alpha - \frac{x}{y}\right| > \frac{c}{y^k}$$

for all integers x and y and for some suitably chosen $k < n$.

If f is the defining polynomial for α, then the equation

$$y^n f\left(\frac{x}{y}\right) = a_0 x^n + a_1 x^{n-1} y + \ldots + a_n y^n = T, \qquad (**)$$

where T is a constant integer, can have only finitely many integer solutions x and y. To prove this suppose $(**)$ has infinitely many solutions $\{x_i, y_i\}$, and let $\beta = \min |\alpha_i - \alpha_j|$ where $i \neq j$ and $\alpha_1, \ldots, \alpha_n$ are the roots of $f(x) = 0$. (We need to assume that these roots are distinct, but this is always true for the defining polynomial of an algebraic number, see Problem 2, Chapter 5.) Now one of the roots α_i, call it α_1, is a limit point of the sequence $\{x_n/y_n\}$, for otherwise

$$y^n f\left(\frac{x}{y}\right) = a_0 y^n \prod_{i=1}^{n} \left(\frac{x}{y} - \alpha_i\right)$$

would be unbounded as y tends to infinity. Hence there are infinitely many solutions $\{x, y\}$ of $(**)$ with $|\alpha_1 - x/y| < \beta/2$. But for each of these we have

$$\left|\alpha_1 - \frac{x}{y}\right| = \frac{T}{a_0 y^n \prod_{i=2}^{n} |x/y - \alpha_i|} \leq \frac{T}{a_0(\beta/2)^{n-1}} \frac{1}{y^n}$$

and this contradicts our conjecture. Hence $(**)$ can have only finitely many solutions if the conjecture holds.

Thue proved this conjecture with $k = n/2 + 1$. Improvements by Siegel, Dyson, and finally Roth have shown that it holds with $k = 2 + \varepsilon$ for $\varepsilon > 0$ (this is best possible by Hurwitz's theorem). Proofs of these results are given in Le Veque (1955) and Baker (1975). It follows that certain classes of two variable polynomial equations have only a finite number of integer solutions, for example equations of the form $f(x, y) = T$ where f is an irreducible polynomial, without multiple roots and of degree at least three, and T is a constant. Recently Faltings has greatly extended these results using some deep theorems from algebraic geometry, a brief discussion will be given in Chapter 15.

8.2 The Hermite and Lindamann theorems

The main method for establishing the transcendence of a number α begins by assuming that α satisfies a polynomial equation, then a formula F is constructed using the coefficients of this polynomial, two properties of F are derived which contradict one another, and the transcendence of α follows. Hermite, in 1873, was the first to use this method successfully in his proof of the transcendence of e. The proofs in this section, and the next, also use this method; the first two rely upon

LEMMA 2.1 *Let f be the polynomial given by $f(x) = \sum_{j=0}^{m} c_j x^j$, where x is real or complex. Define H by*

$$H(x) = \mathrm{e}^x \sum_{i=0}^{m} f^{(i)}(0) - \sum_{i=0}^{m} f^{(i)}(x),$$

where $f^{(i)}(x)$ denotes the derived polynomial $\frac{\mathrm{d}^i}{\mathrm{d}x^i} f(x)$, then we have the bound

$$|H(x)| \leq \mathrm{e}^{|x|} \sum_{j=0}^{m} |c_j|\, |x|^j.$$

Proof. We have

$$\sum_{i=0}^{m} f^{(i)}(x) = f(x) + \sum_{i=1}^{m} \sum_{j=i}^{m} j(j-1)\ldots(j-i+1)c_j x^{j-i}$$

$$= \sum_{j=0}^{m} c_j j! \sum_{i=0}^{j} \frac{x^{j-i}}{(j-i)!} = \sum_{j=0}^{m} c_j \sum_{i=0}^{j} \frac{j!\,x^i}{i!},$$

note as usual $0! = 1$. So it follows that

$$\sum_{i=0}^{m} f^{(i)}(0) = \sum_{j=0}^{m} c_j j!.$$

Using these identities we have

$$H(x) = \sum_{i=0}^{\infty} \frac{x^i}{i!} \sum_{j=0}^{m} c_j j! - \sum_{j=0}^{m} c_j \sum_{i=0}^{j} \frac{j!\,x^i}{i!} = \sum_{j=0}^{m} c_j \sum_{i=j+1}^{\infty} \frac{j!\,x^i}{i!},$$

hence

$$|H(x)| \le \sum_{j=0}^{m} |c_j| \sum_{i=j+1}^{\infty} \frac{|x|^i}{(i-j)!} = \sum_{j=0}^{m} |c_j|\,|x|^j \sum_{i=1}^{\omega} \frac{|x|^i}{i!}$$

$$\le e^{|x|} \sum_{j=0}^{m} |c_j|\,|x|^j. \qquad \square$$

THEOREM 2.2 (Hermite) *The real number* e *is transcendental.*

Proof. Suppose e satisfies the equation

$$a_0 + a_1 e + \ldots + a_n e^n = 0$$

where $a_0, a_1, \ldots, a_n$ are integers and $a_n \neq 0$. Let

$$f(x) = x^{p-1}(x-1)^p \ldots (x-n)^p$$

for some prime p to be chosen later, let H be as in Lemma 2.1 with this f, and let L be given by: $L = \sum_{j=0}^{n} a_j H(j)$. The polynomial equation for e above gives

$$L = \sum_{j=0}^{n} a_j \left(e^j \sum_{i=0}^{m} f^{(i)}(0) - \sum_{i=0}^{m} f^{(i)}(j) \right) = -\sum_{i=0}^{m} \sum_{j=0}^{n} a_j f^{(i)}(j)$$

where $m = (n+1)p - 1$. Now consider $f^{(i)}(j)$; we have $f^{(i)}(j) = 0$ if $i < p$ and $0 < j \le n$, and if $i < p-1$ and $j = 0$, therefore for all i and j in the sum L, except $i = p-1$ and $j = 0$, $f^{(i)}(j)$ is divisible by $p!$. In the exceptional case we have

$$f^{(p-1)}(0) = (p-1)!\,(-1)^{np}(n!)^p,$$

hence if $p > n$ and $p > a_0$, L is divisible by $(p-1)!$ but not by p. Thus it is non-zero and we have

$$L \ge (p-1)!. \qquad (*)$$

For our second property of L we have by Lemma 2.1 with $m = (n+1)p - 1$

$$|L| \le \sum_{j=0}^{n} |a_j|\,|H(j)| \le \sum_{j=0}^{n} |a_j|\,e^{|j|} \sum_{i=0}^{(n+1)p-1} |c_i|\,|j|^i < b^p \qquad (**)$$

for some b which is not dependent on p, as n and each a_i and c_j are independent of p. Finally choose p so that $(p-1)! > b^p$. But $(*)$ and $(**)$ imply $(p-1)! < b^p$ giving a contradiction, the result follows. $\square$

The proof of our next result uses similar methods but is more complicated. We give Baker's proof which is based on Lindamann's original argument.

THEOREM 2.3 (Lindamann) *Let $\alpha_1, \ldots, \alpha_n$ be distinct algebraic numbers and $a_1, \ldots, a_n$ be non-zero algebraic numbers, then*

$$a_1 e^{\alpha_1} + \ldots + a_n e^{\alpha_n} \neq 0.$$

Proof. Let us suppose

$$a_1 e^{\alpha_1} + \ldots + a_n e^{\alpha_n} = 0. \tag{1}$$

In the first part of the proof we show that it is sufficient to consider a special case of (1) which satisfies the assertions (i), (ii) and (iii) below.

(i) We may assume that each a_i is a rational integer.

Using (1) construct the product

$$\prod (a_{1,i_1} e^{\alpha_1} + \ldots + a_{n,i_n} e^{\alpha_n})$$

where, for each j, a_{j,i_j} ranges over the conjugates of a_j. Expanding this product we have, by (1), an equation of the form

$$b_1 e^{\beta_1} + \ldots + b_k e^{\beta_k} = 0$$

where each β_j is a linear combination of the α_i and each b_j is a symmetric polynomial in $a_1, \ldots, a_n$ and their conjugates. It follows by the symmetric function theorem [see (ii) on p. viii] that each b_j is a rational number and so, multiplying this expression by the LCM of the denominators of the b_j, we can assume that each b_j is a rational integer.

(ii) We may further assume that, for each i, each conjugate $\alpha_{i,j}$ of α_i occurs as an exponent in (1), and if $a_i e^{\alpha_i}$ and $a_j e^{\alpha_j}$ are two terms of (1) and α_j is conjugate to α_i then $a_j = a_i$.

Assume that assertion (i) is satisfied. Let g be the minimum polynomial (with rational coefficients) having roots $\alpha_1, \ldots, \alpha_n$ and suppose g has degree m. Note each conjugate $\alpha_{i,j}$ is a root of g, label these roots $\alpha_1, \ldots, \alpha_m$ $(n \leq m)$; they are all distinct. Finally let $a_{n+1} = \ldots = a_m = 0$. Using (1) form the product

$$\prod (a_1 e^{\alpha_{i1}} + \ldots + a_m e^{\alpha_{im}}) \tag{2}$$

taken over all permutations $i1, \ldots, im$ of $1, \ldots, m$. Expanding we obtain, by (1), an equation of the form

$$c_1 e^{\gamma_1} + \ldots + c_k e^{\gamma_k} = 0. \tag{3}$$

Here each c_i is a polynomial in $a_1, \ldots, a_m$ and so is a rational integer, and each γ_j has the form

$$\gamma_j = t_1 \alpha_1 + \ldots + t_m \alpha_m$$

for integers t_i satisfying $0 \leq t_i \leq m!$ and $\sum_{l=1}^{m} t_i = m!$. Now note that each conjugate of each α_i occurs in (3), for a conjugate of γ_j has the form $t_1\alpha_{1,i1} + \ldots + t_m\alpha_{m,im}$. Also note that (3) is symmetric in each γ_j thus the coefficients of terms [of (3)] with conjugate exponents must be identical.

(iii) Finally we may assume that if (1) satisfies assertions (i) and (ii) then at least one of the coefficients a_i is non-zero.

Introduce an ordering relation $\dot{<}$ on the complex numbers by: if $z_j = x_j + iy_j$ $(j = 1, 2)$ then $z_1 \dot{<} z_2$ if and only if $x_1 < x_2$, or $x_1 = x_2$ and $y_1 < y_2$. Now in (1) each exponent is distinct [by condition (i)] hence in each bracket of the product (2) there is a unique summand $a_k e^{\alpha_{ik}}$ with largest exponent α_{ik}, as measured by the relation $\dot{<}$, having a non-zero coefficient a_k. Thus in the sum (3) there is a unique summand with largest exponent, formed by taking the product of the summands with largest exponents from each of the brackets of the product in (2). This term will have a non-zero coefficient.

The second half of the proof follows closely the argument used in the derivation of Hermite's theorem. Suppose (1) holds and satisfies assertions (i), (ii), and (iii). Let t be a positive integer such that $t\alpha_1, \ldots, t\alpha_n$ are algebraic integers [see note (ii) on p. 73] and define f_i and L_i $(i = 1, \ldots, n)$ by

$$f_i(x) = \frac{t^{np}}{(x - \alpha_i)}[(x - \alpha_1) \ldots (x - \alpha_n)]^p$$

and

$$L_i = \sum_{k=1}^{n} a_k \left[e^{\alpha_k} \sum_{j=0}^{m} f_i^{(j)}(0) - \sum_{j=0}^{m} f_i^{(j)}(\alpha_k) \right] = \sum_{k=1}^{n} a_k H_i(\alpha_k)$$

where $m = np - 1$ and p is a prime to be specified later. We shall consider $|L_1 \ldots L_n|$. Note first that as in the proof of Theorem 2.2 we have, by (1),

$$L_i = -\sum_{j=0}^{m} \sum_{k=1}^{n} a_k f_i^{(j)}(\alpha_k).$$

Now $f_i^{(j)}(\alpha_k)$ is an algebraic integer divisible by $p!$ unless $j = p - 1$ and $k = i$. In this case we have

$$f_i^{(p-1)}(\alpha_i) = t^{np}(p-1)! \prod_{\substack{k=1 \\ k \neq i}}^{n} (\alpha_i - \alpha_k)^p,$$

that is an algebraic integer divisible by $(p - 1)!$ but not by p, provided p is sufficiently large (p must be larger than the height of each f_i). Hence L_i is divisible by $(p - 1)!$ and is non-zero.

Suppose we order the set $\alpha_1, \ldots, \alpha_n$ so that no two of $\alpha_1, \ldots, \alpha_s$ are conjugate, and $\alpha_{s+1}, \ldots, \alpha_n$ are the conjugates (excluding themselves) of $\alpha_1, \ldots, \alpha_s$, then by (ii) we have

$$L_i = -\sum_{j=0}^{m} \sum_{k=1}^{s} a_k[f_i^{(j)}(\alpha_k) + \ldots + f_i^{(j)}(\alpha_{k,u}) + \ldots].$$

As each conjugate of α_k occurs in the inner sum of L_i, it follows that $\sum_{k=1}^{s} a_k[f_i^{(j)}(\alpha_k) + \ldots]$ can be expressed as a polynomial in α_i with rational coefficients independent of i (use the symmetric function theorem). Therefore, using the divisibility property of L_i above, we have $|L_1 \ldots L_n|$ is a rational integer divisible by $((p-1)!)^n$, and so by (iii) $|L_1 \ldots L_n| \geq (p-1)!$.

Finally we have, by Lemma 2.1 and as $m = np - 1$,

$$|L_i| \leq \sum_{k=1}^{n} |a_k|\,|H_i(\alpha_k)|$$

$$\leq \sum_{k=1}^{n} |a_k|\,e^{|\alpha_k|} \sum_{j=1}^{m} |c_{j,i}|\,|\alpha_k|^j \leq d_i^p$$

for some suitably chosen $c_{j,i}$ and d_i independent of p. Now we obtain a contradiction by choosing p to satisfy $(p-1)! > (d_1 \ldots d_n)^p$, and the result follows. □

COROLLARY 2.4 *The following numbers are transcendental:*

π, e^α, $\sin\alpha$, $\cos\alpha$, $\tan\alpha$, $\sinh\alpha$, $\cosh\alpha$, $\arcsin\alpha$, $\arccos\alpha$, *and* $\ln\alpha$,

where α is algebraic, non-zero, and not equal to one in the last two cases.

Proof. We have $e^{i\pi} + e^0 = 0$, hence by Lindamann's theorem $i\pi$ is not algebraic and so π is transcendental. Also if $\sin\alpha = a$ and a is algebraic then we have

$$e^{i\alpha} - e^{-i\alpha} - 2iae^0 = 0$$

so again, by our theorem, $2ia$ is not algebraic; this contradiction shows that $\sin\alpha$ is transcendental. The remaining cases are similar. □

A famous problem first proposed by the ancient Greeks and known as 'squaring the circle' was: construct a square equal in area to a given circle using only a ruler and compass. Lindamann's result shows that this is impossible as $\sqrt{\pi}$ is transcendental (see Problem 16, Chapter 6).

8.3 The Gelfond–Schneider theorem

In 1900, at the International Congress of Mathematicians in Paris, Hilbert proposed a collection of problems for the new century. These included the Riemann hypothesis (see Chapter 13) and Fermat's Last Theorem (see Chapter 14) amongst others. The seventh problem, which had its origin in the work of Euler, asked for a proof of

THEOREM 3.1 *Gelfond–Schneider theorem.*
If α and β are algebraic numbers, $\alpha \neq 0$ or 1, and β is irrational then $\omega = \alpha^\beta$ is transcendental.

This was established by Gelfond and Schneider, independently, in 1934 by extending earlier partial results of Gelfond and Kuzmin. It is of interest to note that Hilbert had thought this was one of the most intractable of his problems even though Lindamann's results were known at the time. We shall present a version of Gelfond's proof as it appears in Siegel (1949), it follows the same general lines as Lindamann's proof but is considerably more complex.

We begin with a sketch of the proof. On the assumption that α, β and $\omega = \alpha^\beta$ belong to some algebraic number field K, we construct a real number λ and obtain a contradiction by deriving conflicting inequalities for the norm of λ in K. To do this we define R by

$$R(x) = \eta_1 e^{\theta_1 x} + \ldots + \eta_t e^{\theta_t x}$$

where t is large and each θ_i has the form $(a + b\beta) \ln \alpha$. The coefficients η_i in this expression are chosen so that, for some integer s^*, the Taylor expansion of $R(x)$ at $x = s^*$ has its first $r - 1$ coefficients zero for some r satisfying $r \geq ct$ where c is a constant. The number λ is given by

$$\lambda = (\ln \alpha)^{-r} R^{(r)}(s^*).$$

We derive lower and upper bounds for $|N(\lambda)|$ which conflict with one another if t is sufficiently large. The proof of the final inequality uses Cauchy's integral formula on a circle, this can be avoided by using finite sums to estimate the absolute value of the integral. We shall not do this as it would further complicate the proof. A completely elementary proof of the real case can be found in Gelfond and Linnik (1965).

Before the main proof we have two lemmas concerning the solution of linear equations over $\mathbb{Z}$ and K.

LEMMA 3.2 *Suppose a, w, $v \in \mathbb{Z}$ where $0 < w < v$ and $a \geq 1$, and suppose $a_{jk} \in \mathbb{Z}$ where $|a_{jk}| \leq a$ for $1 \leq j \leq w$ and $1 \leq k \leq v$. Then rational integers $x_1, \ldots, x_v$, not all zero, can be found to satisfy*

$$a_{j1}x_1 + \ldots + a_{jv}x_v = 0 \quad \textit{for} \quad 1 \leq j \leq w,$$

and

$$|x_k| \leq (av)^{w/(v-w)} \quad \textit{for} \quad 1 \leq k \leq v.$$

Proof. Note first that if $r=[(av)^{w/(v-w)}]$, we have $av<(r+1)^{(v-w)/w}$ and

$$avr+1\le av(r+1)<(r+1)^{v/w}. \tag{1}$$

Secondly, given $x_1,\ldots,x_v$, let

$$y_j=a_{j1}x_1+\ldots+a_{jv}x_v$$

for $1\le j\le w$, and let $-b_j$ and c_j be the sum of the negative and positive coefficients, respectively, in this sum. Hence if

$$0\le x_k\le r \quad\text{for}\quad 1\le k\le v, \tag{2}$$

we have, for all j, using the conditions in the lemma

$$-b_jr\le y_j\le c_jr \quad\text{and}\quad b_j+c_j\le av,$$

and so the number of values that y_j can take is bounded by

$$(b_j+c_j)r+1\le avr+1.$$

Thus the total number of w-tuples $\{y_1,\ldots,y_w\}$ is bounded by $(avr+1)^w$. But the number of v-tuples $\{x_1,\ldots,x_v\}$ satisfying (2) is $(r+1)^v$. Hence, by (1), there are distinct v-tuples $\{x_1',\ldots,x_v'\}$ and $\{x_1'',\ldots,x_v''\}$ corresponding to the same w-tuple $\{y_1,\ldots,y_w\}$. If we set $x_k=x_k'-x_k''$ for $1\le k\le v$ the lemma follows. □

From now on in this section we use the word 'integer' to denote an algebraic integer. Let K be an algebraic number field with degree d and integer basis $\{\gamma_1,\ldots,\gamma_d\}$. If $\alpha\in K$ we define $\overline{|\alpha|}$ by

$$\overline{|\alpha|}=\max_{1\le i\le d}|\alpha_i|$$

where α_i denotes the ith field conjugate of α in K (see p. 75). If α is an integer in K, and so $\alpha=a_1\gamma_1+\ldots+a_d\gamma_d$ where each $a_i\in\mathbb{Z}$ (see Theorem 1.9, Chapter 5), we have, for $1\le i\le d$,

$$|a_i|\le c\,\overline{|\alpha|} \tag{3}$$

where c is a positive constant (see Problem 4, Chapter 5).

LEMMA 3.3 *Suppose $b, m, n\in\mathbb{Z}$ where $0<m<n$ and $b\ge 1$, and suppose α_{jk} are integers in K satisfying $\overline{|\alpha_{jk}|}\le b$ for $1\le j\le m$ and $1\le k\le n$. Then we can find integers $\xi_1,\ldots,\xi_n$ in K, not all zero, to satisfy*

$$\alpha_{j1}\xi_1+\ldots+\alpha_{jn}\xi_n=0 \quad\textit{for}\quad 1\le j\le m \tag{4}$$

and

$$\overline{|\xi_k|}<c'(1+(c'bn)^{m/(n-m)}) \quad\textit{for}\quad 1\le k\le n \tag{5}$$

and for some constant c'.

Proof. If (4) has a solution $\{\xi_1, \ldots, \xi_n\}$, let $\xi_s = x_{s1}\gamma_1 + \ldots + x_{sd}\gamma_d$ for $1 \le s \le n$ where all the coefficients x_{si} belong to $\mathbb{Z}$. Further, let

$$\alpha_{js}\gamma_r = a_{jsr1}\gamma_1 + \ldots + a_{jsrd}\gamma_d, \tag{6}$$

where again all coefficients are rational integers. For $1 \le j \le m$, these give

$$\begin{aligned} 0 = \sum_{s=1}^{n} \alpha_{js}\xi_s &= \sum_{s=1}^{n} \alpha_{js} \sum_{r=1}^{d} x_{sr}\gamma_r \\ &= \sum_{r=1}^{d} \sum_{s=1}^{n} x_{sr} \sum_{u=1}^{d} a_{jsru}\gamma_u \\ &= \sum_{u=1}^{d} \left[\sum_{r=1}^{d} \sum_{s=1}^{n} a_{jsru}x_{sr} \right] \gamma_u. \end{aligned} \tag{7}$$

Hence, as $\{\gamma_1, \ldots, \gamma_d\}$ is a basis for K, (4) is soluble if each term in the square brackets is zero. This gives a set of dm equations (dm because $1 \le j \le m$ and $1 \le u \le d$) in dn variables x_{sr}.

By (3) and (6)

$$|a_{jsru}| \le c\ \overline{|\alpha_{js}\gamma_r|} \le cb \max_{1 \le r \le d} \overline{|\gamma_r|} = c''b$$

for some constant c''. So by Lemma 3.2 (with $w = dm$, $v = dn$, and $a = c''b$), as $0 < dm < dn$, the set of dm equations

$$\sum_{r=1}^{d} \sum_{s=1}^{n} a_{jsru}x_{sr} = 0$$

has a non-trivial solution in $\mathbb{Z}$ satisfying

$$|x_{sr}| < 1 + (c''bdn)^{m/(n-m)}$$

for $1 \le r \le d$ and $1 \le s \le n$. This solves (7) and (4). For (5) note that

$$\begin{aligned} \overline{|\xi_s|} &\le |x_{s1}|\ \overline{|\gamma_1|} + \ldots + |x_{sd}|\ \overline{|\gamma_d|} \\ &\le d\{1 + (c''bdn)^{m/(n-m)}\} \max_{1 \le r \le d} \overline{|\gamma_r|} \\ &< c'(1 + (c'bn)^{m/(n-m)}) \end{aligned}$$

for some suitably chosen constant c'. □

We come now to the proof of the Gelfond–Schneider theorem and begin by setting up the notation. By assumption α, β, and $\omega = \alpha^{\beta}$ belong to some algebraic number field K with degree d. Set $m = 2d + 2$ and let n be a rational integer of the form $t/2m$ where t is a square $t = u^2$, the exact value of n (and so of t and u) will be chosen at the end of the proof. Also let $c_1, c_2, \ldots$ be rational integers, independent of n, to be chosen as the proof progresses. For $i = 1, \ldots, t$, let θ_i be given by: $a, b \in \mathbb{Z}$ and

$$\theta_{au+b} = (a + b\beta) \ln \alpha \quad \text{for } 0 \leq a \leq u-1 \quad \text{and} \quad 1 \leq b \leq u,$$

and define R by

$$R(x) = \eta_1 e^{\theta_1 x} + \ldots + \eta_t e^{\theta_t x} \tag{8}$$

where the coefficients η_i will be determined below.

Consider the set of equations, where $R^{(k)}(x)$ denotes the kth derivative of $R(x)$,

$$(\ln \alpha)^{-k} R^{(k)}(s) = 0 \tag{9}$$

for $0 \leq k \leq n-1$ and $1 \leq s \leq m$. This is a set of mn equations in $t = 2mn$ variables $\eta_1, \ldots, \eta_t$. We show first that the coefficients belong to K. A typical coefficient has the form

$$(\ln \alpha)^{-k} \left(\frac{d^k}{dx^k} e^{\theta_i x} \right)_{x=s} = (\ln \alpha)^{-k} \theta_i^k e^{\theta_i s}.$$

If we replace θ_i by $(a + b\beta) \ln \alpha$ and cancel $(\ln \alpha)^k$ this equals

$$(a + b\beta)^k e^{s(a+b\beta) \ln \alpha} = (a + b\beta)^k \alpha^{sa} \omega^{sb},$$

and so each coefficient in (9) belongs to K. By note (ii) on p. 73 a rational integer c_1 exists with the property

$$c_1\alpha, \ c_1\beta, \text{ and } c_1\omega \text{ are integers in } K, \tag{10}$$

hence if we multiply each equation in (9) by $c_1^{n-1+2mu}$ (for a typical coefficient $k \leq n-1$, $s \leq m$, and $a, b \leq u$) each coefficient in the resulting set of equations is an integer in K.

We can determine $\eta_1, \ldots, \eta_t$ now using Lemma 3.3. The absolute value of a conjugate of one of the new coefficients in (9) is at most

$$c_1^{n-1+2mu}(u + u\,|\overline{\beta}|)^{n-1}\,|\overline{\alpha}|^{mu}\,|\overline{\omega}|^{mu} \leq c_2^n n^{(n-1)/2}$$

as $u = O(\sqrt{n})$. Hence, by Lemma 3.3, the set of equations (9) has a solution $\{\eta_1, \ldots, \eta_t\}$, not all zero, in K satisfying

$$\begin{aligned} |\overline{\eta_k}| &< c'[1 + (c' c_2^n n^{(n-1)/2} t)^{mn/(2mn-mn)}] \\ &\leq c_3^n n^{(n+1)/2} \end{aligned} \tag{11}$$

for $1 \leq k \leq t$. This completes the definition of R. Note that $R(x)$ is not identically zero, for if $R(x) = 0$ for all x then expanding the exponentials in (8) we have

$$\eta_1 \theta_1^k + \ldots + \eta_t \theta_t^k = 0$$

with $k = 0, 1, \ldots$. This implies that $\eta_i = 0$ for all i which is impossible by Lemma 3.3. Hence, by (9), rational integers r and s^* can be found satisfying $r \geq n$, $1 \leq s^* \leq m$,

$$R^{(k)}(s)=0 \quad \text{for} \quad 0 \leq k \leq r-1 \quad \text{and} \quad 1 \leq s \leq m,$$

and

$$R^{(r)}(s^*) \neq 0.$$

Now we define λ by

$$\lambda=(\ln \alpha)^{-r} R^{(r)}(s^*) \neq 0. \tag{12}$$

To complete the proof we derive conflicting inequalities for $N(\lambda)$ in K. By (10) $c_1^{r+2mu}\lambda$ is an integer in K, so (as the norm of a non-zero integer is at least 1)

$$|N(\lambda)| \geq c_1^{-d(r+2mu)} \geq c_4^{-r}. \tag{13}$$

To derive our second inequality we use the fact that

$$|N(\lambda)| \leq |\overline{\lambda}|^{d-1}\, |\lambda| \tag{14}$$

as the field K has degree d. We have

$$\begin{aligned} |\overline{\lambda}| &\leq t \max_{1\leq k\leq t} |\overline{\eta_k}|\, |\overline{e^{s^*\theta_k}}|\, |\overline{\theta_k^r}| \\ &\leq t c_3^n n^{(n+1)/2} c_5^u (c_6 u)^r \leq c_7^r r^{(2r+3)/2}, \end{aligned} \tag{15}$$

using (11) and the definitions of θ_k and s^*, as $r \geq n$ and $u^2=t=O(r)$.

To estimate $|\lambda|$ we use Cauchy's integral formula. For $z \in \mathbb{C}$, let T be given by

$$T(z)=r!\,\frac{R(z)}{(z-s^*)^r} \prod_{\substack{k=1\\k\neq s^*}}^{m} \left(\frac{s^*-k}{z-k}\right)^r.$$

Note that $T(s^*)=(\ln \alpha)^r\lambda$ and, as $R^{(j)}(s^*)=0$ for $j=0, \ldots, r-1$, $T(z)$ has a Taylor series expansion at $z=s^*$ valid for all $z \in \mathbb{C}$; that is $T(z)$ is an entire function. Applying Cauchy's formula on the circle C given by $|z|=m\left(1+\dfrac{r}{u}\right)$, we obtain

$$\lambda=(\ln \alpha)^{-r} T(s^*)=(\ln \alpha)^{-r} \frac{1}{2\pi i} \int_C \frac{T(z)}{z-s^*}\, \mathrm{d}z \tag{16}$$

(note that as $s^* \leq m$, s^* lies inside C). To find $|\lambda|$ we estimate each part of this formula separately. In the estimates (i)–(iv) below, z lies on the circle C.

(i) $$\begin{aligned} |R(z)| &\leq t \max_{k,i} |\eta_k|\, e^{|\theta_i|\,|z|} \\ &\leq t \max_{k} |\eta_k|\, e^{u(1+|\beta|)m((u+r)/u)\ln|\alpha|} \\ &\leq t c_3^n n^{(n+1)/2} c_8^{u+r} && \text{by (11)} \\ &\leq c_9^r r^{(r+3)/2} && \text{as } r \geq n \text{ and } t=O(r). \end{aligned}$$

(ii) $|z-k| \geq |z| - k \geq m\left(1+\frac{r}{u}\right) - m = mr/u$ for $1 \leq k \leq m$.

By (ii), as $s^* \leq m$, we have

$$\text{(iii)} \quad \left|(z-s^*)^{-r} \prod_{\substack{k=1 \\ k \neq s^*}}^{m} \left(\frac{s^*-k}{z-k}\right)^r\right| \leq (mr/u)^{-r} \prod_{k=1}^{m} \left(\frac{u}{mr} m\right)^r \leq c_{10}(u/r)^{mr}.$$

Combining (i) and (iii) we have using the definition of T and $u = O(\sqrt{r})$

$$\text{(iv)} \quad |T(z)| \leq r!\, c_9^r r^{(r+3)/2} c_{10}(u/r)^{mr}$$
$$\leq c_{11}^r r^r r^{(r+3)/2} r^{-mr/2} = c_{11}^r r^{(r(3-m)+3)/2}.$$

Now using (16) we obtain

$$|\lambda| \leq |\ln \alpha|^{-r} \frac{1}{2\pi} \left| \int_C \frac{T(z)}{z-s^*} \mathrm{d}z \right|$$
$$\leq |\ln \alpha|^{-r} m\left(1+\frac{r}{u}\right) \frac{u}{mr} c_{11}^r r^{(r(3-m)+3)/2} \qquad \text{by (ii) and (iv)}$$
$$\leq c_{12}^r r^{(r(3-m)+3)/2}.$$

Combining this with (14) and (15) gives

$$|N(\lambda)| \leq c_7^{(d-1)r} r^{(d-1)(2r+3)/2} c_{12}^r r^{(r(3-m)+3)/2}$$

and so, as $m = 2d+2$,

$$|N(\lambda)| \leq c_{13}^r r^{((d-1)(2r+3)+r(3-2d-2)+3)/2}$$
$$= c_{13}^r r^{(3d-r)/2}.$$

Hence, by (13),

$$c_4^{-r} \leq c_{13}^r r^{(3d-r)/2}$$

or

$$r^{(r-3d)/2} < c_{14}^r.$$

Finally, as $r \geq n$ and c_{14} is independent of n, we can choose n so that this inequality does not hold. Thus our assumption that ω is algebraic is false and the proof is complete. □

8.4 Problems 8

1. Show that the number α with continued fraction expansion $[0, a, a^{2!}, \ldots, a^{n!}, \ldots]$ where $a > 1$ is transcendental.

2. Prove that there are uncountably many real numbers which can be shown to be transcendental using the Liouville method.

3. Give Cantor's argument to show that almost all real numbers are transcendental.

4. Show that a real number γ is transcendental if, for each $t>0$, we can find a positive integer n such that

$$0<|x_0\gamma^n+\ldots+x_n|<h^{-t}$$

has infinitely many integer solutions $\{x_0, \ldots, x_n\}$ where h is the height of the polynomial between the modulus signs. Note that if γ is a Liouville number then $n=1$ for all t. [To prove this assume γ satisfies $f(x)=a_0x^v+\ldots+a_v=0$, where $(a_0, \ldots, a_v)=1$ and $a_0>0$, and let the conjugates be $\gamma_1, \ldots, \gamma_v$. Let $g(y)=x_0y^n+\ldots+x_n$ (with $x_0>0$) be such that its zeros $\beta_1, \ldots, \beta_n$ are distinct from $\gamma_1, \ldots, \gamma_v$. Show that

$$0<\left|\frac{1}{a_0^n}\prod_{i=1}^{n}f(\beta_i)\right|=\left|\frac{1}{x_0^n}\prod_{j=1}^{v}g(\gamma_j)\right|,$$

and hence find an estimate for $|g(\gamma_1)|$ using the symmetric function theorem.]

5. A consequence of the Thue–Siegel–Roth theorem is: the polynomial equation $f(x, y)=T$ has only finitely many solutions if f has degree at least 3, is irreducible, and does not have multiple roots. Show that each of these conditions is necessary.

6. Prove that a symmetric polynomial function with rational coefficients in the roots of a polynomial equation $f(x)=0$ represents a rational number provided the coefficients of f are rational.

7. Let f be a polynomial with rational coefficients and roots $\alpha_1, \ldots, \alpha_n$. Show that a symmetric polynomial function in $\alpha_2, \ldots, \alpha_n$ with rational coefficients can be represented by a polynomial function in the remaining root α_1 with rational coefficients.

8. Using Theorem 2.2 only show that $\sinh\gamma$ is transcendental if γ is a rational number.

☆9. A second proof of the transcendence of π.

(i) Assume that $i\pi$ is algebraic and its conjugates are $\beta_1, \ldots, \beta_c$. Show that $(1+e^{\beta_1})\ldots(1+e^{\beta_c})=0$ and obtain the equation

$$2^c-n+e^{\alpha_1}+\ldots+e^{\alpha_n}=0$$

where each α_i is a linear combination of the β_j and is non-zero.

(ii) Let t be the leading coefficient of the minimum polynomial (with integer coefficients) having roots $\beta_1, \ldots, \beta_c$. Further let p be a large prime to be chosen later, let f be the polynomial given by

$$f(x)=t^{np}x^{p-1}(x-\alpha_1)^p\ldots(x-\alpha_n)^p$$

and let H and L be defined as in the proof of Theorem 2.2 using this f. Show that

$$L\geq(p-1)!.$$

(iii) Use Lemma 2.1 to show that π is transcendental.

☆10. Hilbert's proof of the transcendence of e.

(i) Use integration by parts to show that

$$\int_0^\infty x^m f(x) \mathrm{e}^{-x} \,\mathrm{d}x \equiv m!\, f(0) \pmod{(m+1)!}$$

where f is a polynomial with integer coefficients. [Hint. Consider $\int_0^\infty x^k \mathrm{e}^{-x}\,\mathrm{d}x$ first.]

Suppose rational integers $a_0, \ldots, a_n$ exist, with $a_0 \neq 0 \neq a_n$, so that $a_0 + a_1\mathrm{e} + \ldots + a_n\mathrm{e}^n = 0$. Given r (to be chosen later) let

$$I_a^b = \int_a^b x^r[(x-1)\ldots(x-n)]^{r+1}\,\mathrm{e}^{-x}\,\mathrm{d}x \qquad 0 \leq a < b \leq \infty$$

and let

$$P_1 = a_0 I_0^\infty + a_1\mathrm{e} I_1^\infty + \ldots + a_n\mathrm{e}^n I_n^\infty, \qquad P_2 = a_1\mathrm{e} I_0^1 + \ldots + a_n\mathrm{e}^n I_0^n.$$

(ii) By the change of variable $x - k \to y$ evaluate $a_k\mathrm{e}^k I_k^\infty$, hence using (i) show that, for a suitable choice of r, P_1 is a non-zero integer multiple of $r!$.

(iii) Use standard techniques to find a bound for $|a_k I_0^k|$, and thus show that $|P_2| < r!$ for sufficiently large r.

(iv) Deduce that e is transcendental.

11. Where was the condition 'β is irrational' used in the proof of the Gelfond–Schneider theorem?

12. Use Theorem 3.1 to show that, if α, β, γ, and δ are non-zero algebraic numbers and no linear relation with rational coefficients exists between $\ln\beta$ and $\ln\delta$, then

$$\alpha \ln\beta + \gamma \ln\delta \neq 0.$$

Note that Baker's theorem gives a similar result with n rather than two summands.

9

QUADRATIC FORMS

Gauss was the first to study quadratic forms in a systematic way, his famous book published in 1801 and called *Disquisitiones Arithmeticae* (see Gauss, 1965), is largely taken up with this topic. Although he brought together a number of results from earlier mathematicians, many of the basic ideas and theorems are due to him and were first presented in *Disquisitiones Arithmeticae.* For example he realized the fundamental importance of the equivalence of forms (Section 1) and introduced the notion of composition which is central to the theory of binary forms (Chapter 10).

Quadratic forms play an important role in the theory of Diophantine equations. In particular there are many results concerning equations of the type

$$f(x_1, \ldots, x_n) = k$$

where f is a quadratic form. The binary case is the best developed, although even here problems remain. We shall consider this relationship in more detail in Chapter 15.

In this and the following chapter we investigate the arithmetic theory of quadratic forms over the rational integers. There are other approaches. For example quadratic forms over the rational numbers and over the p-adic numbers have been extensively studied. These more general approaches have much to commend them but are outside the scope of this book; a good account is given in Cassels (1978).

The basic concepts in quadratic form theory are equivalence of forms and the determinant (or discriminant) of a form both of which are considered in Section 1. If two forms are equivalent then they have essentially the same properties and the same determinant. Under this relation there are only finitely many equivalence classes (Theorem 1.2) for each determinant and so only finitely many cases need to be considered. The determinant (discriminant in the binary case) is an invariant associated with a form which codifies the essential properties of the form.

In Section 2 we use the finiteness theorem to solve the three-square problem. The remaining sections and Chapter 10 deal exclusively with binary forms. Section 3 lays down the basic framework and in Section 4 we give a method for finding a representative form in each equivalence class. Further results are given in Chapter 10 using both algebraic and analytic methods.

9.1 Equivalence of forms

We begin by giving the basic definitions; note that throughout this and the following chapter all arguments and coefficients of forms are rational integers.

DEFINITION (i) Let a_{ij}, $1 \le i, j \le n$, be integers where $a_{ij} = a_{ji}$. A *quadratic form* (or *form*) of degree n is a function $F: \mathbb{Z}^n \to \mathbb{Z}$ given by

$$F(x_1, \ldots, x_n) = \sum_{i,j=1}^{n} a_{ij} x_i x_j.$$

Forms are called *binary* if $n = 2$, *ternary* if $n = 3$, and so on. Using matrix notation the form F can be represented by

$$F(\mathbf{x}) = \mathbf{x} A \mathbf{x}'$$

where $\mathbf{x}$ is the vector $(x_1, \ldots, x_n)$, A is the symmetric matrix (a_{ij}), and the prime denotes the transpose.

(ii) The *determinant* of F, $D(f)$, is the determinant of the matrix A, this is the most important quantity associated with F.

(iii) A form F is called *positive* (*negative*) *definite* if $F(\mathbf{x}) > 0$ ($F(\mathbf{x}) < 0$, respectively) for all vectors $\mathbf{x}$ except the zero vector $\mathbf{0}$. Otherwise F is called *indefinite*.

(iv) A matrix T is called *unimodular* if it has integer coefficients and $\det(T) = 1$. (Note that T^{-1} is also unimodular.) Two quadratic forms F and G of degree n with matrices A and B respectively are called *equivalent*, denoted by $F \sim G$, if F can be transformed into G using a unimodular matrix; that is if a unimodular matrix T exists such that $B = TAT'$ and then $G(\mathbf{x}) = \mathbf{x}TAT'\mathbf{x}'$. If $\det(T) = -1$ then F and G are said to be *improperly equivalent*.

(v) A quadratic form F is said to *represent* an integer m if integers $x_1, \ldots, x_n$ can be found to satisfy $F(x_1, \ldots, x_n) = m$.

(vi) In the binary case we shall also consider forms of the type

$$f(x, y) = ax^2 + bxy + cy^2$$

where b is even or odd. In this case we define the *discriminant* of f, $d(f)$, to be $b^2 - 4ac$; note that $d(f) = -4D(f)$ and the matrix of f is $\begin{pmatrix} a & b/2 \\ b/2 & c \end{pmatrix}$.

Note that if $D \neq 0$ and F is indefinite then F takes both positive and negative values. [This is not necessarily true if $D = 0$, consider the form

$(ax+by)^2$.] The matrix A of F is symmetric, and so can be expressed as PZP' where P is unimodular and Z is a diagonal matrix. If the diagonal entries are λ_i for $i=1,\ldots,n$ (the eigenvalues of A), they are non zero (as $D\neq 0$) and do not all have the same sign (as F is indefinite). Hence as F is equivalent to the form $\lambda_1 x_1^2+\ldots+\lambda_n x_n^2$ it takes both positive and negative values.

Our first result brings together the basic properties of equivalence.

THEOREM 1.1(i) *If $F\sim G$ then $D(F)=D(G)$.*

(ii) *The relation $\sim$ is an equivalence relation.*

(iii) *For $i=1,\ldots,n$, the diagonal elements of TAT' are $F(t_{i1},\ldots,t_{in})$ where $(t_{i1},\ldots,t_{in})$ is the ith row of T.*

(iv) *If $F\sim G$ then F and G represent the same integers.*

Proof. Suppose T maps F to G. (i) follows as $\det T=\det T'=1$ and (ii) follows as $\det T^{-1}=1$ and $\det TU=1$ if U is also unimodular. (iii) Using (i) in the definition on p. 154, if T_i is the ith row of T then T_iAT_i' is the ith entry on the diagonal of TAT'. (iv) Suppose $G(\mathbf{x})=\mathbf{x}TAT'\mathbf{x}'$ and $F(\mathbf{y})=\mathbf{y}A\mathbf{y}'$, clearly $F(\mathbf{y})=m$ is soluble if and only if $G(\mathbf{x})=m$ is soluble using the transformation $\mathbf{y}=\mathbf{x}T$. □

Part (iv) is very useful for solving quadratic Diophantine equations; if we wish to show that $F(\mathbf{x})=m$ is soluble we need only solve $G(\mathbf{y})=m$ for some G equivalent to F (see Section 2).

The main result of this section is

THEOREM 1.2 *Assume that $D\neq 0$. There are only finitely many equivalence classes of quadratic forms for each degree n and determinant D.*

We shall prove this result by showing that each class contains a form whose coefficients are bounded in terms of n and D (see Problem 1 for the case $D=0$). First we need two lemmas.

LEMMA 1.3 *Given integers $a_1,\ldots,a_n$ where $(a_1,\ldots,a_n)=1$, we can find a unimodular matrix A whose first column is $(a_1,\ldots,a_n)'$.*

Proof. By induction on the degree n. For $n=1$ there is nothing to prove. If $n=2$ choose integers c_1 and c_2 so that $a_1c_2-a_2c_1=1$, and then let $A=\begin{pmatrix} a_1 & c_1 \\ a_2 & c_2\end{pmatrix}$. Now suppose $n>2$ and the result holds for all $k<n$. Let $(a_2,\ldots,a_n)=b$, then $(a_1,b)=1$ and, if $bc_i=a_i$ $(i=2,\ldots,n)$, $(c_2,\ldots,c_n)=1$. By the inductive hypothesis there is a 2×2 matrix B and an $n-1\times n-1$ matrix C such that $\det(B)=\det(C)=1$, the first column of B is $(a_1,b)'$, and the first column of C is $(c_2,\ldots,c_n)'$. Let A,

an $n \times n$ matrix, be given by

$$A = \begin{pmatrix} 1 & 0 \\ 0 & C \end{pmatrix} \begin{pmatrix} B & 0 \\ 0 & I_{n-2} \end{pmatrix} = \begin{pmatrix} 1 & 0 & 0 & \cdot\cdot \\ 0 & c_2 & * & \cdot\cdot \\ 0 & c_3 & * & \cdot\cdot \\ \cdot & \cdot & \cdot & \cdot\cdot \end{pmatrix} \begin{pmatrix} a_1 & * & 0 & \cdot\cdot \\ b & * & 0 & \cdot\cdot \\ 0 & 0 & 1 & \cdot\cdot \\ \cdot & \cdot & \cdot & \cdot\cdot \end{pmatrix}$$

where I_{n-2} is the $n-2 \times n-2$ identity matrix. Clearly $\det(A) = 1$ and multiplying out we see that the first column of A is $(a_1, \ldots, a_n)'$. □

LEMMA 1.4 *Suppose the quadratic form F has degree n, is not negative definite, and has a non-zero determinant D. A vector* **a** *exists satisfying*

$$0 < F(\mathbf{a}) \le \theta\, |D|^{1/n}$$

where $\theta = 3^{s/n}(4/3)^{(n-1)/2}$ *for some s such that* $0 \le s \le (n-1)(n-2)/2$.

Proof. By induction on the degree n. The result is clearly true if $n = 1$ with $\theta = 1$ and $\mathbf{a} = (1)$. Now assume that the result holds for all forms of degree less than n. Let m be the least positive integer represented by F, hence a vector $\mathbf{a} = (a_1, \ldots, a_n)$ exists satisfying $F(a_1, \ldots, a_n) = m$; we have $(a_1, \ldots, a_n) = 1$ by the minimality of m. Now transform F into an equivalent form G using the matrix with first column $\mathbf{a}'$ given by Lemma 1.3. By Theorem 1.1(iii) we see that $G(1, 0, \ldots, 0) = m$, and so to finish the proof we need to find an upper bound for m.

Completing the square, the form H (independent of x_1) is given by

$$mG(\mathbf{x}) = (mx_1 + g_{12}x_2 + \ldots + g_{1n}x_n)^2 + H(x_2, \ldots, x_n) \qquad (1)$$

where g_{ij} is the (i, j)th element of the matrix of G. We have

$$D(H) = m^{n-2}D(G).$$

[This follows because the determinant of mG is $m^n D(G)$ and, by a suitable transformation with a triangular matrix having determinant m, the right-hand side of (1) can be transformed into $x^2 + H(x_2, \ldots, x_n)$ and so has determinant $m^2 D(H)$.] Now we can apply the induction hypothesis to H; there are two cases, see note below (vi) on p. 154.

Case 1. H is not negative definite. So there is a **b** and θ_1 such that

$$0 < H(\mathbf{b}) \le \theta_1\, |m^{n-2}D|^{1/(n-1)} \qquad (2)$$

where $\mathbf{b} = (b_2, \ldots, b_n)$. By taking the absolute least residue modulo m, choose b_1 to satisfy

$$|mb_1 + g_{12}b_2 + \ldots + g_{1n}b_n| \le m/2. \qquad (3)$$

Hence if $\mathbf{b}^* = (b_1, \ldots, b_n)$, $G(\mathbf{b}^*) > 0$ by (1) and (2) and so, by the minimality of m, $G(\mathbf{b}^*) \geq m$. But by (1), (2) and (3)

$$mG(\mathbf{b}^*) \leq m^2/4 + \theta_1 \, |m^{n-2}D|^{1/(n-1)},$$

and so

$$3m^2/4 \leq \theta_1 \, |m^{n-2}D|^{1/(n-1)}.$$

Thus, as $2-(n-2)/(n-1) = n/(n-1)$,

$$m^{n/(n-1)} \leq (4\theta_1/3) \, |D|^{1/(n-1)},$$

or

$$m \leq (4\theta_1/3)^{(n-1)/n} \, |D|^{1/n}.$$

Hence in this case $\theta = (4\theta_1/3)^{(n-1)/n}$. (4)

Case 2. H is negative definite. Let H' be the form obtained from H by changing the sign of each coefficient of H; for all $x_2, \ldots, x_n$, $H'(x_2, \ldots, x_n) = -H(x_2, \ldots, x_n)$ and H' is positive definite. We can apply the inductive hypothesis to H', so we have $\mathbf{c}$ and θ_2 satisfying

$$0 < H'(\mathbf{c}) \leq \theta_2 \, |m^{n-2}D|^{1/(n-1)}$$

where $\mathbf{c} = (c_2, \ldots, c_n)$. Choose c_1 to satisfy

$$m/2 \leq |mc_1 + g_{12}c_2 + \ldots + g_{1n}c_n| \leq m$$

and, by (1) above, this gives $G(\mathbf{c}^*) < m$, where $\mathbf{c}^* = (c_1, c_2, \ldots, c_n)$. Hence by the minimality of m, $G(\mathbf{c}^*) \leq 0$, and so

$$(mc_1 + g_{12}c_2 + \ldots + g_{1n}c_n)^2 - H'(\mathbf{c}) \leq 0.$$

Combining these inequalities we obtain

$$m^2/4 \leq (mc_1 + g_{12}c_2 + \ldots + g_{1n}c_n)^2 \leq H'(\mathbf{c}) \leq \theta_2 \, |m^{n-2}D|^{1/(n-1)}$$

and, as in Case 1, this gives

$$m \leq (4\theta_2)^{(n-1)/n} \, |D|^{1/n}.$$

Hence in this case $\theta = (4\theta_2)^{(n-1)/n}$. (5)

The result follows by induction. For all forms F the number θ can be constructed using (4) and (5) repeatedly as $\theta = 1$ when $n = 1$. □

Proof of Theorem 1.2. We use induction on the degree n, there is a single class if $n = 1$. As above we have for a form F with minimum positive value m (maximum if F is negative definite)

$$mF(\mathbf{x}) = (mx_1 + f_{12}x_2 + \ldots + f_{1n}x_n)^2 + H(x_2, \ldots, x_n).$$

By the inductive hypothesis we may assume that H is equivalent to one of a finite set of forms. Transform F into an equivalent form G by

$$x_1 = y_1 + c_2y_2 + \ldots + c_ny_n, \text{ and } x_i = y_i \text{ if } i > 1.$$

This has a unimodular matrix for all choices of c_i, it does not affect H, and it gives

$$mG(\mathbf{y}) = (my_1 + (f_{12} + c_2 m)y_2 + \ldots + (f_{1n} + c_n m)y_n)^2 + H(y_2, \ldots, y_n).$$

Finally choose c_i so that $|f_{1i} + c_i m| \leq m/2$ for $i = 2, \ldots, n$. Hence, as m is bounded in terms of n and D (Lemma 1.4), the inductive hypothesis implies that there are only finitely many choices for the coefficients of G; this completes the proof. □

9.2 Sums of three squares

In Chapter 6 we considered the problem of representing an integer as a sum of squares and, using elementary methods, we were able to solve the two-square and four-square problems. The three-square problem is more difficult because there is no product identity in this case. But, using the methods of the previous section, we can solve this problem completely. In fact the two- and four-square theorems can also be deduced from this theory; see Problems 8 and 9.

The method is as follows. Given m, we look for a positive definite ternary quadratic form with determinant 1 which represents m and then apply Theorem 1.1 as all forms of this type are equivalent. The proof uses Dirichlet's theorem on primes in arithmetic progressions which states that a prime (in fact infinitely many primes) exists having the form $ak + b$ for some positive integer k provided $(a, b) = 1$. We shall prove this theorem in Chapter 13 using methods independent of the results of this chapter.

We begin with a general lemma about positive definite forms with determinant 1.

LEMMA 2.1 *If F is a positive definite quadratic form in n variables such that $n < 6$ and $D(F) = 1$, then F is equivalent to the form $x_1^2 + \ldots + x_n^2$.*

Proof. We use Lemma 1.4 with $s = 0$. This is valid because Case 2 in the proof of Lemma 1.4 does not arise with a positive definite form. Thus a form F' equivalent to F and a non-zero vector $\mathbf{a}$ exist satisfying

$$0 < F'(\mathbf{a}) \leq (4/3)^{(n-1)/2}.$$

But $n < 6$, so $(4/3)^{(n-1)/2} < 2$ and hence $F'(\mathbf{a}) = 1$. Now as in the proof of Lemma 1.4 we can assume that the leading coefficient of F' is 1 and

$$F'(\mathbf{x}) = (x_1 + a_{12}x_2 + \ldots + a_{1n}x_n)^2 + G(x_2, \ldots, x_n) \qquad (*)$$

for some quadratic form G. Further, the right-hand side of $(*)$ can be transformed into $x_1^2 + G(x_2, \ldots, x_n)$ using a unimodular matrix. As G has determinant 1 we can apply this process to G, and the result follows using $n - 1$ repetitions of this process. □

With more precise inequalities this result can also be proved when $n=6$ and 7, but it is false when $n \geq 8$. For example the form

$$2\left\{\sum_{i=1}^{8} x_i^2 - x_1x_3 - x_2x_4 - \sum_{i=1}^{5} x_{i+2}x_{i+3}\right\}$$

is positive definite and has determinant 1 but is not equivalent to a sum of squares (see Problem 7). More details are given in Cassels (1978) and Serre (1973).

We come now to the main result of this section.

THEOREM 2.2 *A positive integer can be expressed as a sum of three squares if and only if it is not of the form* $4^a(8b+7)$.

Proof. Let $F(\mathbf{x}) = \sum_{i,j=1}^{3} a_{ij}x_ix_j$ where $\det(a_{ij}) = 1$. We shall show first that F is positive definite if and only if $a_{11} > 0$, and $c = a_{11}a_{22} - a_{12}^2 > 0$.

Suppose first $a_{11} > 0$ and $c > 0$; we have, completing the squares,

$$F(x_1, x_2, x_3)$$
$$= \frac{1}{a_{11}c}\{c(a_{11}x_1 + a_{12}x_2 + a_{13}x_3)^2 + (cx_2 + (a_{11}a_{23} - a_{12}a_{13})x_3)^2 + a_{11}x_3^2\},$$

this is clearly always positive unless $x_1 = x_2 = x_3 = 0$. Conversely if $a_{11} \leq 0$ then $F(1, 0, 0) \leq 0$, and if $a_{11} > 0$ and $c \leq 0$ we have $F(x_1, x_2, 0) = \{(a_{11}x_1 + a_{12}x_2)^2 + cx_2^2\}/a_{11}$, and this has some non-trivial, non-positive values; see note below (vi) on p. 154.

Secondly we have

(i) $4m$ can be expressed as a sum of three squares if and only if this is also true for m, and

(ii) if $m \equiv 7 \pmod 8$ then m cannot be expressed as a sum of three squares.

For (i) note that if $4m = x^2 + y^2 + z^2$, then x, y, and z are all even because a square is congruent to 0, 1, or 4 modulo 8 (as 4 is a square the converse is obvious); (ii) follows using the same result as 7 is not a sum of three squares. Hence by (i) and (ii) we only need to consider the cases (a) $m \equiv 2$ or $6 \pmod 8$, and (b) $m \equiv 1$, 3, or $5 \pmod 8$.

To prove the theorem we must find a ternary form with matrix (a_{ij}) and values t_1, t_2, and t_3 such that

$$F(t_1, t_2, t_3) = m,\ a_{11} > 0,\ c = a_{11}a_{22} - a_{12}^2 > 0, \text{ and } D(F) = 1.$$

We shall give a simple solution of this, there are many others. Let $t_1 = t_2 = 0$, $t_3 = 1$, $a_{13} = 1$, $a_{23} = 0$, and $a_{33} = m$, then the first condition is satisfied and the remaining three can be replaced by

$$a_{11} > 0,\ c = a_{11}a_{22} - a_{12}^2 > 0, \text{ and } D = cm - a_{22} = 1$$

If we assume that $m > 1$ (the result clearly holds if $m = 1$) then the condition $a_{11} > 0$ is implied by the remaining two as $0 < c \leq a_{11}(cm - 1)$. Finally, as $a_{22} = cm - 1$, we see that it is sufficient to find an integer c such that

$$c > 0 \text{ and } -c \text{ is a quadratic residue modulo } cm - 1,$$

the solution of the quadratic residue being a_{12}.

Case (a). $m \equiv 2$ or $6 \pmod 8$.
In this case $(4m, m - 1) = 1$, so by Dirichlet's theorem (Chapter 13) there is a prime satisfying $p = 4mt + m - 1$ where $t \geq 0$. Let $c = 4t + 1$ (and so $c > 0$) and then $p = cm - 1$. Now as $p \equiv 1 \pmod 4$ and c is odd we have by quadratic reciprocity (using both the Legendre and Jacobi symbols, see Chapter 4)

$$(-c/p) = (-1/p)(c/p) = (p/c) = (cm - 1/c) = (-1/c) = 1,$$

as $c \equiv 1 \pmod 4$. Thus c has the required properties in this case.

Case (b). $m \equiv 1$, 3, or $5 \pmod 8$.
Let $e = 1$ if $m \equiv 3 \pmod 8$ and $e = 3$ if $m \equiv 1$ or $5 \pmod 8$, then in each case $(em - 1)/2$ is an odd integer and $(4m, (em - 1)/2) = 1$. So by Dirichlet's theorem again there is a prime p such that

$$p = 4mu + (em - 1)/2 = ((8u + e)m - 1)/2$$

and $u \geq 1$. Let $c = 8u + e$, then $c > 0$ and $2p = cm - 1$. We have

$$\begin{aligned}
&\text{if } m \equiv 1 \pmod 8 \text{ then } e \equiv c \equiv 3 \pmod 4 \text{ and } p \equiv 1 \pmod 4,\\
&\text{if } m \equiv 3 \pmod 8 \text{ then } e \equiv c \equiv 1 \pmod 4 \text{ and } p \equiv 1 \pmod 4,\\
&\text{if } m \equiv 5 \pmod 8 \text{ then } e \equiv c \equiv 3 \pmod 4 \text{ and } p \equiv 3 \pmod 4,
\end{aligned}$$

and in each case $(-2/c) = 1$. Finally by quadratic reciprocity we have

$$\begin{aligned}
(-c/p) &= (-1/p)(-1)^{(c-1)(p-1)/4}(p/c)\\
&= (-1)^{(c+1)(p-1)/4}(p/c) = (p/c) \qquad \text{see cases above},\\
&= (p/c)(-2/c) = (-2p/c) = (1 - cm/c) = (1/c) = 1.
\end{aligned}$$

Thus $-c$ is a quadratic residue modulo p, but c is odd so $-c \equiv 1^2 \pmod 2$ and hence $-c$ is a quadratic residue $(\text{mod } 2p)$, that is modulo $cm - 1$. □

Further proofs of this result have appeared in the literature. Gauss gave a formula for the number of representations, $\psi(m)$, of m as a sum of three squares showing that if m is square-free,

$$\psi(m) = 12H(m) \text{ if } m \equiv 1 \text{ or } 2 \pmod 4,$$

and

$$\psi(m) = 24H(m) \text{ if } m \equiv 3 \pmod 8$$

where $H(m)$ is the number of equivalence classes of binary quadratic forms with discriminant $-4m$ in the first case, and $-m$ in the second case; see for example Taussky (1982). Another proof of the main result using the Hasse principle is given in Serre (1973).

9.3 Representation by binary forms

From now on we shall only consider binary forms f of the type

$$f(\mathbf{x}) = f(x, y) = ax^2 + bxy + cy^2$$

where b is even or odd.

In this section we describe the basic method for solving the equation

$$f(x, y) = k. \qquad (*)$$

This will make use of the results on Pell's equation given in Chapter 7. We begin with a

DEFINITION (a) The *discriminant* d of f is $b^2 - 4ac$.

(b) The form f is called *primitive* if $(a, b, c) = 1$, and *imprimitive* otherwise.

We have immediately

(i) $d = -4D$, where D is the determinant of f.

(ii) $d \equiv 0, 1 \pmod 4$, and $d \equiv b \pmod 2$.

(iii) If $a < 0$ and $d < 0$ then f is negative definite.

This last proposition follows because $4af(\mathbf{x}) = (2ax + by)^2 - dy^2$; we shall exclude this case as $-f$ is positive definite. We shall also exclude the case d is a square for we have

LEMMA 3.1 *The binary form f is a product of linear factors if and only if its discriminant d is a square.*

Proof. If $a = 0$ then $d = b^2$ and $f(x, y) = y(bx + cy)$, so we may assume that $a \neq 0$. If $d = e^2$ we have

$$4af(x, y) = (2ax + (b - e)y)(2ax + (b + e)y),$$

but Problem 22 in Chapter 3 gives $4a \mid (2a, b - e)(2a, b + e)$, and so f is a product of linear factors with integer coefficients. Conversely if

$$f(x, y) = (rx + sy)(tx + uy)$$

then

$$d = (ru + st)^2 - 4rstu = (ru - st)^2. \qquad \square$$

Note, our assumption that d is not a square implies

(iv) $a \neq 0 \neq c$,

(v) if $f(x, y) = k$ has a non-trivial solution then $k \neq 0$.

In Section 1 we defined equivalence of forms, the next few lemmas develop this in the binary case.

LEMMA 3.2 *Let A be the matrix of f, let $T = \begin{pmatrix} r & s \\ t & u \end{pmatrix}$ with rows* **s** *and* **u** *and where $ru - st = 1$, and let $K = TAT' = \begin{pmatrix} k & m/2 \\ m/2 & n \end{pmatrix}$. Then*

(i) *$f(\mathbf{s}) = k$, $f(\mathbf{u}) = n$, and $2\mathbf{s}A\mathbf{u}' = m$,*
(ii) *given* **s**, *we can choose* **u** *so that*

$$m^2 \equiv d \pmod{4k} \text{ and } 0 \le m < 2k.$$

Proof. (i) See Theorem 1.1. (ii) We have by (i)

$$\begin{aligned} m^2 &= [(2ar + bs)t + (br + 2cs)u]^2 \\ &= 4(ar^2 + brs + cs^2)(at^2 + btu + cu^2) + (b^2 - 4ac)(ru - st)^2, \end{aligned}$$

and so

$$m^2 \equiv d \pmod{4k}.$$

Now t and u can be any solution of $ru - st = 1$, so let $t = t_0 + rj$ and $u = u_0 + sj$; then

$$m = (2ar + bs)t_0 + (br + 2cs)u_0 + 2jf(\mathbf{s}).$$

By (i) we can choose j so that m satisfies the inequality in the lemma. □

DEFINITION If A is the matrix of the form f and the matrix T satisfies $TAT' = A$ then T is called an *automorph* of f.

THEOREM 3.3 *If T is an automorph of a primitive form f then*

$$T = \begin{pmatrix} \dfrac{v - bw}{2} & aw \\ -cw & \dfrac{v + bw}{2} \end{pmatrix}.$$

for some integers v and w satisfying $v^2 - dw^2 = 4$.

Proof. By Lemma 3.2 with $K = A$ we have

$$ar^2 + brs + cs^2 = a,$$

$$ru - st = 1,$$

$$(2ar + bs)t + (br + 2cs)u = b.$$

Solving these linear equations for t and u we have, using the first equation,

$$au = ar + bs \quad \text{and} \quad at = -cs.$$

Now $a \mid s$ [for if not, suppose $a' \mid a$, $a' \mid c$, but $a' \nmid s$; then as $a(u - r) = bs$

we have $a'|b$ but this is impossible as f is primitive.] So let $s = aw$, and using the first equation above we have

$$4(ar^2 + abrw + a^2cw^2) = 4a$$

or

$$(2r + bw)^2 - dw^2 = 4.$$

Hence if $v = 2r + bw$, that is $r = (v - bw)/2$, $t = -cw$ and $u = (v + bw)/2$, then T has the form given in the lemma. Now see Problem 16. □

COROLLARY 3.4 *If $f(\mathbf{x}_1) = k$, T is an automorph of f, and $\mathbf{x}_2 = \mathbf{x}_1 T$, then $f(\mathbf{x}_2) = k$.*

Proof. We have $f(\mathbf{x}_2) = \mathbf{x}_2 A \mathbf{x}_2' = \mathbf{x}_1 TAT'\mathbf{x}_1' = \mathbf{x}_1 A \mathbf{x}_1' = f(\mathbf{x}_1) = k$. □

DEFINITION A solution $\mathbf{x} = \{x, y\}$ of the equation $(*)$ is called *proper* if x and y have no common factors, otherwise it is called *improper*.

Clearly if t divides both x and y then $t^2|k$ and we can replace $(*)$ by the equation $f(x, y) = k/t^2$. Hence we shall assume that all solutions are proper unless specifically stated otherwise.

The converse of Corollary 3.4 is given by

THEOREM 3.5 *Let f be a primitive form. Given two proper solutions $\mathbf{x}_1 = (x_1, y_1)$ and $\mathbf{x}_2 = (x_2, y_2)$ of the equation $f(\mathbf{x}) = k$, a matrix T of the type above can be found to satisfy $\mathbf{x}_1 = \mathbf{x}_2 T$, provided $\mathbf{x}_1 A \mathbf{s}_1' = \mathbf{x}_2 A \mathbf{s}_2'$ for some suitably chosen $\mathbf{s}_i = (r_i, s_i)$ such that $x_i s_i - y_i r_i = 1$ $(i = 1, 2)$.*

Proof. Suppose $f(\mathbf{x}_1) = f(\mathbf{x}_2) = k$ and $2\mathbf{x}_1 A \mathbf{s}_1' = m$. Define v and w by

$$\begin{aligned} 2akv &= (2ax_1 + by_1)(2ax_2 + by_2) - dy_1y_2 \\ kw &= x_2y_1 - x_1y_2. \end{aligned}$$

We show first that w is an integer. We have

$$\begin{aligned} 2ax_1 + by_1 &= (2ax_1 + by_1)(x_1s_1 - y_1r_1) \\ &= (2ax_1 + by_1)x_1s_1 - my_1 + (bx_1 + 2cy_1)y_1s_1 \\ &= 2ks_1 - my_1 \\ &\equiv -my_1 \pmod{2k}. \end{aligned}$$

Using the condition in the lemma we also have

$$2ax_2 + by_2 \equiv -my_2 \pmod{2k},$$

and so, multiplying by $-y_2$ and y_1 and then by x_2 and $-x_1$,

$$2a(x_2y_1 - x_1y_2) \equiv 0 \quad \text{and} \quad (b + m)(x_2y_1 - x_1y_2) \equiv 0 \pmod{2k}.$$

A similar argument gives (start with $bx_2 + 2cy_1$)

$$2c(x_2y_1 - x_1y_2) \equiv 0 \quad \text{and} \quad (b - m)(x_2y_1 - x_1y_2) \equiv 0 \pmod{2k}.$$

But, as $(a, b, c) = 1$,

$$(2a, b+m, b-m, 2c) = (2a, 2b, 2c, b+m) \leq 2,$$

so it follows that $k | x_2y_1 - x_1y_2$ and w is an integer.

Secondly we show that $v^2 - dw^2 = 4$. We have

$$\begin{aligned} 4a^2k^2(v^2 - dw^2) &= (2ax_1 + by_1)^2(2ax_2 + by_2)^2 + (b^2 - 4ac)^2y_1^2y_2^2 \\ &\quad - 2(b^2 - 4ac)y_1y_2[4a^2x_1x_2 + 2ab(x_2y_1 + x_1y_2) + b^2y_1y_2] \\ &\quad - 4a^2(b^2 - 4ac)(x_2y_1 - x_1y_2)^2 \\ &= 16a^2f(\mathbf{x}_1)f(\mathbf{x}_2). \end{aligned}$$

(Check first that the terms involving $b^4y_1^2y_2^2$ cancel, so $16a^2$ is a factor.) The result follows as $a \neq 0 \neq k$.

Thirdly, note that v is an integer because v is rational by definition and v^2 is integral as it satisfies the equation $v^2 = 4 + dw^2$.

Let

$$T = \begin{pmatrix} \dfrac{v - bw}{2} & aw \\ -cw & \dfrac{v + bw}{2} \end{pmatrix}.$$

It follows that $\mathbf{x}_1 = \mathbf{x}_2T$. For

$$\mathbf{x}_2T = \frac{1}{2}(x_2v - (2cy_2 + bx_2)w, (2ax_2 + by_2)w + y_2v)$$

and

$$\begin{aligned} &\frac{1}{2}(x_2v - (2cy_2 + bx_2)w) \\ &\quad = \frac{1}{4ak}[x_2(4a^2x_1x_2 + 2ab(x_1y_2 + x_2y_1) + 4acy_1y_2) \\ &\qquad - 2a(2cx_2y_1y_2 - 2cx_1y_2^2 + bx_2^2y_1 - bx_1x_2y_2)] \\ &\quad = x_14af(\mathbf{x}_2)/4ak = x_1. \end{aligned}$$

Similarly $\frac{1}{2}((2ax_2 + by_2)w + y_2v) = y_1$ and so the result follows. □

Representation of integers by binary forms

Using the preliminary results established above we shall consider the equation

$$f(\mathbf{x}) = f(x, y) = k, \qquad (*)$$

where $k > 0$ and f is primitive.

To each solution $\mathbf{x}$ of $(*)$ we can attach a unique number m, called the *attached number*, which satisfies the congruence relation, involving k and d, given in Lemma 3.2(ii) (with $\mathbf{x}$ replacing $\mathbf{s}$). By Theorem 3.5 two proper solutions $\mathbf{x}_1$ and $\mathbf{x}_2$ of $(*)$ with the same attached number m are related by the equation $\mathbf{x}_1 = \mathbf{x}_2 T$ where T is defined in terms of a solution $\{v, w\}$ of the equation

$$v^2 - dw^2 = 4, \qquad (**)$$

see Theorem 3.3. If T_1 and T_2 are two matrices of this type then the product T_1T_2 is also of this type. [If v_1, w_1 and v_2, w_2 correspond to T_1 and T_2, respectively, then $v_1v_2 + dw_1w_2$, $v_1w_2 - v_2w_1$ corresponds to T_1T_2.] Hence by Theorem 3.5 and Corollary 3.4 all proper solutions $\mathbf{x}$ of $(*)$ with attached number m will be generated from any single fixed solution $\mathbf{x}_0$ by $\mathbf{x} = \mathbf{x}_0 T$ for some automorph T.

Example Consider the equation (see Theorem 1.3, Chapter 6)

$$x^2 - xy + y^2 = 7.$$

Here $k = 7$, $d = -3$, and $m = 5$ or 9. The solutions $\{2, -1\}$, $\{-2, 1\}$, $\{1, 3\}$, $\{-1, -3\}$, $\{3, 2\}$, and $\{-3, -2\}$ are attached to 5, and the solutions $\{1, -2\}$, $\{-1, 2\}$, $\{3, 1\}$, $\{-3, -1\}$, $\{2, 3\}$, and $\{-2, -3\}$ are attached to 9. The reader should check this.

In the indefinite case $(d > 0)$ the equation $(**)$ has infinitely many solutions generated by a fundamental solution $\{v_0, w_0\}$ (see Problem 18, Chapter 7) and so each matrix T will have the form $T_0^n (n \in \mathbb{Z})$ where T_0 is built up using v_0 and w_0. In the positive definite case $(**)$ has only finitely many solutions and so this also holds for the equation $(*)$.

Definition In the indefinite case we choose a fixed proper solution $\mathbf{x}_0$ with attached number m (as above) and call it the *primary* solution of the equation $(*)$. In the positive definite case all proper solutions of $(*)$ are called *primary*.

As noted on p. 163 each improper solution $\{x_1, y_1\}$ of $(*)$ arises from a proper solution $\{x_2, y_2\}$ of the equation $f(x, y) = k/t^2$ where $(x_1, y_1) = t$, $x_2 = x_1/t$, and $y_2 = y_1/t$. We call the solution $\{x_1, y_1\}$ an *improper primary* solution of $(*)$ when $\{x_2, y_2\}$ is a primary solution of the second equation.

For the positive definite case we have

Lemma 3.6 *Let f be a primitive binary form with discriminant $d < 0$, let m be as given in Lemma 3.2, and suppose $\mathbf{x}$ is a proper solution of $(*)$ with attached number m, then for each m the equation $(*)$ has z proper solutions where $z = 6$ if $d = -3$, $z = 4$ if $d = -4$, and $z = 2$ if $d < -4$. (See example above.)*

Proof. This follows immediately from Theorem 3.5 as the equation $v^2 - dw^2 = 4$ has z solutions. They are $\{\pm 2, 0\}$ for all d, and also $\{\pm 1, \pm 1\}$ if $d = -3$ and $\{0, \pm 1\}$ if $d = -4$. □

One method for finding primary solutions is as follows. Multiplying by $4a$, the equation $(*)$ can be written as

$$u^2 - dy^2 = 4ak$$

where $u = 2ax \pm by$. In Chapter 7 we gave a criterion for solving equations of this type. Note that although $u \in \mathbb{Z}$ it does not necessarily follow that $x \in \mathbb{Z}$, see example below.

Example Consider the equation

$$21x^2 + 11xy + y^2 = 3 \tag{1}$$

It has discriminant 37 and attached numbers $m = 1$ and 5, as $m^2 \equiv 37$ (mod 12) with $0 \le m < 6$ in this example. The corresponding equation is

$$u^2 - 37y^2 = 252. \tag{2}$$

(i) Equation (2) has the solution $u = 17$, $y = \pm 1$, and x satisfies $17 = 42x \pm 11$. In this case there is no corresponding integer solution to (1) as $x \notin \mathbb{Z}$.

(ii) Secondly equation (2) has the solution $u = 20$, $y = \pm 2$, and x satisfies $20 = 42x \pm 22$. In this case (1) has the primary solution $x = 1$, $y = -2$ with attached number 1. Note that the improper solution of (2) gave rise to a proper solution of (1).

(iii) Thirdly equation (2) has the solution $u = 57$, $y = \pm 9$, and x satisfies $57 = 42x \pm 99$. In this case (1) has the primary solution $x = -1$, $y = 9$ with attached number 5.

(iv) To find the general solution of (1) we note that $\{146, 24\}$ is the fundamental solution of the equation $v^2 - 37w^2 = 4$, and this gives the automorph matrix

$$T_0 = \begin{pmatrix} -59 & 504 \\ -24 & 205 \end{pmatrix}.$$

Hence the solutions of (1) are given by $(1, -2)T_0^n$ and $(-1, 9)T_0^n$ for $n \in \mathbb{Z}$. For instance $\{-11, 94\} = (1, -2)T_0$ and $\{11, -27\} = (-1, 9)T_0^{-1}$ are further solutions of equation (1).

The number of representations

From now on we shall use the notation $\langle a, b, c \rangle$ to denote the form f where $f(x, y) = ax^2 + bxy + cy^2$.

In Section 1 we showed that the number of equivalence classes of forms with discriminant d is finite. Let $h(d)$ denote the number of equivalence classes of primitive forms with discriminant d. [Note that if a class contains a primitive (imprimitive) form then all the forms of the class are primitive (imprimitive, respectively), see Lemma 3.2.] Given a form $f = \langle a, b, c\rangle$ with $(a, b, c) = e > 1$ and discriminant d then the form $f' = \langle a/e, b/e, c/e\rangle$ is primitive and has discriminant d/e^2, so the total number of forms (primitive or not) with discriminant d is given by

$$\sum_{e^2|d, e>0} h(d/e^2).$$

THEOREM 3.7 *Let $k > 0$ and $(d, k) = 1$. Let $\psi_i(k)$ be the number of primary and improper primary solutions of the equation*

$$f_i(x, y) = k, \tag{1}$$

where, for each $i = 1, \ldots, h(d)$, f_i is a representative of the ith equivalence class of primitive forms with discriminant d, and let

$$\psi(k) = \sum_{i=1}^{h(d)} \psi_i(k).$$

We have

$$\psi(k) = z\sum_{n|k} (d/n)$$

using the Kronceker symbol $(d/\cdot)$, and where z is given by Lemma 3.6 if $d < 0$, and $z = 1$ if $d > 0$.

Proof. We use Lemma 3.2. Note first that the number of solutions of the congruence

$$m^2 \equiv d \pmod{4k} \tag{2}$$

with the added condition

$$0 \le m < 2k \tag{3}$$

is exactly half the number of solutions of (2) on its own, for if m is a solution to (2) so is $m \pm 2k$. Hence the number of common solutions of (2) and (3) is

$$\sum_{u|k}{}' (d/u)$$

where this sum is taken over all square-free divisors u of k, see Problem 15 of Chapter 4. Now if m satisfies (2) and (3) then there is an n such that $m^2 - 4kn = d$ and so $\langle k, m, n\rangle$ is a primitive form [as $(d, k) = 1$] with discriminant d which represents k. Hence $\langle k, m, n\rangle$ is equivalent to one of the forms f_i given in the statement of the theorem. Further by Lemma 3.6 we know that the equation $f_i(x, y) = k$ has z primary solutions and so the total number of (proper) primary solutions to (1) is

$$z\sum_{u|k}{}' (d/u)$$

for u square-free. Now the total number of primary and improper primary solutions is

$$\psi(k) = z \sum_{e^2|k, e>0} \sum_{e_1|k/e^2}{}' (d/e_1) = z\sum_{n|k} (d/n)$$

for e_1 square-free. This follows because any integer n can be written in the form e_1e^2 where e_1 is square-free and $e > 0$. Since $e^2|k$ and $e_1|k/e^2$ is equivalent to $n|k$, the second equality above is valid and the result follows. □

It is clear that this result is particularly useful when $h(d) = 1$; an extension to cases when $h(d) \neq 1$ is given in Theorem 1.8, Chapter 10. We shall show in the next section that $h(-4) = 1$ and so Theorem 3.7 gives

COROLLARY 3.8 *The number of solutions of the equation $x^2 + y^2 = k$ for odd k is*

$$4\sum_{n|k} (-1/n) = 4(d_1 - d_3)$$

where d_1 (and d_3) are the number of divisors of k congruent to 1 *(congruent to* 3, *respectively) modulo* 4.

The reader should check that this agrees with the result given in Problem 2 of Chapter 6. A number of further applications are given in the problem section.

9.4 Algorithms for reduced forms

Finally in this chapter we shall give an elementary procedure for choosing a unique representative, called the *reduced form,* for each equivalence class of forms. By counting the total number of reduced forms we can calculate $h(d)$, the number of equivalence classes for discriminant d. Using a different approach a formula for $h(d)$ will be given in the next chapter. Here we shall consider the positive definite and indefinite cases seperately; both algorithms rely upon

LEMMA 4.1 *Every class of binary forms with discriminant d contains a form $\langle a, b, c\rangle$ whose coefficients satisfy*

(i) $|b| \leq |a| \leq |c|$,

(ii) *if* $d > 0$, *then* $ac < 0$ *and* $|a| \leq \frac{1}{2}\sqrt{d}$,

(iii) *if* $d < 0$, *then* c *is positive and* $0 < a \leq \sqrt{(|d|/3)}$.

Proof. (i) This follows in a similar manner to Theorem 1.2. Consider a class K containing the form $\langle a_0, b_0, c_0\rangle$. Let a be the least integer, in absolute terms, represented by a form in K; so integers r and s exist satisfying $a = a_0r^2 + b_0rs + c_0s^2$ with $(r, s) = 1$. Hence, as in Lemma 1.4,

K contains the form $\langle a, b, c\rangle$ for suitably chosen b and c. Now $|c| \geqslant |a|$ by the minimality of a, and $|b| \leq |a|$ follows as in the proof of Theorem 1.2.

(ii) If $d > 0$ then by (i)

$$|ac| \geq b^2 = d + 4ac > 4ac,$$

i.e. $ac < 0$. This gives

$$4a^2 \leq 4\,|ac| = -4ac = d - b^2 \leq d$$

and so (ii) follows.

(iii) If $d < 0$ we may assume that $a > 0$ and $c > 0$ (as we have excluded the negative definite case), then we have

$$d \leq 4a^2 - b^2 \leq 3a^2$$

and this gives the result. □

The algorithm for the positive definite case is given by

THEOREM 4.2 *Each equivalence class of positive definite forms contains exactly one form (the reduced form) of the type* $\langle a, b, c\rangle$ *where*

$$-a < b \leq a < c \quad \text{or} \quad 0 \leq b \leq a = c. \tag{i}$$

Note that this result does not distinguish between primitive and imprimitive forms. The class number equals the number of solutions of these inequalities which satisfy $(a, b, c) = 1$.

Proof. Note first that

$$\langle a, -b, a\rangle \sim \langle a, b, a\rangle \quad \text{and} \quad \langle a, -a, c\rangle \sim \langle a, a, c\rangle. \tag{ii}$$

(For the first equivalence use the matrix $\begin{pmatrix} 0 & 1 \\ -1 & 0 \end{pmatrix}$, and use $\begin{pmatrix} 1 & 1 \\ 0 & 1 \end{pmatrix}$ for the second.) Now as a and c are positive we have by Lemma 4.1

$$-a \leq b \leq a \leq c, \tag{iii}$$

so if $a = c$ we may assume that $b \geq 0$ using the first equivalence of (ii), and if $a < c$ we can assume that $b \neq -a$ using the second part of (ii).

Conversely we need to show that if $\langle a, b, c\rangle \sim \langle a', b', c'\rangle$ and both satisfy (i) then $a = a'$, $b = b'$, and $c = c'$. To begin with we establish the first of these equations. Without loss of generality assume that $a' \leq a$ and the unimodular matrix $\begin{pmatrix} r & s \\ t & u \end{pmatrix}$ transforms $\langle a, b, c\rangle$ into $\langle a', b', c'\rangle$, so

$$a' = ar^2 + brs + cs^2 \quad \text{and} \quad b' = (2ar + bs)t + (br + 2cs)u. \tag{iv}$$

The first of these gives, using (i),

$$a \geq a' \geq ar^2 - a|rs| + as^2 = a(|r| - |s|)^2 + a|rs| \geq a|rs|. \tag{v}$$

So $|rs| = 0$ or 1 and in either case $a = a'$ follows.

Secondly we show that $b = b'$ (note that $c = c'$ follows as both forms have the same discriminant). To begin with suppose $c > a = a'$. Then $cs^2 > as^2$ if $s \neq 0$, and so $s = 0$ using an argument similar to (v). Hence $ru = 1$ and

$$b' = 2art + b \equiv b \pmod{2a}.$$

But, by (i), both b and b' lie between $-a+1$ and a, and so must be equal. A similar conclusion follows if we assume $c' > a'$. So for the final case we have $a = a' = c = c'$ but, by (i), both b and b' are non-negative and $b^2 = b'^2$, thus $b = b'$. □

A table of positive definite reduced forms for $d \geq -200$ is given on pp. 340–1; here as an example we calculate $h(-15)$. If $\langle a, b, c\rangle$ is a form with discriminant -15 then we have, by Lemma 4.1, $a = 1$ or 2 and $b = \pm 1$ as the discriminant is odd. If $a = 1$ then $b \neq -1$ (by Theorem 4.2) so we have the form $\langle 1, 1, 4\rangle$, and if $a = 2$ then $c = 2$ (because $d = -15$) and so $b \geq 0$ and we have the form $\langle 2, 1, 2\rangle$. Thus $h(-15) = 2$.

The algorithm for the indefinite case is less straightforward, it makes use of some continued fraction results. By Lemma 4.1, given an indefinite form $f = \langle a, b, c\rangle$ we may assume that

$$|b| \leq |a| \leq |c|, \; ac < 0, \text{ and } |a| \leq \tfrac{1}{2}\sqrt{d} \tag{i}$$

where d is the discriminant of f. If $f' = \langle a', b', c'\rangle$ is another form satisfying (i) then each of the equivalences $f \sim f'$, $\langle a, -b, c\rangle \sim \langle a', -b', c'\rangle$, and $\langle -a, -b, -c\rangle \sim \langle -a', -b', -c'\rangle$ implies the remaining two $\left(\text{for if } T = \begin{pmatrix} r & s \\ t & u \end{pmatrix} \text{ gives the first equivalence, it also gives the third one and } \begin{pmatrix} r & -s \\ -t & u \end{pmatrix} \text{ gives the second}\right)$. Hence we may assume

$$0 \leq b \leq a \quad \text{and} \quad a \geq |a'|. \tag{ii}$$

THEOREM 4.3 *Suppose $f \sim f'$, and $f = \langle a, b, c\rangle$ and $f' = \langle a', b', c'\rangle$ satisfy* (i) *and* (ii) *above. There is a matrix T with top row* (r, s) *which transforms f into f' such that the number r/s is a continued fraction convergent to $\alpha = (-b + \sqrt{d})/2a$.*

Proof. By Problem 19 we can choose T so that $r > 0$ and $s > 0$ and we have

$$ar^2 + brs + cs^2 = a'. \tag{iii}$$

Case 1. $a' > 0$. Dividing (iii) by as^2 and using (ii) we derive

$$\left(\frac{r}{s} + \frac{b}{2a}\right)^2 - \frac{d}{4a^2} = \left(\frac{r}{s} - \alpha\right)\left(\frac{r}{s} + \beta\right) = \frac{a'}{as^2} < 1$$

where $\beta = (b + \sqrt{d})/2a$. As r/s and β are positive, $r/s > \alpha > 0$ and so

$$\frac{r}{s} + \beta = \alpha + \frac{\sqrt{d}}{a} > \frac{\sqrt{d}}{a} \geq 2$$

by (i). This gives, using Lemma 4.1 and (ii),

$$0 < \frac{r}{s} - \alpha < \frac{1}{s^2(r/s + \beta)} < \frac{1}{2s^2},$$

that is r/s is a continued fraction convergent to α (see Problem 13, Chapter 7).

Case 2. $a' < 0$. By (i) $c < 0$ as $a > 0$, and so (iii) gives

$$\left(\frac{s}{r} - \frac{1}{\alpha}\right)\left(\frac{s}{r} + \frac{1}{\beta}\right) = \frac{a'}{cr^2} > 0.$$

Using a similar argument to that in Case 1 we have s/r is a convergent to $1/\alpha$ and, by comparing expansions of α and $1/\alpha$, it follows that r/s is a convergent to α. The reader should check the details. □

This result greatly reduces the number of matrices that need be examined to check if $f \sim f'$; the bottom row of the matrix can be calculated easily once the top row is known. The next theorem reduces this number still further, it relies on the fact that a quadratic number has a periodic continued fraction expansion, see Theorem 1.7 of Chapter 7.

THEOREM 4.4 *Using the notation of Section* 1, *Chapter* 7, *suppose* α (*defined in Theorem* 4.3) *has the continued fraction expansion* $[q_0, \ldots, q_{k-1}, \overset{*}{q}_k, \ldots, \overset{*}{q}_{k+r-1}]$, $r/s = a_m/b_m$, *and* $r'/s' = a_{m+r}/b_{m+r}$ *where* $m \geq k$. *There is an automorph* U *of* $f = \langle a, b, c\rangle$ *with the property* $(r', s') = (r, s)U$.

Proof. As α is quadratic (with period r) we have by Theorem 1.1 of Chapter 7

$$\alpha = \frac{r\alpha_m + t}{s\alpha_m + u} = \frac{r'\alpha_m + t'}{s'\alpha_m + u'}$$

where $r = a_{m-1}$ etc. So the vector $(\alpha, 1)$ satisfies

$$(\alpha, 1) = j(\alpha_m, 1)\begin{pmatrix} r & s \\ t & u \end{pmatrix} = j'(\alpha_m, 1)\begin{pmatrix} r' & s' \\ t' & u' \end{pmatrix}$$

for some suitably chosen constants j and j'. But by Problem 2(ii) of Chapter 7 we have

$$\begin{pmatrix} r' & s' \\ t' & u' \end{pmatrix} = \begin{pmatrix} r & s \\ t & u \end{pmatrix} U$$

for some unimodular matrix $U = \begin{pmatrix} r'' & s'' \\ t'' & u'' \end{pmatrix}$. This gives $(\alpha, 1)U = j''(\alpha, 1)$ where $j'' = j/j'$, and so

$$\alpha = \frac{r''\alpha + t''}{s''\alpha + u''}$$

or $s''\alpha^2 + (u'' - r'')\alpha - t'' = 0$. But $a\alpha^2 + b\alpha + c = 0$ and so there is an integer w satisfying

$$s'' = aw, \; u'' - r'' = bw, \text{ and } t'' = -cw.$$

Finally if we let $v = (r'' + bw)/2$ then $r'' = (v - bw)/2$ and $u'' = (v + bw)/2$ and it follows, by Theorem 3.3, that U is an automorph of f. □

Using these two results we can give a finite list of possible top rows of matrices which will establish $f \sim f'$ if this is true, provided f and f' satisfy (i) and (ii) of Lemma 4.1. So, by Lemma 3.2(i), if $f(r, s) \neq a'$ for all convergents r/s in this list then $f \not\sim f'$, and if $f(r, s) = a'$ for some convergent r/s it is a relatively simple matter to find, where possible, the remaining entries of the transformation matrix as it must be unimodular, see Problem 24. This is illustrated in the following example and in Problem 23. A list of reduced indefinite forms with discriminants not more than 200 is given on pp. 342–3.

Example Consider forms $f = \langle a, b, c \rangle$ with discriminant $d = 48$.

By Lemma 4.1 $|a| \leq \frac{1}{2}\sqrt{48}$ that is $|a| \leq 3$ and $|b| = 0$ or 2 as d is even. If $|b| = 2$ then $48 = 4 - 4ac$, so $-ac = 11$ and $|a| = 1$ (as $|a| \leq |c|$). This is impossible as $|b| \leq |a|$; hence $b = 0$ and $-ac = 12$. This gives $|a| = 1$ and $|c| = 12$, or $|a| = 2$ and $|c| = 6$, or $|a| = 3$ and $|c| = 4$ but the second case must be discounted as the form is not primitive. Consider $\langle 1, 0, -12 \rangle$ with $\alpha = \frac{1}{2}\sqrt{48} = \sqrt{12} = [3, \overset{*}{2}, \overset{*}{6}]$ (see Table 3 on p. 339). Theorem 4.3 gives the possible top rows $(3, 1)$ and $(7, 2)$. As $3^2 - 12 \cdot 1^2 = -3$ suppose

$$T = \begin{pmatrix} 3 & 1 \\ 3t - 1 & t \end{pmatrix} \quad \text{with} \quad T\begin{pmatrix} 1 & 0 \\ 0 & -12 \end{pmatrix}T' = \begin{pmatrix} -3 & 0 \\ 0 & 4 \end{pmatrix}.$$

Comparing the bottom right-hand entries we have $(3t - 1)^2 - 12t^2 = 4$ or $t = -1$, and with this value of t the other entries agree, hence $\langle 1, 0, -12 \rangle \sim \langle -3, 0, 4 \rangle$. Using the same matrix we also have $\langle -1, 0, 12 \rangle \sim \langle 3, 0, -4 \rangle$. Finally is $\langle 1, 0, -12 \rangle \sim \langle -1, 0, 12 \rangle$? The above list of possible solutions shows that 1 and -3 are represented by the form $\langle 1, 0, -12 \rangle$; hence, as -1 is not included, $\langle 1, 0, -12 \rangle \not\sim \langle -1, 0, 12 \rangle$ and $h(48) = 2$.

9.5 Problems 9

1. Let F be a form in n variables where $F(\mathbf{x}) = \mathbf{x}A\mathbf{x}'$ and $\det A = 0$. Show that $F(\mathbf{x}) = \mathbf{y}B\mathbf{y}'$ for some $r \times r$ matrix B where $r < n$, $\det B \neq 0$ (unless all the

coefficients of A are zero), and each coordinate in $\mathbf{y}$ is a linear combination of the coordinates of $\mathbf{x}$.

2. Give another proof of Lemma 1.3 as follows: let $(a_1, a_2) = b_1$, $(a_{i+2}, b_i) = b_{i+1}$, $i = 1, \ldots, n-2$, and so $b_{n-1} = 1$. Choose c_i and d_i to satisfy

$$a_1 d_1 - a_2 c_1 = b_1, \qquad b_i d_{i+1} + (-1)^i a_{i+2} c_{i+1} = b_{i+1}.$$

Note $(c_i, d_i) = 1$ for all i. Now construct a matrix A with $\det A = 1$, first column $(a_1, \ldots, a_n)'$, diagonal $(a_1, d_1, \ldots, d_{n-1})$ and zeros in all other entries below this diagonal.

3. Which of the following forms are equivalent?

$$2x^2 - 3y^2, \quad 2x^2 + 3y^2, \quad x^2 + 6y^2, \quad -2x^2 + 3y^2, \quad 2x^2 + 4xy + 5y^2.$$

4. How many non-equivalent binary forms are there with determinant 2?

5. Show that the forms

$$F(x, y) = 3x^2 + 2xy + 23y^2, \qquad G(x, y) = 7x^2 + 6xy + 11y^2$$

have the same determinant but are not equivalent. Further show that the forms $z^2 - F(x, y)$ and $z^2 - G(x, y)$ are equivalent. [Hint. Use the transformations $x = x$, $y = x + s$, $z = 5x + t$, and $u = t - 2s$, $v = 3s - t$, $w = x - 14s + 5t$. See also Cassels (1978. pp. 131–2).]

6. Let $ax^2 + by^2 + cz^2$ and $a'x^2 + b'y^2 + c'z^2$ be two positive definite forms. Show that they are equivalent if and only if $\{a', b', c'\}$ is a permutation of $\{a, b, c\}$.

7. Check that the 8-dimensional form given on p. 159 has determinant 1 and is positive definite.

8. Use the quadratic form method of Section 2 to prove that if $p \equiv 1 \pmod 4$ then p can be expressed as a sum of two squares.

9. Use Theorem 2.2 to show that
 (i) every positive integer is a sum of four squares,
 (ii) every positive integer is a sum of three triangular numbers, that is a sum of numbers of the form $n(n+1)/2$.

10. Show that

$$x_1^2 + x_2^2 + x_3^2 + x_4^2 = n, \qquad (x_1, x_2, x_3, x_4) = 1$$

are jointly soluble if and only if $n > 0$ and $8 \nmid n$.

11. Show that the form F equivalent to $x^2 + y^2 + z^2$, defined in the proof of Theorem 2.2, is given by

$$F(x, y, z) = \frac{a^2 + c}{cm - 1} x^2 + 2axy + (cm - 1)y^2 + 2xz + mz^2,$$

where a satisfies $a^2 \equiv -c \pmod{cm - 1}$. By considering the matrix which transforms F into a sum of squares, show that

$$x_1^2+x_2^2+x_3^2=m, \qquad (x_1, x_2, x_3)=1$$

are jointly soluble if and only if $m\equiv 1$, 2, 3, 5, or 6 (mod 8).

12. Use the identity $9(x^2+y^2)=(2x-y)^2+(2x+2y)^2+(2y-x)^2$ to show that a square-free integer n congruent to 2 (mod 3) and where $n>17$ and if $k|n$ then $k\equiv 1$ (mod 4), can be expressed as a sum of three distinct positive squares.

☆13. Using the method of proof of Theorem 2.2 show that the equation

$$x^2+y^2+2z^2=m$$

is soluble if and only if m is not of the form $4^a(16b+14)$.

☆14. Show that there are exactly two equivalence classes of positive definite ternary forms with determinant 3 and give a member of each class. Further show that the equation

$$x^2+y^2+3z^2=m$$

is soluble if $m\equiv 5$ or 13 (mod 24). [It can be shown that this equation is soluble if and only if m is not of the form $9^a(9b+6)$.]

15. Let $t=4n-m^2$. Show that the equations

$$n=x_1^2+x_2^2+x_3^2+x_4^2, \quad m=x_1+x_2+x_3+x_4$$

are jointly soluble if and only if either (i) $t=0$ and $4|m$, or (ii) $t>0$, $m\equiv n$ (mod 2) and $t\neq 4^a(8b+7)$.

16. Show that the elements in the automorph matrix T given in Theorem 3.3 are integers.

17. Solve, where possible,
(i) $3x^2+xy+3y^2=2$,
(ii) $3x^2+xy+3y^2=5$,
(iii) $x^2+xy-5y^2=1$,
(iv) $x^2+xy-7y^2=5$.

18. Let $e_1(n)$ be the difference between the number of divisors of n of the form $3k+1$ and the number of divisors of n of the form $3k+2$, let $e_2(n)$ be the difference between the number of divisors of n congruent to 1, 2, or 4 modulo 7 and the number of divisors of n congruent to 3, 5, or 6 modulo 7, and let $N(f(x, y)=0)$ denote the number of solutions of the equation $f(x, y)=0$. Show that

(i) $N(x^2+xy+y^2=n)=6e_1(n)$,
(ii) if m is odd, $N'(x^2+3y^2=4m)=e_1(m)$, where N' only counts positive odd solutions,
(iii) if $k>0$, $N(x^2+xy+2y^2=k)=2e_2(k)$,
(iv) if t is positive and odd, $N(x^2+7y^2=2t)=0$ and $N(x^2+7y^2=t)=2e_2(t)$.

19. Suppose f is an indefinite form satisfying the conditions of Lemma 4.1 and T is an automorph of f built up using v and w where $v^2-dw^2=4$.

(i) Show that the signs of v and w can be chosen so that the elements of T are positive integers.

(ii) Let R be a unimodular matrix with top row (r, s). Show that an automorph T can be chosen so that the top row elements r' and s' of the product matrix RT are both positive. [Hint. If $rs < 0$, write $r's' = rs + X$ and show that X can be chosen large and positive.]

20. Show that the following forms are equivalent where t and u are integers: $\langle a, b, c\rangle$, $\langle c, -b, a\rangle$, $\langle a, b+2at, at^2+bt+c\rangle$, $\langle a+bu+cu^2,\ \ b+2cu,\ \ c\rangle$. Deduce that $\langle a, ka, c\rangle$ is equivalent to $\langle a, 0, c'\rangle$ or $\langle a, a, c'\rangle$ for some suitably chosen c'.

21. Which of the following forms are equivalent: $\langle 3, 0, 7\rangle$, $\langle 2, 2, 11\rangle$, $\langle 34, 86, 55\rangle$, $\langle 5, -4, 5\rangle$, $\langle 2, -2, 11\rangle$, $\langle 1, -6, 30\rangle$?

22. Using Problem 19 of Chapter 7, prove that, if p is prime,

$$p \equiv 1 \pmod 4 \text{ if and only if } \langle 1, 0, -p\rangle \sim \langle -1, 0, p\rangle.$$

23. Find $h(-67)$, $h(-68)$, $h(29)$, and $h(69)$ and list the reduced forms.

24. Suppose $\langle a, b, c\rangle$ and $\langle a', \pm b', c'\rangle$ are primitive forms with discriminant d and suppose we can find r and s to satisfy $(r, s) = 1$ and $ar^2 + brs + cs^2 = a'$. Show that if either

(i) a' is square-free and b' is zero, or

(ii) $(a, a') = 1$ and a' is prime,

then the two forms are equivalent for some choice of the sign of b'. Note that this can be used to simplify equivalence calculations in the indefinite case once the continued fraction expansions have been found.

10

GENERA AND THE CLASS GROUP

Two quadratic forms f and f' are equivalent if f can be transformed into f' using a unimodular matrix T, that is if

$$f(\mathbf{x}T)=f'(\mathbf{x}).$$

In the previous chapter we showed that this is an equivalence relation with a finite number of equivalence classes. Here we shall investigate the binary case in more detail using both algebraic and analytic methods.

We begin by introducing equivalence of forms modulo a prime (that is, studying forms locally, see Section 3 of Chapter 3), this gives a new class structure with classes called genera. Secondly, we show that the set of equivalence classes (in the original sense) forms a group, called the class group, under the operation of composition of forms defined below. The genera are cosets of this group and this fact enables us to derive some arithmetical properties of the class group. Finally we give a formula for $h(d)$, the order of the class group for discriminant d.

There is a close connection between the class group and the theory of ideals in quadratic number fields. A correspondence exists between binary forms with discriminant d and ideals in the field $\mathbb{Q}(\sqrt{d})$, and between the class group for d and the ideal class group for $\mathbb{Q}(\sqrt{d})$. We shall not develop this as the theory of ideals will not be treated in this book, the reader can find details in, for example, Cohen (1962), Hua (1982), or Jones (1950).

10.1 The genus of a form

We begin by giving a number of definitions and lemmas which will lead to the notion of the genus of a form. Most of the development is straightforward relying on the results of Chapter 4; consequently we shall leave some of the details to the reader.

DEFINITION Let p be a prime and let $\mathbf{x}$ be the vector (x, y). Given two binary forms f_1 and f_2 then we say *f_1 is equivalent to f_2 modulo p*, denoted by $f_1 \sim f_2 \pmod p$, if and only if for all $\mathbf{x}$

$$f_1(\mathbf{x}) \equiv f_2(\mathbf{x}T) \pmod p$$

where T is a 2×2 matrix with integer coefficients, independent of $\mathbf{x}$, satisfying $(\det T, p) = 1$.

Note that this relation is an equivalence relation (see Problem 1), and if f_1 and f_2 have discriminants d_1 and d_2, then

$$d_2 \equiv (\det T)^2 d_1 \pmod{p}.$$

LEMMA 1.1 *If p is an odd prime and $\langle a, b, c\rangle$ is a primitive form, then integers a' and c' can be found to satisfy*

$$\langle a, b, c\rangle \sim \langle a', 0, c'\rangle \pmod{p},$$

and if $p \nmid a$ we can take $a' = a$.

Proof. Suppose first $p \nmid a$. Choose a_1 so that $4aa_1 \equiv 1 \pmod{p}$ then, if d is the discriminant of the form, we have, multiplying by $4aa_1$,

$$\begin{aligned} ax^2 + bxy + cy^2 &\equiv a_1(2ax + by)^2 - a_1 dy^2 \\ &\equiv a(2a_1(2ax + by))^2 - a_1 dy^2 \\ &\equiv a(x + 2a_1 by)^2 - a_1 dy^2 \pmod{p} \end{aligned}$$

This gives $\langle a, b, c\rangle \sim \langle a, 0, a_1 d\rangle \pmod{p}$ using the transformation $x' = x + 2a_1 by$, $y' = y$. A similar argument can be used if $p \nmid c$. If p divides both a and c then $p \nmid b$, and we have $\langle a, b, c\rangle \sim \langle 0, b, 0\rangle \sim \langle b, 0, -b\rangle \pmod{p}$ using the transformation $x' = x + y$, $y' = -x + y$. □

LEMMA 1.2 *Suppose p is an odd prime, d is the discriminant of the form $\langle a, b, c\rangle$, $p \nmid d$, and q is a quadratic non-residue modulo p. Then*

(i) $\langle a, b, c\rangle \sim \langle 1, 0, -1\rangle \pmod{p}$ *if* $(d/p) = 1$,
(ii) $\langle a, b, c\rangle \sim \langle 1, 0, -q\rangle \pmod{p}$ *if* $(d/p) = -1$.

Proof. Note first that if $p \nmid ac$ then integers r and s can be found to satisfy

$$ar^2 + cs^2 \equiv 1 \pmod{p}.$$

For as r runs over the set $P = \{0, 1, \ldots, p-1\}$, ar^2 takes $(p+1)/2$ distinct values, this is also true for $-1 - cs$ as s runs over P. The overlap gives a solution to the congruence. Now suppose $p \nmid a$ (the proof is similar if $p \nmid c$). Choose t and u so that $p \nmid ru - st$ and let

$$T = \begin{pmatrix} r & s \\ t & u \end{pmatrix}, \qquad b_1 = 2art + 2c'su, \qquad c_1 = at^2 + c'u^2$$

where c' is given by Lemma 1.1. We have

$$\langle a, b, c\rangle \sim \langle a, 0, c'\rangle \sim \langle 1, b_1, c_1\rangle \sim \langle 1, 0, -d\rangle \pmod{p}$$

using T and Lemma 1.1 again for the last equivalence.

If $(d/p)=1$ we can find an integer e satisfying $e^2 \equiv d \pmod p$, and the transformation $x'=x$, $y'=ey$ establishes the result in this case as $p \nmid e$. The other case is similar. □

LEMMA 1.3 *If $\langle a, b, c\rangle$ has an odd discriminant then it is equivalent to either $\langle 1, 1, 1\rangle$ or $\langle 0, 1, 0\rangle \pmod 2$.*

Proof. We have $2 \nmid b$ as $2 \nmid d$. If a and c are both odd then $\langle a, b, c\rangle \sim \langle 1, 1, 1\rangle \pmod 2$. If $2|a$ then $\langle a, b, c\rangle \sim \langle 0, 1, c\rangle \pmod 2$ and the transformation $x'=x+cy$, $y'=y$ gives the result. □

These lemmas give

THEOREM 1.4 *If two forms have the same discriminant d and $p \nmid d$ then they are equivalent modulo p.*

Now we consider the case $p|d$. First we need

LEMMA 1.5 *Let f be a primitive form. For all n integers x and y can be found satisfying $(x, y)=1$ and $(f(x, y), n)=1$.*

Proof. Suppose first n is a prime p. Choose $x=1$ and $y=0$ if $p \nmid a$, $x=0$ and $y=1$ if $p \nmid c$, and $x=y=1$ if $p|a$ and $p|c$ (in this case $p \nmid b$ as f is primitive); for each choice we have $p \nmid f(x, y)$. The general case follows using the Chinese remainder theorem (1.8, Chapter 3). □

THEOREM 1.6 *Let $p>2$ and suppose, for $i=1$ and 2, $\langle a_i, b_i, c_i\rangle$ is a primitive form with discriminant d_i which represents an integer k_i where $p|d_i$ and $(k_i, d_i)=1$. We have*

$$\langle a_1, b_1, c_1\rangle \sim \langle a_2, b_2, c_2\rangle \pmod p \text{ if and only if } (k_1/p)=(k_2/p).$$

Proof. Assume that $p \nmid a_i$ $(i=1, 2)$. [If $p|a_i$ then $p \nmid c_i$, as f_i is primitive, and we can repeat the argument below with c_i in place of a_i.] We have by Lemma 1.1

$$\langle a_i, b_i, c_i\rangle \sim \langle a_i, 0, c_i'\rangle \sim \langle a_i, 0, 0\rangle \pmod p \qquad (*)$$

as the second form has discriminant $-4a_ic_i'$ which is divisible by p. Now by Lemma 1.5, integers x and y exist satisfying

$$(k_i, p)=1 \quad \text{and} \quad a_ix^2+b_ixy+c_iy^2 \equiv k_i \pmod p, \qquad (**)$$

and so we have using $(*)$

$$(k_1/p)=(a_1x^2+b_1xy+c_1y^2/p)=(a_1x^2/p)=(a_1/p).$$

Hence $(k_1/p)=(a_1/p)$ for any k_1 satisfying $(**)$. But we can take a_2 for k_1, and so $(k_1/p)=(a_2/p)=(k_2/p)$.

Conversely if $(k_1/p)=(k_2/p)$ then $(a_1/p)=(a_2/p)$ and so, by Problem 3 of Chapter 4, $a_1 \equiv a_2z^2 \pmod p$ for some integer z. Hence using $(*)$

twice we have

$$\langle a_1, b_1, c_1\rangle \sim \langle a_1, 0, 0\rangle \sim \langle a_2, 0, 0\rangle \sim \langle a_2, b_2, c_2\rangle \pmod p. \qquad \square$$

Finally we consider the case $p=2$ and $4|d$; we shall write d^* for $d/4$.

Definition Let k be an odd integer represented by a form f with discriminant $4d^*$. Using Jacobi's symbol let

$$\alpha(k) = (-1/k) \text{ if and only if } d^* \equiv 0 \text{ or } 3 \pmod 4$$
$$\beta(k) = (2/k) \text{ if and only if } d^* \equiv 0 \text{ or } 2 \pmod 8$$
$$\gamma(k) = (-2/k) \text{ if and only if } d^* \equiv 0 \text{ or } 6 \pmod 8.$$

Theorem 1.7 *For $i=1$ and 2, let f_i be a primitive form with discriminant $4d_i^*$ which represents an odd integer k_i.*

(i) *If $d_1^* \equiv d_2^* \equiv 0$ or 3 (mod 4) then*

$$f_1 \sim f_2 \pmod 4 \text{ if and only if } \alpha(k_1) = \alpha(k_2).$$

(ii) *If $d_1^* \equiv d_2^* \equiv 2$ (mod 8) then*

$$f_1 \sim f_2 \pmod 8 \text{ if and only if } \beta(k_1) = \beta(k_2).$$

(iii) *If $d_1^* \equiv d_2^* \equiv 6$ (mod 8) then*

$$f_1 \sim f_2 \pmod 8 \text{ if and only if } \gamma(k_1) = \gamma(k_2).$$

(iv) *If $d_1^* \equiv d_2^* \equiv 0$ (mod 8) then*

$$f_1 \sim f_2 \pmod 8 \text{ if and only if } \alpha(k_1) = \alpha(k_2) \text{ and } \beta(k_1) = \beta(k_2).$$

The remaining cases are treated in Problem 3.

Proof. We shall prove the first part of (i), all the remaining parts are similar (see Problem 2).

Let $f = \langle a, 2b, c\rangle$. Note that a and c cannot both be even, we assume a is odd (if not, interchange x and y) and so $d^* = b^2 - ac \equiv 0 \pmod 4$. Let $f(x, y) \equiv k \pmod 4$.

Case 1. $2|b$. So $ac \equiv 0 \pmod 4$ and $c \equiv 0 \pmod 4$, this gives $k \equiv ax^2 \equiv a \pmod 4$ as x must have the same parity as k.

Case 2. $2 \nmid b$. So $1 - ac \equiv 0 \pmod 4$, and this gives $a \equiv c \pmod 4$ which implies that x and y have different parities as k is odd. Hence k is congruent to ax^2 or cy^2 modulo 4, either of which imply $k \equiv a \pmod 4$.

Thus $k \equiv a \pmod 4$ in both cases and the first part of our result follows by the method we used to complete the proof of Theorem 1.6. $\square$

With these results we can introduce the genus of a form.

Definition Suppose f is a primitive form with discriminant d, k is properly represented by f, and $(k, 2d) = 1$. Let $p_1, \ldots, p_m$ be the odd prime divisors of d. The *character system* of f is the set

$$(k/p_1), \ldots, (k/p_m), \ \alpha(k), \ \beta(k), \ \gamma(k)$$

where the last three entries are included only if d satisfies the corresponding conditions given prior to Theorem 1.7.

Note that, by Theorems 1.4, 1.6, and 1.7 and as k is odd, each element of a character system is a well-defined Jacobi symbol whose value is independent of k. Hence we can make the following

DEFINITION Two forms with discriminant d and the same character system are said to belong to the same *genus* (plural, *genera*).

For example the character system for forms with discriminant -136 is $(k/17)$, $\gamma(k)$ and this gives two genera of reduced forms: $\langle 1, 0, 34\rangle$, $\langle 2, 0, 17\rangle$ representing 1 and 19 respectively with character system 1, 1, and $\langle 5, 2, 7\rangle$, $\langle 5, -2, 7\rangle$ both of which represent 5 and have the character system -1, -1.
The tables on pp. 340–3 list all genera of reduced forms with discriminant d satisfying $|d| \leq 200$.

Clearly if two forms are equivalent then they are equivalent modulo p for all primes p, so each genus consists of a number of equivalence classes of forms. We shall use this fact in the next section to derive the structure of the sets of equivalence classes. Here we use it to extend the scope of Theorem 3.7, Chapter 9; for as it stands it can only be used effectively when $h(d) = 1$.

THEOREM 1.8 *Suppose each genus of forms with discriminant d contains exactly one reduced form. Let* $k > 0$, $(k, 2d) = 1$ *and let f be a form with discriminant d. An algorithm (given below) exists to determine the solubility of the equation*

$$f(x, y) = k.$$

Proof. From our discussion above we see that if f_i and f_j are forms with the same discriminant but different character systems then the equations

$$f_i(x, y) = k, \qquad f_j(x', y') = k$$

cannot both be soluble. But by Theorem 3.7 of Chapter 9 and the condition above there are $\psi(k)$ primary or improper primary solutions to the equations

$$f_i(x, y) = k, \qquad i = 1, \ldots, h(d),$$

where each f_i belongs to a different genus. Hence if $\psi(k) > 0$ one of these equations has $\psi(k)$ solutions and the remaining equations are insoluble. □

We illustrate the method of solution in the following examples.

EXAMPLE 1 Let $d=40$ and suppose $k>0$ and $(k, 10)=1$.

There are two reduced forms $\langle 1,0,-10\rangle$ and $\langle 2,0,-5\rangle$, and the character system is $(k/5)$, $\beta(k)$. The first form has the character system $1,1$ (take $k=1$) and the second has the character system $-1,-1$ (take $k=3$), so the condition of Theorem 1.8 applies. As $k\equiv 1$, 9, 31, or 39 (mod 40) if $(k/5)=\beta(k)=1$, and $k\equiv 3$, 13, 27, or 37 (mod 40) if $(k/5)=\beta(k)=-1$, we have

(i) if $k\equiv 1$, 9, 31, or 39 (mod 40) and $f\sim\langle 1,0,-10\rangle$ then $f(x,y)=k$ has $\psi(k)$ primary or improper primary solutions,

(ii) if $k\equiv 3$, 13, 27, or 37 (mod 40) and $f\sim\langle 2,0,-5\rangle$ then $f(x,y)=k$ has $\psi(k)$ primary or improper primary solutions,

(iii) in all other cases subject to the conditions in this example, the equation $f(x,y)=k$ is insoluble.

EXAMPLE 2 Let $d=-36$ and suppose $k>0$ and $(k,6)=1$.

The character system is $(k/3)$, $\alpha(k)$ and this is cyclic modulo 12. The reduced forms are $\langle 1,0,9\rangle$ with character system $1,1$ (take $k=1$) and $\langle 2,2,5\rangle$ with character system $-1,1$ (take $k=5$). So we have, using Lemma 3.6 and Theorem 3.7, Chapter 9,

(i) if $k\equiv 1 \pmod{12}$, $x^2+9y^2=k$ has $2\sum_{n\mid k}(-1/n)$ solutions

(ii) if $k\equiv 5 \pmod{12}$, $2x^2+2xy+5y^2=k$ has $2\sum_{n\mid k}(-1/n)$ solutions, and there are no solutions in the other cases.

Note that the congruence conditions do not necessarily imply the solubility of the corresponding equation because $\psi(k)$ could be zero. For instance in Example 1, the equation $2x^2-5y^2=77$ is insoluble because $\psi(77)=\sum_{n\mid 77}(40/n)=0$.

10.2 Composition and the class group

We let $\langle\langle a,b,c\rangle\rangle$ denote the class of forms equivalent to $\langle a,b,c\rangle$. As an example to introduce composition, consider forms with discriminant -63. There are four equivalence classes which can be written as

$$\langle\langle 2,1,8\rangle\rangle,\ \langle\langle 4,1,4\rangle\rangle,\ \langle\langle 8,1,2\rangle\rangle,\ \langle\langle 16,1,1\rangle\rangle,$$

using the table on p. 340 and Problem 20, Chapter 9. This suggests a 'product', or 'composition', of classes given by

$$\langle\langle a_1,b,c_1\rangle\rangle * \langle\langle a_2,b,c_2\rangle\rangle = \langle\langle a_1a_2,b,e\rangle\rangle$$

where e is chosen so that the discriminants agree, that is $e=c_2/a_1=c_1/a_2$. Thus

$$\langle\langle 2,1,8\rangle\rangle * \langle\langle 4,1,4\rangle\rangle = \langle\langle 8,1,2\rangle\rangle.$$

Also the class $\langle\langle 1, 1, 16\rangle\rangle$, which is identical to $\langle\langle 16, 1, 1\rangle\rangle$, will act as an identity. But how do we define the composition of $\langle\langle 4, 1, 4\rangle\rangle$ and $\langle\langle 8, 1, 2\rangle\rangle$? Using Problem 20, Chapter 9, again, we have

$$\langle 2, 1, 8\rangle \sim \langle 8, -1, 2\rangle,\ \langle 4, 1, 4\rangle \sim \langle 4, -1, 4\rangle,\ \langle 8, 1, 2\rangle \sim \langle 2, -1, 8\rangle,$$

and then

$$\begin{aligned}\langle\langle 4, 1, 4\rangle\rangle * \langle\langle 8, 1, 2\rangle\rangle &= \langle\langle 4, -1, 4\rangle\rangle * \langle\langle 2, -1, 8\rangle\rangle\\ &= \langle\langle 8, -1, 2\rangle\rangle\\ &= \langle\langle 2, 1, 8\rangle\rangle.\end{aligned}$$

We shall see below that the set of classes of forms having discriminant -63 is a cyclic group under the operation $*$.

Gauss was the first to define composition, but Dirichlet introduced some considerable simplifications usually followed now, see Venkov (1970). We have based this section on Cassels (1978), and further results may be found there. Note that we have excluded square discriminants and negative definite and imprimitive forms from our discussion.

LEMMA 2.1 *Suppose $\langle a_1, b, c_1\rangle$ and $\langle a_2, b, c_2\rangle$ are primitive equivalent forms and m is an integer satisfying*

$$m|c_1, \qquad m|c_2, \qquad (a_1, a_2, m) = 1.$$

We have

$$\langle ma_1, b, c_1/m\rangle \sim \langle ma_2, b, c_2/m\rangle.$$

Proof. Let T be the unimodular matrix given in the lemma, so if $T = \begin{pmatrix} r & s\\ t & u\end{pmatrix}$ we have, multiplying by T^{-1},

$$\begin{pmatrix} a_1 & b/2\\ b/2 & c_1\end{pmatrix}\begin{pmatrix} r & t\\ s & u\end{pmatrix} = \begin{pmatrix} u & -s\\ -t & r\end{pmatrix}\begin{pmatrix} a_2 & b/2\\ b/2 & c_2\end{pmatrix}.$$

Equating the off-diagonal entries we obtain

$$2a_1t + bu = bu - 2c_2s, \qquad br + 2c_1s = -2a_2t + br$$

or

$$a_1t + c_2s = 0, \qquad a_2t + c_1s = 0.$$

Hence, by the hypothesis, $m|t$, and so the matrix

$$T_1 = \begin{pmatrix} r & sm\\ t/m & u\end{pmatrix}$$

is unimodular. It gives the required equivalence as the reader can check easily. □

DEFINITION (i) Primitive forms $f_j = \langle a_j, b_j, c_j \rangle$ $(j = 1, 2)$, with the same discriminant, are called *concordant* if (i1) $a_1 a_2 \neq 0$, (i2) $b_1 = b_2 = b$, and (i3) $a_2 | c_1$.

(ii) If f_1 and f_2 are concordant then the form $f_3 = \langle a_1 a_2, b, c_1/a_2 \rangle$ is called the *composition* of f_1 and f_2 and we write $f_3 = f_1 * f_2$. Note that, as f_1 and f_2 have the same discriminant d, we have $c_2/a_1 = c_1/a_2$ and f_3 also has discriminant d. Further if $(a_1, a_2) = 1$ then (i3) in the definition above is redundant for $(d - b^2)/4 = a_1 c_1 = a_2 c_2$ and so $a_2 | c_1$.

LEMMA 2.2 *Let C_1 and C_2 be equivalence classes of primitive forms with discriminant d. For $j = 1$ and 2, we can find $f_j = \langle a_j, b_j, c_j \rangle \in C_j$ so that f_1 and f_2 are concordant and, given m, we can choose a_1 and a_2 so that $(a_1, a_2) = (a_1, m) = (a_2, m) = 1$.*

Proof. By Lemma 1.5, C_j represents integers a_j $(j = 1, 2)$ where $(a_1, m) = 1$ and $(a_2, a_1 m) = 1$. Using the unimodular matrices $\begin{pmatrix} 1 & 0 \\ n_j & 1 \end{pmatrix}$ we see that $\langle a_j, b_j, c_j \rangle$ is equivalent to a form with central coefficient $b_j' = 2a_j n_j + b_j$; and, as $b_j \equiv b_j' \pmod 2$ and $(a_1, a_2) = 1$, we can choose n_1 and n_2 so that $b_1' = b_2'$. □

The third entry c in the form $\langle a, b, c \rangle$ will play only a minor role in what follows (note that it must be integral and is determined by a, b and the discriminant d), so we shall often write $\langle a, b, - \rangle$ for this form.

LEMMA 2.3 *Let C_1 and C_2 be classes of primitive forms with discriminant d. If $f_i \in C_i$ $(i = 1, 2)$, let C_3 be the class containing the form $f_1 * f_2$. This is a well-defined operation on the set of classes of primitive forms with discriminant d.*

We call C_3 the *composition* of C_1 and C_2, denoted by $C_3 = C_1 * C_2$.

Proof. Suppose we have four forms $\langle a_j, b, - \rangle$, $\langle a_j', b', - \rangle$ $(j = 1, 2)$ where

$$\begin{aligned} &\langle a_1, b, - \rangle \sim \langle a_1', b', - \rangle \in C_1, \\ &\langle a_2, b, - \rangle \sim \langle a_2', b', - \rangle \in C_2, \\ &\langle a_1, b, - \rangle \text{ and } \langle a_2, b, - \rangle \text{ are concordant}, \\ &\langle a_1', b', - \rangle \text{ and } \langle a_2', b', - \rangle \text{ are concordant}. \end{aligned} \tag{i}$$

We shall show that $\langle a_1 a_2, b, - \rangle \sim \langle a_1' a_2', b', - \rangle$; this gives the result. Let $m = a_1 a_2 a_1' a_2'$ then, by Lemma 2.2, we can find integers A_1, A_2, and B so that

$$\begin{aligned} &\langle A_1, B, - \rangle \in C_1, \qquad \langle A_2, B, - \rangle \in C_2, \\ &\langle A_1, B, - \rangle \text{ and } \langle A_2, B, - \rangle \text{ is a concordant pair} \end{aligned} \tag{ii}$$

and where

$$(A_1, A_2) = (A_1A_2, m) = 1. \tag{iii}$$

We shall deduce $\langle a_1a_2, b, -\rangle \sim \langle A_1A_2, B, -\rangle$, a similar argument gives $\langle a_1'a_2', b', -\rangle \sim \langle A_1A_2, B, -\rangle$ completing the proof.

As $(a_1a_2, A_1A_2) = 1$ and b and B have the same parity we can find b^* to satisfy

$$b^* \begin{cases} \equiv b \pmod{2a_1a_2} \\ \equiv B \pmod{2A_1A_2}. \end{cases}$$

Hence

$$\langle a_j, b, -\rangle \sim \langle a_j, b^*, -\rangle \quad (j = 1, 2), \quad \langle a_1a_2, b, -\rangle \sim \langle a_1a_2, b^*, -\rangle,$$
$$\langle A_j, B, -\rangle \sim \langle A_j, b^*, -\rangle \quad (j = 1, 2), \quad \langle A_1A_2, B, -\rangle \sim \langle A_1A_2, b^*, -\rangle. \tag{iv}$$

We have

$$\begin{aligned} \langle A_1, b^*, -\rangle &\sim \langle A_1, B, -\rangle && \text{by (iv)} \\ &\sim \langle a_1, b, -\rangle && \text{by (i) and (ii)} \\ &\sim \langle a_1, b^*, -\rangle && \text{by (iv).} \end{aligned} \tag{v}$$

The third entry of $\langle a_1, b^*, -\rangle$ is $(d - b^{*2})/4a_1$, this is divisible by a_2. Also the third entry of $\langle A_1, b^*, -\rangle$ is $(d - b^{*2})/4A_1$, this too is divisible by a_2 [use (iii)]. Finally $(a_1, A_1, a_2) = 1$, hence the conditions of Lemma 2.1 are satisfied and we have by (v)

$$\langle a_1a_2, b^*, -\rangle \sim \langle A_1a_2, b^*, -\rangle.$$

A similar argument shows that

$$\langle A_1a_2, b^*, -\rangle \sim \langle A_1A_2, b^*, -\rangle,$$

and these equivalences and (iv) give the result. □

The class group is given by

THEOREM 2.4 *Let $\mathcal{G}$ be the set of classes of primitive forms having discriminant d with the operation of composition defined in Lemma 2.3, then $\mathcal{G}$ is a finite Abelian group.*

This group $\mathcal{G}$ is called the class group for discriminant d.

Proof. By Lemma 2.3, $\mathcal{G}$ is closed under composition. Secondly let

$$C_0 \begin{cases} = \langle\langle 1, 0, -d/4\rangle\rangle & \text{if } d \equiv 0 \pmod 4, \\ = \langle\langle 1, 1, (1-d)/4\rangle\rangle & \text{if } d \equiv 1 \pmod 4, \end{cases}$$

then C_0 is the identity of $\mathcal{G}$. This follows because for any b the forms $\langle 1, b, -\rangle$ and $\langle a, b, -\rangle$ are concordant, $\langle\langle 1, b, -\rangle\rangle * \langle\langle a, b, -\rangle\rangle = \langle\langle a, b, -\rangle\rangle$, and a form represents 1 if and only if it belongs to C_0 (see

Problem 10). Thirdly as $\langle a, b, c\rangle$, $\langle c, b, a\rangle$ is a concordant pair ($ac \neq 0$ as d is not square) and

$$\langle\langle a, b, c\rangle\rangle * \langle\langle c, b, a\rangle\rangle = \langle\langle ac, b, 1\rangle\rangle = C_0,$$

each $C \in \mathcal{G}$ has an inverse in $\mathcal{G}$. Finally associativity is proved by a method similar that used in the proof of the first part of Lemma 2.3 (see Problem 8). Hence $\mathcal{G}$ is a group; it is Abelian by definition and finite by Theorem 1.2, Chapter 9. □

We shall use the following lemma in our discussion of genera.

LEMMA 2.5 *A class $C \in \mathcal{G}$ of forms with discriminant d properly represents an integer z^2 where $(z, 2d) = 1$, if and only if $C = C_1^2$ for some class $C_1 \in \mathcal{G}$.*

Proof. Suppose first C represents z^2, so by Lemma 3.2, Chapter 9, there is a form $\langle z^2, b, c\rangle \in C$. Now $d = b^2 - 4z^2c$, therefore $(b, z) = 1$ [as $(z, d) = 1$] and the form $\langle z, b, zc\rangle$ is primitive. Hence $C_1^2 = C$ if $C_1 = \langle\langle z, b, zc\rangle\rangle$. Conversely suppose $C_1^2 = C$. By Lemma 1.5, C_1 represents an integer z where $(z, 2d) = 1$ and so there is a form $\langle z, b, c\rangle \in C_1$ with $(b, z) = 1$. The unimodular matrix $\begin{pmatrix} 1 & 0 \\ t & 1 \end{pmatrix}$ transforms $\langle z, b, c\rangle$ into the form $\langle z, b', c'\rangle$ where $c' = zt^2 + bt + c$. We can choose t so that $z \mid c'$, that is $c' = zc''$. Hence $\langle z, b', zc''\rangle \in C_1$, therefore $\langle z^2, b', c''\rangle \in C$ and the result follows. □

Genera and the class group

Properties of the class group can be used to give some information about genera.

DEFINITION The genus which contains the identity class C_0 of the class group $\mathcal{G}$ is called the *principal genus* $\mathcal{G}_0$.

Note that, as a form in C_0 represents $1, f \in \mathcal{G}_0$ if and only if each character in the character system for f has the value $+1$. We shall show that $\mathcal{G}_0$ is a subgroup of $\mathcal{G}$ and the genera are the cosets of $\mathcal{G}_0$ in $\mathcal{G}$. It follows that each genus contains the same number of classes; for some examples see the tables on pp. 340–3. We begin with

LEMMA 2.6 *If f is a form in the principal genus $\mathcal{G}_0$ and has discriminant d, then f represents a square z^2 for some integer z satisfying $(z, 2d) = 1$.*

Proof. We use Legendre's theorem (given in Chapter 14): the equation

$$w^2 = a_1y^2 + b_1z^2,$$

is soluble in integers w, y, and z, with z odd and $(w, y, z) = 1$, if b_1 is odd, $(a_1, b_1) = 1$, a_1Rb_1, and b_1Ra_1. (aRb stands for the proposition: the congruence $x^2 \equiv a \pmod{b}$ is soluble, see Problem 3, Chapter 14.) The proof of Legendre's theorem is independent of the work in this chapter.

Let $f = \langle a, b, c\rangle$ where $(a, 2d) = 1$ (Problem 9). We need to solve the equation

$$ax^2 + bxy + cy^2 = z^2 \tag{i}$$

or, multiplying by $4a$,

$$w^2 = (2ax + by)^2 = dy^2 + 4az^2. \tag{ii}$$

Case 1. $d \equiv 1 \pmod 4$. By Legendre's theorem we can find w_1, y_1, and z_1 satisfying

$$w_1^2 = dy_1^2 + az_1^2,$$

if (a) dRa and (b) aRd. (a) follows immediately as $b^2 \equiv d \pmod{a}$. For (b) we have, as f belongs to $\mathscr{G}_0$ and f represents an odd integer a, the character system for f is

$$(a/p) = 1 \text{ for each } p \mid d,\ p \text{ a prime.}$$

So using the methods of Section 2, Chapter 3, we can solve $x^2 \equiv a \pmod{d}$, that is aRd holds. Finally a solution to (ii) is given by $w = 2w_1$, $y = 2y_1$, and $z = z_1$; z is odd as a is odd.

Case 2. $d \equiv 0 \pmod 4$. Let $d = 4d^*$; again using Legendre's theorem we have

$$w_2^2 = d^*y_2^2 + az_2^2$$

is soluble if d^*Ra and aRd^*. This follows as in Case 1 except that now d^* may be even. Let t be the exponent of 2 in d^*, we need to show that $aR2^t$. If $t = 0$ there is nothing to prove. If $t = 1$ we have $1^2 \equiv a \pmod 2$ as a is odd. If $t = 2$ then the character α is defined and $\alpha(a) = 1$, so $a \equiv 1 \pmod 4$ and we have $1^2 \equiv a \pmod 4$. If $t = 3$ then α and β are defined and give $a \equiv 1 \pmod 8$, hence $1^2 \equiv a \pmod 8$. If $t > 3$ we can extend the previous case. Now a solution to (ii) is given by $w = 2w_2$, $y = y_2$, and $z = z_2$, and again z is odd.

Thus in both cases we can find a solution to (ii) where a divides w, y, and z (multiply the solutions above by a). This gives a solution to (i) if we set

$$2ax + by = w \quad \text{or} \quad x = (w - by)/2a$$

[check that x is an integer using $d \equiv b \pmod 2$]. By Problem 3, Chapter 14, we can choose the solution of (ii) so that z is odd (as a is odd) and $(w, d) = 1$. Hence $(z, d) = 1$. Using this problem again, note also that it is not necessary for a or d to be square-free. □

THEOREM 2.7 (Gauss) *If $C \in \mathcal{G}$ then $C \in \mathcal{G}_0$ if and only if $C = C_1^2$ for some class $C_1 \in \mathcal{G}$, and hence $\mathcal{G}_0$ is a subgroup of $\mathcal{G}$.*

Proof. By Lemma 2.6, if $f \in C$ where $C \in \mathcal{G}_0$, then f represents z^2 with $(z, 2d) = 1$; so by Lemma 2.5 we can find C_1 to satisfy $C = C_1^2$. Conversely, if $C = C_1^2$, then using Lemma 2.5 again, a form in C represents a square z^2 where $(z, 2d) = 1$. Hence the character system for C is: (z^2/p) for $p \mid d$, p odd, and $\alpha(z^2)$, $\beta(z^2)$, $\gamma(z^2)$ where defined. These are all $+1$ and so $f \in \mathcal{G}_0$. □

THEOREM 2.8 *The genera in the class group $\mathcal{G}$ are the cosets of $\mathcal{G}_0$ in $\mathcal{G}$. In particular each genus contains the same number of classes.*

Proof. If f belongs to the genus $\mathcal{G}_n$ and $g \in \mathcal{G}_0$, then $f * g$ also belongs to $\mathcal{G}_n$. For if f represents z and g represents w^2, where z and w are prime to $2d$, then $f * g$ represents zw^2, and so the character systems of f and $f * g$ are identical. Now use Theorem 2.7. □

Finally we consider the number of genera in the class group.

DEFINITION A class C in the class group $\mathcal{G}$ is called *ambiguous* if $C = C^{-1}$; the set of ambiguous classes is denoted by $\mathcal{A}$.

It is a simple matter to check that $\mathcal{A}$ is a subgroup of $\mathcal{G}$, also each ambiguous class contains forms of the type $\langle a, 0, c \rangle$ or $\langle a, a, c \rangle$; see Problem 13.

THEOREM 2.9 (Gauss) (i) *In the class group $\mathcal{G}$ the number of ambiguous classes equals the number of genera.*

(ii) *If $\mathcal{G}$ has only one genus then the order of $\mathcal{G}$ is odd.*

Proof. (i) As $\mathcal{A}$ is the set of classes C such that $C^2 = C_0$, $\mathcal{A}$ is the kernel of the natural homomorphism of $\mathcal{G}$ onto $\mathcal{G}_0$. So if g is the order of $\mathcal{G}_0$, m is the number of genera and a is the order of $\mathcal{A}$ then, by Theorem 2.8, $gm/g = a$, that is $m = a$.

(ii) If $m = 1$, there is only one ambiguous class C_0. So if $C \neq C_0$, $C \neq C^{-1}$. Hence the elements of $\mathcal{G}$ are C_0 and a number of pairs of distinct classes $\{C, C^{-1}\}$, that is the order of $\mathcal{G}$ is odd. □

The formula for m (the number of genera) is given by: if t is the number of odd prime divisors of the discriminant d and $d = 4d^*$, if d is even, then $m = 2^{t-1}$ if d is odd or if d is even and $d^* \equiv 1 \pmod 4$; $m = 2^t$ if d is even and $d^* \equiv 2, 3, 4, 6$, or $7 \pmod 8$; and $m = 2^{t+1}$ if $32 \mid d$. For the derivation of this formula see Problems 13, 14, and 15. As an example consider the cases $d = 188$, 189 and 192 (see Table 5 on p. 343).

(i) $d = 188 = 4 \cdot 47$, so $d^* = 47 \equiv 7 \pmod 8$ and $t = 1$. Hence we have $m = 2^t = 2$.

(ii) $d = 189 = 3^3 \cdot 7$, so d is odd and $t = 2$. Hence $m = 2^{t-1} = 2$.

(iii) $d = 192 = 2^6 \cdot 3$, so $32 \mid d$ and $t = 1$. Hence $m = 2^{t+1} = 4$.

10.3 A formula for the class number

In the previous sections we obtained some data on the set of equivalence classes of binary forms (the class group). This set is the disjoint union of a number m of genera and we have shown that m is determined by the discriminant d. We also know that each genus contains the same number g of classes, but we have little information on the value of g; this is a difficult problem which has been only partially solved. We shall give a formula for $h(d)(=mg)$, the number of classes in the class group for d, but it is not easy to interpret. For example, we can ask for which d is $h(d)=1$? If $d<0$, Heilbronn and Stark have shown that $h(d)=1$ for only 13 values of d, the smallest being -163 (see tables); there is no corresponding result when $d>0$ although it has been conjectured that there are infinitely many values of d in this case. Similar results have been proved when $h(d)=2$, here the smallest discriminant is -427. Gauss conjectured that $h(d)$ is unbounded as $|d|\to\infty$. This has been shown to hold if $d<0$ (see p. 229) but only a partial result is known if $d>0$. Further details are given in Baker (1975) and Hua (1982).

As noted earlier there is a close correspondence between quadratic forms and quadratic number fields. The class group for forms with discriminant d is isomorphic to the ideal class group of $\mathbb{Q}(\sqrt{(d/4)})$, if $d\equiv 0 \pmod 4$, and $\mathbb{Q}(\sqrt{d})$, if $d\equiv 1 \pmod 4$; and so the number $h(d)$ is an important constant associated with the field. For instance $h(d)=1$ if and only if the ring of integers of the field has unique factorization. The formula also has some applications in quadratic residue theory, in the proof of Dirichlet's theorem (see p. 239), and to Pell's equation.

The original derivation of the formula for $h(d)$ was given by Dirichlet, our derivation is based on Landau (1958). We begin by considering the series

$$K(d)=\sum_{n=1}^{\infty}\frac{(d/n)}{n}$$

where $(d/\cdot)$ denotes the Kronecker symbol, and we derive a relationship between $K(d)$ and $h(d)$. We then use Gauss sums to evaluate $K(d)$. The derivation requires four lemmas from unrelated mathematical areas; the first concerns lattice points in conics.

LEMMA 3.1 *Let E denote the interior of an ellipse with centre the origin, or the interior of the region bounded by a branch of a hyperbola, centred at the origin, together with two lines passing through the origin which intersect this hyperbola* (*see Figure* 2 *on p.* 191). *Let e denote the area of E. Now transform E into E' by the expansion $x\to x\sqrt{t}$, $y\to y\sqrt{t}$ $(t\geq 1)$. Given $n>0$, let $T(t)$ be the number of lattice points* (*points with integer coordinates*) *in E' whose coordinates x, y satisfy*

$$x\equiv x_0 \pmod n, \qquad y\equiv y_0 \pmod n,$$

then

$$\lim_{t\to\infty} T(t)/t = e/n^2.$$

Proof. In the plane of the curve E draw a grid of lines

$$x = (x_0 + rn)/\sqrt{t}, \qquad y = (y_0 + sn)/\sqrt{t}.$$

where $r, s \in \mathbb{Z}$. Note that each resulting square has area n^2/t. Let $U(t)$ equal n^2/t times the number of those squares whose top left-hand vertices lie in E. We have $U(t) = T(t)$. Now using elementary area estimates we obtain

$$e = \lim_{t\to\infty} \frac{n^2}{t} U(t). \qquad \square$$

The second lemma concerns the series $K(d)$. By Theorem 3.5, Chapter 4, the Kronecker symbol $(d/\cdot)$ is a real character modulo $|d|$ and so it follows, by Theorem 3.4, Chapter 5, and Abel's lemma, that the series for $K(d)$ converges. Further we have

LEMMA 3.2

$$\lim_{t\to\infty} \frac{1}{t} \sum_{\substack{x\le t\\(x,d)=1}} \sum_{n|x} (d/n) = \frac{\phi(|d|)}{|d|} K(d).$$

Proof. If $(d, n) = 1$ let $J(t, d, n)$ denote the cardinality of the set of integers $\{x : 1 \le x \le t/n \text{ and } (x, d) = 1\}$, that is

$$J(t, d, n) = \sum_{\substack{x\le t/n\\(x,d)=1}} 1 = \sum_{\substack{x\le t\\(x,d)=1, n|x}} 1.$$

Hence using DSI (see p. 23) we have, as $(d/n) = 0$ if $(d, n) > 1$,

$$X(t, d) = \frac{1}{t} \sum_{\substack{x\le t\\(x,d)=1}} \sum_{n|x} (d/n) = \frac{1}{t} \sum_{n=1}^{t} (d/n) \sum_{\substack{x\le t\\(x,d)=1, n|x}} 1$$

$$= \sum_{n=1}^{\infty} \frac{(d/n) J(t, d, n)}{t} \qquad (*)$$

because $J(t, d, n) = 0$ if $n > t$. Now as $J(t, d, n') \le J(t, d, n)$ if $n' \ge n$, and $J(t, d, n)/t \le 1/n$, the Weierstrass M-test shows that the series on the right-hand side of $(*)$ is uniformly convergent in t. Also for fixed n

$$\lim_{t\to\infty} \frac{J(t, d, n)}{t} = \phi(|d|)/n|d|.$$

Hence

$$\lim_{t\to\infty} X(t, d) = \sum_{n=1}^{\infty} (d/n) \lim_{t\to\infty} \frac{J(t, d, n)}{t} = \frac{\phi(|d|)}{|d|} K(d). \qquad \square$$

The third lemma concerns fundamental discriminants. A discriminant d is called *fundamental* if it has no odd prime squared factor, and if it is even and $d = 4d^*$ then $d^* \equiv 2$ or 3 (mod 4). Every discriminant d can be expressed as $d = d_0 e^2$ where d_0 is fundamental. For example note that -84 is fundamental but -44 is not; see Problem 20, Chapter 1.

LEMMA 3.3 *Let $d = d_0 e^2$ where d_0 is a fundamental discriminant then*

$$K(d) = \prod_{p \mid e} \left(1 - \frac{(d_0/p)}{p}\right) K(d_0).$$

Proof. Using the properties of the Kronecker symbol, we have

$$K(d) = \sum_{n=1}^{\infty} \frac{(d_0 e^2/n)}{n} = \sum_{\substack{n=1 \\ (e,n)=1}}^{\infty} \frac{(d_0/n)}{n}.$$

Suppose the prime factorization of e is $\prod_{i=1}^{k} p_i^{\alpha_i}$, then, by Problem 10, Chapter 1, we have

$$\begin{aligned} K(d) &= K(d_0) - \sum_{p_i \mid e} \frac{(e/p_i)}{p_i} K(d_0) + \sum_{\substack{p_i \mid e, p_j \mid e \\ i \neq j}} \frac{(e/p_i p_j)}{p_i p_j} K(d_0) - \dots \\ &= \prod_{p \mid e} \left(1 - \frac{(d_0/p)}{p}\right) K(d_0). \end{aligned}$$

□

The final lemma is an elementary result about trigonometrical sums.

LEMMA 3.4 *If $0 < \theta < 2\pi$ then*

(i) $\displaystyle\sum_{n=1}^{\infty} \frac{\sin n\theta}{n} = \frac{\pi}{2} - \frac{\theta}{2}$,

(ii) $\displaystyle\sum_{n=1}^{\infty} \frac{\cos n\theta}{n} = -\ln\left(2 \sin \frac{\theta}{2}\right)$.

Proof. Using the standard expansion for $\ln(1 - z)$ we have if $0 < \theta < 2\pi$,

$$\begin{aligned} \sum_{n=1}^{\infty} \frac{e^{in\theta}}{n} &= -\ln(1 - e^{i\theta}) = -\ln\left(2 \sin \frac{\theta}{2}\right) + i \arctan\left(\cot \frac{\theta}{2}\right) \\ &= -\ln\left(2 \sin \frac{\theta}{2}\right) + i\left(\frac{\pi}{2} - \frac{\theta}{2}\right). \end{aligned}$$

Now take real and imaginary parts. □

The main derivation

We begin by showing that $h(d) = XK(d)$ for some non-zero factor X defined in terms of the area of a conic. As t increases the primary and improper primary solutions of the equation $f(x, y) = k$, $1 \leq k \leq t$, can be represented by lattice points in the conic $f(x, y) \le t$ and, using Lemma 3.1, we can relate the number of these lattice points to the area of a region of the plane bounded by the conic $f(x, y) = 1$. This region is the interior of an ellipse if $d < 0$ and a segment of a hyperbola if $d > 0$.

Suppose

$$f = \langle a, b, c \rangle \tag{i}$$

is a form with discriminant $d > 0$. As we may replace f by an equivalent form, we may assume that $a > 0$ and so the hyperbola $f(x, y) = k$ $(k > 0)$ cuts the x-axis. Now by Theorem 3.5, Chapter 9, if $\{x', y'\}$ is a solution of the equation $f(x, y) = k$, so is $(x', y')T$ where T is an automorph for f. Hence this equation has a unique solution in the segment of the halfplane $x \ge 0$ lying between the lines $y = 0$ and $2awx - (v - bw)y = 0$, as this second line is $y = 0$ transformed by T. Thus we need to calculate the area shaded in Figure 2.

LEMMA 3.5 *Let $f = \langle a, b, c \rangle$ with discriminant d, and let $x, y \in \mathbb{R}$.*

(a) *If $d < 0$ then $f(x, y) = 1$ represents an ellipse with area $2\pi/\sqrt{-d}$.*

(b) *If $d > 0$ let $\delta = (v + w\sqrt{d})/2$ where $\{v, w\}$ is the fundamental solution of the equation $v^2 - dw^2 = 4$. The area of the curvilinear triangle in the halfplane $x \ge 0$ bounded by the lines $y = 0$ and $2awx - (v - bw)y = 0$ and the hyperbola $ax^2 + bxy + cy^2 = 1$ is $\ln \delta/\sqrt{d}$.*

Proof. (a) This is a standard result from elementary calculus. (b) As noted above we may assume that $a > 0$. We apply the transformation

$$2ax + (b + \sqrt{d})y = r, \qquad 2ax + (b - \sqrt{d})y = s, \tag{ii}$$

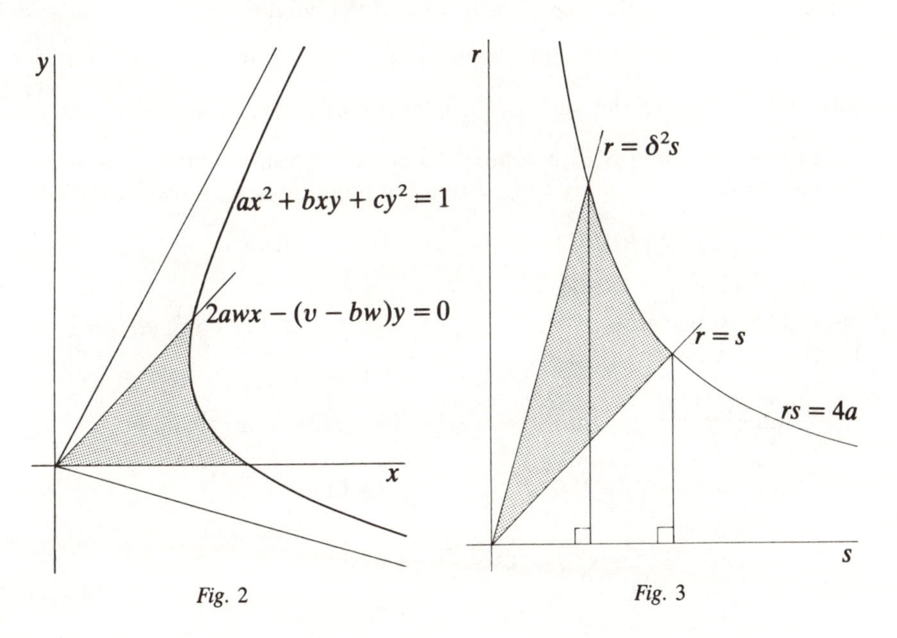

Fig. 2 *Fig. 3*

which maps the asymptotes of the original hyperbola onto the r and s axes. It has Jacobian $4a\sqrt{d}$. The transformation (ii) maps $y=0$ onto $r=s$ with $s\geq 0$, $2awx-(v-bw)y=0$ onto $r=\delta^2 s$, and the original hyperbola becomes $rs=4a$. This is illustrated in Figures 2 and 3. As the two right-angled triangles in Figure 3 have the same area, the area of the shaded curvilinear triangle in Figure 3 is given by

$$\left|4a\int_{2\sqrt{a}}^{2\sqrt{a/\delta}} \mathrm{d}t/t\right| = 4a \ln \delta.$$

Hence, using the Jacobian, the area shaded in Figure 2 is $\ln \delta/\sqrt{d}$. □

Let f be a form with discriminant d, let $\psi(k,f)$ be the number of primary and improper primary representations of k by the form f and let

$$S(t,f)=\sum_{\substack{k\leq t\\(k,d)=1}} \psi(k,f).$$

LEMMA 3.6 *For δ as in Lemma 3.5, we have*

$$\lim_{t\to\infty}\frac{S(t,f)}{t}=\begin{cases}2\pi\phi(|d|)/|d|^{3/2} & \text{if } d<0\\ \phi(d)\ln\delta/d^{3/2} & \text{if } d>0.\end{cases}$$

Proof. Let $R(t,f,x,y)$ denote the number of solutions $\{x',y'\}$ of

$$0\leq f(x',y')\leq t,\qquad x'\equiv x \pmod{|d|},\qquad y'\equiv y \pmod{|d|},$$

where in the indefinite case ($d>0$) we add the conditions

$$y'\geq 0 \quad\text{and}\quad 2awx'-(v-bw)y'\leq 0.$$

Also let $\sum_{x,y}^{f}$ denote the sum taken over all pairs $\{x,y\}$ where x and y range independently over a complete residue system modulo $|d|$ subject to the condition $(f(x,y),d)=1$. From the definitions we have

$$\sum_{x,y}^{f} R(t,f,x,y)=\sum_{\substack{k\leq t\\(k,d)=1}}\psi(k,f)=S(t,f).$$

Now if A is the area of the conic or segment of the conic defined in Lemma 3.5 we have

$$\begin{aligned}\lim_{t\to\infty}\frac{S(t,f)}{t} &= \sum_{x,y}^{f}\lim_{t\to\infty}\frac{R(t,f,x,y)}{t} && \text{interchanging } \textstyle\sum \text{ and lim}\\ &= \sum_{x,y}^{f} A/|d|^2 && \text{by Lemma 3.1}\\ &= \frac{A}{|d|^2}\sum_{x,y}^{f}1=\frac{A\phi(|d|)}{|d|},\end{aligned}$$

using Problem 9, Chapter 3, which gives the value of $\sum_{x,y}^{f} 1$. The result follows by Lemma 3.5. □

Our main result is

THEOREM 3.7 *We have*

$$h(d)=\begin{cases}\dfrac{zK(d)\sqrt{-d}}{2\pi} & \text{if } d<0\\[2ex] \dfrac{K(d)\sqrt{d}}{\ln\delta} & \text{if } d>0,\end{cases}$$

where z is given in Lemma 3.6, Chapter 9, and δ in Lemma 3.5 above.

Proof. If $f_1, \ldots, f_{h(d)}$ is a set of representatives of equivalence classes for discriminant d we have

$$\begin{aligned}\sum_{i=1}^{h(d)} S(t, f_i) &= \sum_{\substack{k\le t\\(k,d)=1}} \sum_{i=1}^{h(d)} \psi(k, f_i)\\ &= \sum_{\substack{k\le t\\(k,d)=1}} \psi(k) \qquad \text{by definition of } \psi\\ &= z \sum_{\substack{k\le t\\(k,d)=1}} \sum_{n\mid k} (d/n)\end{aligned}$$

by Theorem 3.7, Chapter 9. Hence by Lemma 3.2

$$\frac{|d|}{\phi(|d|)} \lim_{t\to\infty} \sum_{i=1}^{h(d)} \frac{S(t, f_i)}{t} = \frac{z\,|d|}{\phi(|d|)} \lim_{t\to\infty} \frac{1}{t} \sum_{\substack{k\le t\\(k,d)=1}} \sum_{n\mid k} (d/n) = zK(d).$$

Finally we use Lemma 3.6. As the right-hand side of the equation in Lemma 3.6 is independent of f we have

$$\frac{|d|}{\phi(|d|)} \sum_{i=1}^{h(d)} \lim_{t\to\infty} \frac{S(t, f_i)}{t} = \begin{cases}2\pi h(d)/\sqrt{-d} & \text{if } d<0\\ h(d)\ln\delta/\sqrt{d} & \text{if } d>0,\end{cases}$$

and the result follows from these two equations. □

Note that in the positive definite and indefinite cases both primary and improper primary solutions have been counted.

In the second part of our derivation we shall evaluate $K(d)$. By Lemma 3.3 we may restrict our attention to fundamental discriminants. The evaluation relies upon Gauss sums. Let $p \equiv 1 \pmod 4$, p a prime, we have shown (Theorem 3.1, Chapter 6) that if ψ is the Legendre symbol $(\cdot/p)$ then

$$g(\psi) = \sum_{t=1}^{p-1} (t/p)e_p(t) = \overset{+}{\sqrt{p}}$$

where $e_a(t) = e^{2\pi i t/a}$. A straightforward extension of this result gives (see Problem 18, Chapter 6)

$$\sum_{y=1}^{|d|-1} (d/t)e_{|d|}(nt) = (d/n)\sqrt{d} \tag{i}$$

for fundamental discriminants d, where we use the convention $\sqrt{d} = \overset{+}{\sqrt{d}}$ if $d > 0$ and $\sqrt{d} = i\overset{+}{\sqrt{-d}}$ if $d < 0$. This gives

$$\begin{aligned} K(d)\sqrt{d} &= \sum_{n=1}^{\infty} \frac{(d/n)\sqrt{d}}{n} && \text{by definition} \\ &= \sum_{n=1}^{\infty} \frac{1}{n} \sum_{t=1}^{|d|-1} (d/t)e_{|d|}(nt) && \text{by (i)} \\ &= \sum_{t=1}^{|d|-1} (d/t) \sum_{n=1}^{\infty} \frac{e_{|d|}(nt)}{n}. && \text{(ii)} \end{aligned}$$

We now use Lemma 3.4.

THEOREM 3.8 *If d is a fundamental discriminant and if $d > 0$ then*

$$h(d) = -\frac{1}{\ln \delta} \sum_{t=1}^{d-1} (d/t) \ln \sin(\pi t/d),$$

where δ is given in Lemma 3.5.

Proof. By (ii) above and Lemma 3.4 we have, as $K(d)\sqrt{d}$ is real,

$$\begin{aligned} K(d)\sqrt{d} &= \sum_{t=1}^{d-1} (d/t) \sum_{n=1}^{\infty} \frac{\cos(2\pi nt/d)}{n} \\ &= -\sum_{t=1}^{d-1} (d/t)\ln(2\sin(\pi t/d)) \\ &= -\sum_{t=1}^{d-1} (d/t)\ln(\sin(\pi t/d)) - \ln 2 \sum_{t=1}^{d-1} (d/t). \end{aligned}$$

The result follows by Theorem 3.4, Chapter 5, and Theorem 3.7. □

Finally we consider the positive definite case. Note that in this case the formula for $h(d)$ is a finite sum of integer-valued terms.

THEOREM 3.9 *If d is a fundamental discriminant, $d < 0$ and $d' = [-d/2]$, then*

$$h(d) = \frac{z}{2(2-(d/2))} \sum_{t=1}^{d'} (d/t).$$

Proof. We have using (ii)

$$K(d)\sqrt{d} = \sum_{t=1}^{-d-1} (d/t) \sum_{n=1}^{\infty} \frac{e_{-d}(nt)}{n},$$

and so, as $K(d)\sqrt{d}$ is imaginary and using our convention,

$$K(d)\sqrt{-d} = \sum_{t=1}^{-d-1} (d/t) \sum_{n=1}^{\infty} \frac{\sin(-2\pi nt/d)}{n}$$

$$= \sum_{t=1}^{-d-1} (d/t)(\pi/2 + \pi t/d) = \frac{\pi}{d} \sum_{t=1}^{-d-1} (d/t)t, \qquad \text{(iii)}$$

by Lemma 3.4 and Lemma 2.1, Chapter 6. We derive a similar formula for $(d/2)K(d)\sqrt{-d}$; the result will follow by (iii). Note that by Lemma 3.4 if $2\pi < \theta < 4\pi$

$$\sum_{n=1}^{\infty} \frac{\sin n\theta}{n} = \sum_{n=1}^{\infty} \frac{\sin n(\theta - 2\pi)}{n} = \frac{\pi}{2} - \frac{\theta - 2\pi}{2} = (3\pi - \theta)/2. \qquad \text{(iv)}$$

Hence using the same argument as above we have

$$(d/2)K(d)\sqrt{d} = \sum_{n=1}^{\infty} \frac{(d/2n)\sqrt{d}}{n} = \sum_{n=1}^{\infty} \frac{1}{n} \sum_{t=1}^{-d-1} (d/t)\mathrm{e}_{-d}(2nt),$$

and, as the left-hand side is imaginary and using our convention,

$$(d/2)K(d)\sqrt{-d} = \sum_{t=1}^{-d-1} (d/t) \sum_{n=1}^{\infty} \frac{\sin(-4\pi nt/d)}{n}$$

$$= \sum_{1 \le t \le -d/2} (d/t)(\pi/2 + 2\pi t/d)$$

$$+ \sum_{-d/2 < t < -d} (d/t)(\pi/2 + 2\pi t/d + \pi)$$

by (iv). Note that if $t = -d/2$ then $(d/t) = 0$. Now using Lemma 2.1, Chapter 6 again, we have

$$(d/2)K(d)\sqrt{-d} = \frac{2\pi}{d} \sum_{t=1}^{-d-1} (d/t)t + \pi \sum_{-d/2 < t < -d} (d/t)$$

$$= 2K(d)\sqrt{-d} - \pi \sum_{t=1}^{d'} (d/t)$$

by (iii). Hence

$$(2 - (d/2))K(d)\sqrt{-d} = \pi \sum_{t=1}^{d'} (d/t)$$

and the result follows by Theorem 3.7. □

Some applications of these results are given in the problem section. Here we consider two examples.

(i) Discriminant $d = -31$. In this case $(d/2) = 1$ as $d \equiv 1 \pmod 8$ and so

$$h(-31) = \sum_{t=1}^{15} (-31/t) = 3,$$

for $(-31/t)=-1$ if $t=3$, 6, 11, 12, 13, and 15, and $(-31/t)=1$ otherwise.

(ii) Discriminant $d=-63$. In this case -63 is not fundamental as $-63=-7\cdot 3^2$. Hence, by Theorem 3.7 and Lemma 3.3,

$$h(-63)=\frac{\sqrt{63}}{\pi}K(-63)=\frac{3\sqrt{7}}{\pi}\frac{4}{3}K(-7)=4h(-7)$$

$$=4\sum_{t=1}^{3}(-7/t)=4.$$

10.4 Problems 10

1. Show that equivalence of forms modulo p (a prime) is an equivalence relation.

2. Give proofs of the remaining parts of Theorem 1.7.

3. Let f and g be forms with discriminant $4d^*$.
   (i) Show that if $d^*\equiv 2 \pmod 4$ then f and g are equivalent modulo 4.
   (ii) Show that if $d^*\equiv 1 \pmod 4$ then f and g are also equivalent modulo 4, and that they are equivalent modulo 8 to either x^2+3y^2 or x^2+7y^2.

4. Let f be a reduced form with discriminant 21, find all the primary solutions of $f(x, y)=k$ for $1\le k\le 42$ and $(k, 42)=1$.

5. Find all the solutions of the equation $3x^2+10y^2=k$ for $1\le k\le 240$ and $(k, 30)=1$.

6. Investigate the solubility of the equation $f(x, y)=k$, where $(k, 2d)=1$, for reduced forms with discriminants d equal to (i) -20, (ii) -84, and (iii) 96.

7. Check that Theorem 1.8 agrees with our results on Pell's equation given in Chapter 7.

8. Complete the proof of Theorem 2.4 by showing that composition is associative.

9. Show that a form f properly represents z if and only if f is equivalent to a form with leading coefficient z. Further show that, if $(z, 2d)=1$ where d is the discriminat of f, then f is equivalent to a form with leading coefficient z.

10. Prove that $f\in C_0$ if and only if f represents 1 where C_0 is the identity class of the class group.

11. Find C_1 when $C=C_1^2$ and C equals (i) $\langle\langle 3,2,4\rangle\rangle$, (ii) $\langle\langle 3,1,12\rangle\rangle$, and (iii) $\langle\langle -1,0,34\rangle\rangle$.

12. Use Problem 7, Chapter 1, to show that if $n>1$ and C belongs to the class group $\mathcal{G}$, then $C=C_1^n$ for some $C_1\in\mathcal{G}$ if and only if C represents z^n for some integer z satisfying $(z, 2d)=1$.

13. A form of the type $\langle a, ka, c\rangle$ is called an *ambiguous* form. This problem shows that a class is ambiguous if and only if it contains an ambiguous form.

Prove:

(i) A form belongs to an ambiguous class if and only if it is improperly equivalent (i.e. the determinant of the transformation is -1) to itself.

(ii) An ambiguous form is equivalent to a form of the type $\langle a, 0, c\rangle$ or $\langle a, a, c\rangle$.

(iii) An ambiguous form belongs to an ambiguous class.

(iv) If S is the matrix $\begin{pmatrix} r & s \\ t & u \end{pmatrix}$ with determinant -1 which satisfies $S^2 = I$, show that $u = -r$ and $t = (1 - r^2)/s$, if $s \neq 0$.

(v) Let f belong to an ambiguous class and let S be as given in (iv) with $s \neq 0$. Find a unimodular matrix T such that the top right-hand entry of TST^{-1} is zero. Now show that if f is transformed into g by T then g is an ambiguous form equivalent to f, that is each ambiguous class contains an ambiguous form.

(vi) Apply the above to show that $\langle 4, 1, 4\rangle \sim \langle 9, 9, 4\rangle$.

14. Let t be the number of distinct odd prime factors of a positive discriminant d. Show that the number of ambiguous forms (with discriminant d) of the type $\langle a, 0, c\rangle$ or $\langle a, a, c\rangle$ is 2^{t+1} if d is odd, or if $d = 4d^*$ and $d^* \equiv 1 \pmod 4$; 2^{t+2} if $d = 4d^*$ and $d^* \equiv 2, 3, 4, 6$, or $7 \pmod 8$; and 2^{t+3} if $32 \mid d$. Further show that if $d < 0$ the number of ambiguous forms in an ambiguous class is exactly half that of the positive case.

15. Apply the results on automorphs given in Section 3, Chapter 9, to show that the number of ambiguous forms of the type $\langle a, 0, c\rangle$ or $\langle a, a, c\rangle$ in an ambiguous class is 4 if the discriminant is positive and 2 if it is negative. Finally derive the formula for the number of genera m for each discriminant given on p. 187, and check that it agrees with the tables on pp. 340–3.

16. Given a class group $\mathcal{G}$ of order h which contains m genera show that

(i) h is odd if and only if $m = 1$,

(ii) if $h = 2$ or 6 then $m = 2$,

(iii) if $h = 4$, then $\mathcal{G}$ is cyclic if and only if $m = 2$.

17. Suppose $d < 0$ and p is a prime such that $(d/p) = 1$. Let g be a form with discriminant d which represents p and let h be the order of the class group for d. Show that p^h is represented by the identity form f_0 and so deduce

$$1 + h \geq \ln((-d - 1)/4)/\ln p$$

18. Let $n > 1$ and c be odd. By considering the form $f_0 = \langle 1, 0, k\rangle$ with discriminant $d < 0$, where $e^2 + k = c^n$ and e and c are to be chosen suitably, use Problem 12 to find a form g which represents c. Hence show that there is a class group which contains an element of order n.

19. Use the results of Section 3 to calculate $h(12)$, $h(-23)$, and $h(-575)$.

20. Given a positive integer n, use Lemma 3.3 and Theorem 3.7 to show that there is a discriminant d such that $h(d) > n$.

21. Let p be a prime congruent to 3 modulo 4, using Theorem 3.9 show that there are more quadratic residues than non-residues in the interval $(0, p/2)$.

☆22. Let $d > 0$ and $d \equiv 0$ or 1 (mod 4). This problem gives an upper bound for the fundamental solution of the equation

$$x^2 - dy^2 = 4. \tag{i}$$

Suppose x_0, y_0, and δ satisfy $x_0 > 0$, $y_0 > 0$, $\delta = (x_0 + y_0\sqrt{d})/2$, and $\{x_0, y_0\}$ is the solution of (i) with least δ. Prove that

$$\delta < d^{\sqrt{d}} \tag{ii}$$

using the following method. Let

$$S(n) = \sum_{i=1}^{n} \sum_{t=1}^{i} (d/t),$$

you may assume the result

$$|S(n)| \leq n\sqrt{d}. \tag{iii}$$

Show that

$$K(d) = \left(\sum_{n=1}^{k-1} + \sum_{n=k}^{\infty}\right) \frac{2S(n)}{n(n+1)(n+2)} = T_1 + T_2,$$

estimate T_1 using the logarithm function and T_2 using (iii), and by making a suitable choice for k deduce that $2K(d) < \ln d + 2$ if $d \geq 5$. Now prove (ii); for further details see Hua (1982, pp. 326–8).

11
PARTITIONS

Every positive integer can be expressed as a sum of smaller integers in a number of ways. If $n > 0$ and

$$n = a_1 + a_2 + \ldots + a_r$$

where each a_i is a positive integer, the set $\{a_1, a_2, \ldots, a_r\}$ is called a *partition* of n and the individual terms $a_1, a_2, \ldots$ are called *parts*; the order of the parts is unimportant although it is usual to write a partition in decreasing order. The total number of partitions of n is denoted by $p(n)$. For example $p(5) = 7$ as

$$5, 4+3, 3+2, 3+1+1, 2+2+1, 2+1+1+1, 1+1+1+1+1$$

is the set of partitions of 5. (Note that as the primes play no role in this chapter, we can use the symbol p to denote the partition function.) Partition theory deals with additive problems, in Chapter 6 we studied sums of squares which is another branch of additive number theory; some similarities exist between these topics, for example see Problem 17.

There are many relationships between subsets of the set of partitions of an integer n. The most famous, due to Euler, states that the number of partitions of n with odd parts equals the number of partitions of n with distinct parts. In the example above the first three partitions have distinct parts whilst the first, fourth, and seventh partitions are the only ones with exclusively odd parts. A number of other properties of this type will be discussed in this chapter.

Euler's result is remarkable because there is no obvious connection between the two sets, but it becomes almost a triviality once it has been expressed in terms of generating functions (Theorem 1.2). For example we shall show that

$$\prod_{n=1}^{\infty} (1 - q^n)^{-1} = \sum_{n=1}^{\infty} p(n) q^n$$

(Theorem 1.1), and the expression on the left-hand side is called the *generating function* for p. Identities amongst these generating functions enable us to deduce many partition results. It is of interest to note that some of these identities arose first in other branches of mathematics, for example the Jacobi identity (Theorem 2.1) and the use of q as the bound variable both came from elliptic function theory.

In this chapter we derive the basic results in the first section, prove the Jacobi identity and use it to deduce some further partition results in Section 2, and give some bounds for the partition function in the last section. Andrew's book (1976), *The Theory of Partitions* is the only one devoted exclusively to partition theory, it contains many further ideas and results including a 'general theory of partition identities'. Knopp's book (1970) describes in detail the connections with modular function theory, and some further results are given in Hardy and Wright (1954).

As infinite products will be used extensively in this chapter we state here the basic definitions and results. An infinite product $\prod_{n=1}^{\infty} f(n)$ is said to *converge* if $\lim_{m\to\infty} \prod_{n=1}^{m} f(n)$ tends to a *non-zero* finite limit. It can be shown that if each a_i is a real number satisfying $0 < a_i < 1$, then the products $\prod_{i=1}^{\infty} (1 + a_i)$ and $\prod_{i=1}^{\infty} (1 - a_i)$ converge if and only if $\sum_{i=1}^{\infty} a_i$ converges. In all cases the convergence is absolute, that is the value of the sum or products is unaltered by a permutation of their terms.

11.1 Elementary properties

We begin with some notation. Let N denote the set of positive integers; if $H \subset N$ let $\hat{H}$ denote the set of all partitions with parts in H, and let $\hat{H}_d$ denote the set of those partitions in $\hat{H}$ in which no part is used more than d times. So $\hat{N}$ denotes the set of all partitions, and $\hat{N}_1$ denotes the set of all partitions with distinct parts. Further let $p(\hat{H}, n)$ denote the number of partitions of a positive integer n contained in $\hat{H}$, we write $p(n)$ for $p(\hat{N}, n)$. Finally let $p(0) = 1$ and $p(m) = 0$ if $m < 0$.

The basic tools in partition theory are generating functions, they are given by

DEFINITION Let $\{a_0, a_1, \ldots\}$ be an infinite sequence of integers. The power series

$$f(q) = \sum_{n=0}^{\infty} a_n q^n$$

is called the *generating function* for this sequence.

Note that the sequence $\{a_0, a_1, \ldots\}$ is determined completely by f. We shall usually treat these power series formally but in most cases they are absolutely convergent if we assume that $0 < q < 1$.

THEOREM 1.1 *Suppose $H \subset N$, d is a positive integer, and f and f_d are given by*

$$f(q) = \sum_{n=0}^{\infty} p(\hat{H}, n) q^n, \qquad f_d(q) = \sum_{n=0}^{\infty} p(\hat{H}_d, n) q^n,$$

then

$$f(q)=\prod_{n\in H}(1-q^n)^{-1} \quad \textit{and} \quad f_d(q)=\prod_{n\in H}\frac{1-q^{(d+1)n}}{1-q^n}.$$

Proof. Leaving aside convergence questions for the moment, we have

$$\prod_{n\in H}(1-q^n)^{-1}=\prod_{n\in H}(1+q^n+\ldots+q^{kn}+\ldots)=\Sigma^* q^{x_1h_1+\ldots+x_kh_k}$$

where $h_1, h_2, \ldots$ is an enumeration (in increasing order) of the elements of H, and the sum Σ^* is taken over all finite sequences of non-negative integers $\{x_1, \ldots, x_k\}$. Now for each n, q^n occurs in this sum each time the equation

$$n=x_1h_1+\ldots+x_kh_k$$

has a solution $\{x_1, \ldots, x_k\}$, in which case n has a partition in $\hat{H}$. Conversely each partition of a positive integer using parts from H will occur as an exponent in the sum Σ^*. Hence we have

$$\prod_{n\in H}(1-q^n)^{-1}=\sum_{n=0}^{\infty}p(\hat{H}, n)q^n.$$

The second part follows similarly using finite geometric series.

We show now that the sums and products above converge if $0<q<1$. Note first that $\prod_{i=1}^{n}(1+q^{h_i}+\ldots+q^{kh_i}+\ldots)$ is a finite product of absolutely convergent series and so is itself absolutely convergent. Also

$$S_m=\sum_{n=0}^{h_m}p(\hat{H}, n)q^n\le\prod_{i=1}^{m}(1-q^{h_i})^{-1}<\prod_{i=1}^{\infty}(1-q^{h_i})^{-1}<\infty.$$

The last inequality holds as $\sum_{i=1}^{\infty}q^{h_i}$ is convergent (see note on infinite products on the previous page). Hence S_m is an increasing bounded sequence and thus tends to a limit. Further

$$\sum_{n=0}^{\infty}p(\hat{H}, n)q^n\ge\prod_{i=1}^{m}(1-q^{h_i})^{-1}\to\prod_{i=1}^{\infty}(1-q^{h_i})^{-1},$$

as $m\to\infty$. This completes the proof. □

THEOREM 1.2 (*Euler*) *Let O denote the set of odd positive integers, then*

$$p(\hat{O}, n)=p(\hat{N}_1, n).$$

Proof. By the result above we have

$$\sum_{n=0}^{\infty} p(\hat{O}, n)q^n = \prod_{n=1}^{\infty} (1-q^{2n-1})^{-1} \quad \text{and} \quad \sum_{n=0}^{\infty} p(\hat{N}_1, n)q^n = \prod_{n=1}^{\infty} (1+q^n).$$

But

$$\prod_{n=1}^{\infty} (1+q^n) = \prod_{n=1}^{\infty} \frac{(1-q^{2n})}{(1-q^n)} = \prod_{n=1}^{\infty} \frac{1}{(1-q^{2n-1})}$$

by cancellation. The result follows because identical functions have identical power series expansions. □

The reader should try a few numerical examples to illustrate this theorem. Some similar results are given in the problem section.

Graphical representation

Consider for example $n = 13$ with the partition $13 = 5 + 3 + 3 + 2$. We can represent this by the dot pattern

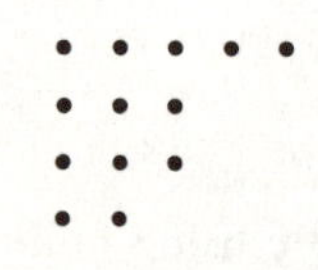

that is the four parts are represented by horizontal rows of dots. Clearly the pattern can be 'read' vertically, that is 13 also has the partition $4+4+3+1+1$, these two partitions are called conjugate. This simple representation has some useful consequences.

DEFINITION Given n and partitions **a** and **b** of n

$$n = a_1 + a_2 + \ldots + a_k = b_1 + b_2 + \ldots + b_m,$$

where $a_{i+1} \leq a_i$ and $b_{i+1} \leq b_i$ for all i, we represent the partition **a** by the set of points (dots) in the plane

$$(0, 0), (1, 0), \ldots, (a_1 - 1, 0); (0, -1), (1, -1), \ldots, (a_2 - 1, -1); \\ \ldots; (0, -k+1), (1, -k+1), \ldots, (a_k - 1, -k+1)$$

with a similar pattern for **b**. Then **a** and **b** are called *conjugate* if the dot pattern for **b** is the mirror image in the line $y = -x$ of the dot pattern for the partition **a**.

We have immediately

THEOREM 1.3 *The number of partitions of n with at most m parts equals the number of partitions of n in which no part exceeds m.*

Proof. Use the fact that conjugacy defines a bijection between the two classes of partitions. □

For example if $n = 5$ and $m = 2$ the partitions are 5, 4 + 1, 3 + 2, and 2 + 2 + 1, 2 + 1 + 1 + 1, 1 + 1 + 1 + 1 + 1.

We come now to Euler's pentagonal number theorem which provides an efficient algorithm for evaluating $p(n)$. Note that a pentagonal number is one of the form $m(3m + 1)/2$ or $m(3m - 1)/2$.

THEOREM 1.4 *Let $p_1(\hat{H}, n)$ ($p_2(\hat{H}, n)$) denote the number of partitions of n in $\hat{H}$ with an odd (even, respectively) number of parts. We have*

$$p_2(\hat{N}_1, n) - p_1(\hat{N}_1, n) = \begin{cases} (-1)^r & \text{if } n = r(3r \pm 1)/2 \\ 0 & \text{otherwise.} \end{cases}$$

We shall give an elementary graphical proof due to Franklin; for an analytical proof see p. 207.

Proof. Let $\mathbf{a}$ be a partition of n into r distinct parts. So $n = a_1 + \ldots + a_r$ and $a_i > a_{i+1}$ for all i. Let $s(\mathbf{a}) = a_r$ (the smallest part), let $t(\mathbf{a})$ be the largest integer c such that $a_1 = a_2 + 1$, $a_2 = a_3 + 1, \ldots, a_{c-1} = a_c + 1$, and let $t(\mathbf{a}) = 1$ if $a_1 \neq a_2 + 1$. Note $t(\mathbf{a}) \leq r$.

We define a correspondence between the partitions of n enumerated by $p_1(\hat{N}_1, n)$ and $p_2(\hat{N}_1, n)$; this is illustrated in the example in Figure 4.

Case 1 $s(\mathbf{a}) \leq t(\mathbf{a})$

Take the partition $\mathbf{a}$ (of n), add one to the first $s(\mathbf{a})$ parts and delete the last part; so the new partition (of n) is

$$(a_1 + 1) + \ldots + (a_{s(a)} + 1) + a_{s(a)+1} + \ldots + a_{r-1}.$$

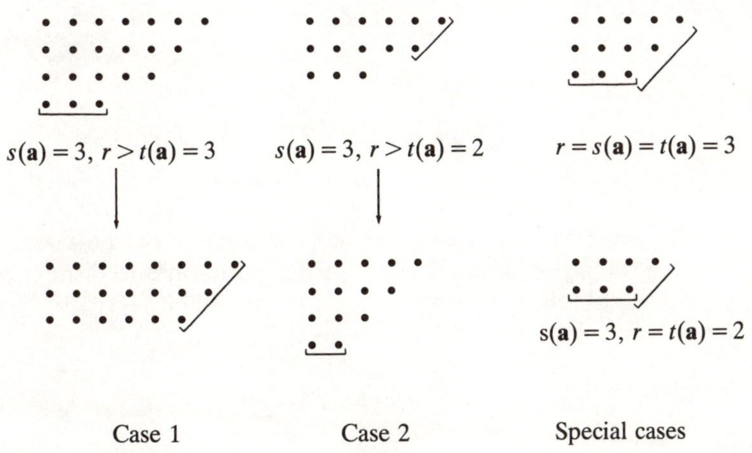

$s(\mathbf{a}) = 3$, $r > t(\mathbf{a}) = 3$

$s(\mathbf{a}) = 3$, $r > t(\mathbf{a}) = 2$

$r = s(\mathbf{a}) = t(\mathbf{a}) = 3$

$s(\mathbf{a}) = 3$, $r = t(\mathbf{a}) = 2$

Case 1 Case 2 Special cases

Fig. 4

Case 2 $s(\mathbf{a}) > t(\mathbf{a})$
Take the partition $\mathbf{a}$ (of n), subtract one from each of the first $t(\mathbf{a})$ parts and add the new part $t(\mathbf{a})$; so the new partition (of n) is

$$(a_1 - 1) + \ldots + (a_{t(a)} - 1) + a_{t(a)+1} + \ldots + a_r + t(a).$$

As $a_{t(a)} - 1 > a_{t(a)+1}$ and $a_r > t(a)$, this is a partition into distinct parts. These two cases provide a one-to-one correspondence between the partitions enumerated by $p_1(\hat{N}_1, n)$ and $p_2(\hat{N}_1, n)$ except in the following special cases. First consider the example $n = 12$ with partition $\mathbf{c}$: $12 = 5 + 4 + 3$, so $s(\mathbf{c}) = t(\mathbf{c}) = 3$; the correspondence fails in this case, see Figure 4. The special cases are (i) $s(\mathbf{a}) = t(\mathbf{a}) = r$ when

$$n = (2r-1) + (2r-2) + \ldots + r = r(3r-1)/2,$$

and (ii) $s(\mathbf{a}) = t(\mathbf{a}) + 1 = r + 1$ when

$$n = 2r + (2r-1) + \ldots + (r+1) = r(3r+1)/2.$$

The term $(-1)^r$ counts these special cases and so the result follows. □

THEOREM 1.5 (*Euler's pentagonal number theorem*) *For* $0 < q < 1$

$$\prod_{n=1}^{\infty} (1 - q^n) = 1 + \sum_{m=1}^{\infty} (-1)^m q^{m(3m-1)/2}(1 + q^m) = \sum_{m=-\infty}^{\infty} (-1)^m q^{m(3m-1)/2}.$$

Proof. We have

$$\sum_{m=-\infty}^{\infty} (-1)^m q^{m(3m-1)/2} = 1 + \left\{ \sum_{m=1}^{\infty} + \sum_{m=-1}^{-\infty} \right\} (-1)^m q^{m(3m-1)/2}$$

$$= 1 + \sum_{m=1}^{\infty} (-1)^m \{ q^{m(3m-1)/2} + q^{m(3m+1)/2} \}$$

$$= \sum_{n=0}^{\infty} \{ p_2(\hat{N}_1, n) - p_1(\hat{N}_1, n) \} q^n,$$

by Theorem 1.4. Now

$$\prod_{n=1}^{\infty} (1 - q^n) = \prod_{n=1}^{\infty} (1 + (-1)q^n + 0 \cdot q^{2n} + \ldots)$$

$$= \Sigma^* (-1)^{a_1 + a_2 + \ldots} q^{a_1 + 2a_2 + 3a_3 + \ldots}$$

where the sum Σ^* is taken over all finite sequences of zeros and ones $\{a_1, a_2, \ldots\}$. Hence $a_1 + 2a_2 + 3a_3 + \ldots$ is a partition into distinct parts, and it has an odd (even) number of parts if and only if $(-1)^{a_1 + \ldots + a_k} = -1$ (1, respectively). Hence

$$\prod_{n=1}^{\infty} (1 - q^n) = \sum_{n=0}^{\infty} \{ p_2(\hat{N}_1, n) - p_1(\hat{N}_1, n) \} q^n,$$

and combining this with the above the result follows. □

THEOREM 1.6 *If $n > 0$ then*

$$p(n) - p(n-1) - p(n-2) + \ldots + (-1)^m p(n - m(3m-1)/2) + (-1)^m p(n - m(3m+1)/2) + \ldots = 0.$$

Note that $p(0) = 1$ and $p(n) = 0$ if $n < 0$.

Proof. Let $P(n)$ denote the sum on the left-hand side of the equation above. If we collect together those powers of q in $\sum_{n=0}^{\infty} P(n)q^n$ which have the same coefficient $p(n)$, we have

$$\begin{aligned}\sum_{n=0}^{\infty} P(n)q^n &= \sum_{n=0}^{\infty} p(n)q^n\{1 - q - q^2 + q^5 + q^7 - \ldots\} \\ &= \sum_{n=0}^{\infty} p(n)q^n\left\{1 + \sum_{m=1}^{\infty} (-1)^m (q^{m(3m-1)/2} + q^{m(3m+1)/2})\right\} \\ &= \sum_{n=0}^{\infty} p(n)q^n\left[\prod_{m=1}^{\infty} (1 - q^m)\right] = 1\end{aligned}$$

by Theorems 1.1 and 1.5. Hence $P(n) = 0$ if $n > 0$. □

Clearly this result provides a recursive procedure for calculating $p(n)$; for large n it is very efficient. For example

$$p(5) = p(4) + p(3) - p(0) = 2p(3) + p(2) - 1 = 3p(2) + 2p(1) - 1 = 7.$$

11.2 Jacobi's identity

Many identities exist between the generating functions defined in the previous section; we begin this section by deriving one of the most famous—Jacobi's identity. Although it arose in the theory of elliptic functions, a number of partition results can be deduced by considering special cases. We shall write $\Pi_i^n f(i)$ for $\prod_{i=1}^{n} f(i)$ throughout this section.

THEOREM 2.1 (*Jacobi's identity*) *If $|q| < 1$ and $z \neq 0$ then*

$$\begin{aligned}\Pi_n^{\infty}(1 - q^{2n})(1 + q^{2n-1}z)(1 + q^{2n-1}z^{-1}) &= 1 + \sum_{n=1}^{\infty} q^{n^2}(z^n + z^{-n}). \\ &= \sum_{n=-\infty}^{\infty} z^n q^{n^2}.\end{aligned}$$

Note that the second and third expressions are virtually identical.

Proof. We begin by deriving a recursive relation. For a fixed m let

$$P_m(z)=\Pi_n^m(1+q^{2n-1}z)(1+q^{2n-1}z^{-1}). \tag{1}$$

As $P_m(z)=P_m(z^{-1})$ we can write this in the form

$$P_m(z)=Q_0+(z+z^{-1})Q_1+\ldots+(z^m+z^{-m})Q_m, \tag{2}$$

where each Q_i is independent of z. Substituting q^2z for z in (1) we have

$$P_m(q^2z)=\Pi_n^m(1+q^{2n+1}z)(1+q^{2n-3}z^{-1}) \tag{3}$$

and, as (1) and (3) only differ by terms at each end of their products,

$$(1+qz)(1+q^{2m-1}z^{-1})P_m(q^2z)=(1+q^{2m+1}z)(1+q^{-1}z^{-1})P_m(z).$$

Now multiplying by qz, cancelling $(1+qz)$, and using (2) we have

$$\begin{aligned}(qz+q^{2m})[Q_0+(q^2z+q^{-2}z^{-1})Q_1+\ldots+(q^{2m}z^m+q^{-2m}z^{-m})Q_m]\\ =(1+q^{2m+1}z)[Q_0+(z+z^{-1})Q_1+\ldots+(z^m+z^{-m})Q_m],\end{aligned}$$

and equating the coefficients of z^{1-n}, for $1\leq n\leq m$, we obtain

$$q^{1-2n}Q_n+q^{2m-2n+2}Q_{n-1}=Q_{n-1}+q^{2m+1}Q_n,$$

or

$$Q_n=q^{2n-1}(1-q^{2m-2n+2})Q_{n-1}/(1-q^{2m+2n}).$$

As $\Sigma_i^n(2i-1)=n^2$, this recursive relation gives, for $1\leq n\leq m$,

$$Q_n=q^{n^2}\frac{\Pi_i^n(1-q^{2m-2n+2i})}{\Pi_i^n(1-q^{2m+2i})}Q_0. \tag{4}$$

Now Q_m, the coefficient of z^m in (1), is given by

$$Q_m=q^{1+3+\ldots+(2m-1)}=q^{m^2},$$

and so, using (4), we see that

$$Q_0=\Pi_i^m(1-q^{2m+2i})/\Pi_i^m(1-q^{2i}). \tag{5}$$

Combining (4) and (5) we have, for $0\leq n\leq m$,

$$\begin{aligned}Q_n&=q^{n^2}\Pi_i^{m+n}(1-q^{2m-2n+2i})/\Pi_i^{m+n}(1-q^{2i})\\ &=q^{n^2}Q_n'/\Pi_i^m(1-q^{2i}),\end{aligned}$$

where

$$Q_n'=\Pi_i^n(1-q^{2m-2n+2i})\Pi_i^{m-n}(1-q^{2m+2n+2i}).$$

Using (2) this gives

$$\Pi_i^m(1-q^{2i})P_m(z)=Q_0'+\sum_{n=1}^m q^{n^2}(z^n+z^{-n})Q_n' \tag{6}$$

Now, for each n, $Q_n'\to 1$ as $m\to\infty$, and so the result will follow if we can show that the series in (6), with ∞ replacing m, is uniformly convergent.

To do this let

$$a(m, n) = \begin{cases} Q_0 & \text{if } m = 0 \\ q^{n^2}(z^n + z^{-n})Q_n'/\Pi_i^m(1 - q^{2i}) & \text{if } 1 \le n \le m \\ 0 & \text{if } n > m. \end{cases}$$

By the note on products in the introduction to this chapter we have

$$|Q_n'| < \Pi_k^\infty(1 + |q|^{2k}) = c_1 \quad \text{and} \quad |\Pi_i^m(1 - q^{2i})^{-1}| < \Pi_k^\infty(1 - |q|^{2k})^{-1} = c_2,$$

where c_1 and c_2 are constants, hence

$$|a(m, n)| \le c_1 c_2 \, |q|^{n^2}(|z|^n + |z|^{-n}) = b_n.$$

Note that b_n is independent of m and so, as $|q| < 1$,

$$\begin{aligned} \frac{b_{n+1}}{b_n} &= |q|^{2n+1} \, (|z|^{n+1} + |z|^{-n-1})/(|z|^n + |z|^{-n}) \\ &< |q|^{2n+1} \, (|z| + |z|^{-1}) \to 0 \end{aligned}$$

as $n \to \infty$. Therefore the series $\sum_{n=0}^{\infty} b_n$ is convergent. Hence, by the Weierstrass M-test, the series

$$\sum_{m=0}^{\infty} a(m, n)$$

is uniformly convergent for all n, and it follows that the term-by-term limit above is valid. □

We shall consider some applications of this result. First note that if we replace q by $-q^{3/2}$ and z by $q^{1/2}$ we have

$$\begin{aligned} \Pi_n^\infty(1 - q^n) &= \Pi_n^\infty(1 - q^{3n})(1 - q^{3n-1})(1 - q^{3n-2}) \\ &= \sum_{n=-\infty}^{\infty} (-q^{3/2})^{n^2} q^{n/2} = \sum_{n=-\infty}^{\infty} (-1)^n q^{n(3n+1)/2}, \end{aligned}$$

as $(-1)^{n^2} = (-1)^n$. This is a version of Euler's pentagonal number theorem (1.5), and so Theorem 2.1 provides a second proof of Euler's result. Again using Jacobi's identity, we can extend Theorem 1.4 to

THEOREM 2.2 *Let $\hat{N}_{j;k}$ denote the set of partitions with distinct parts in which each part is congruent to $-j$, 0, or j modulo $2k + 1$, then*

$$p_2(\hat{N}_{j;k}, n) - p_1(\hat{N}_{j;k}, n) = \begin{cases} (-1)^m & \textit{if } n = ((2k \mp 1)m(m+1) \pm 2jm)/2 \\ 0 & \textit{otherwise,} \end{cases}$$

where p_1 and p_2 are defined in Theorem 1.4.

Proof. By Theorem 2.1 with $q^{k+1/2}$ replacing q and $-q^{k-j+1/2}$ replacing z we have

$$\sum_{n=-\infty}^{\infty} (-1)^n q^{((2k+1)n(n+1)-2jn)/2}$$
$$= \Pi_n^{\infty}(1-q^{(2k+1)n})(1-q^{(2k+1)n-j})(1-q^{(2k+1)(n-1)+j}).$$

Now as in the proof of Theorem 1.4 we can rewrite the right-hand side as a sum of powers of q with typical term

$$(-q^{(2k+1)x_1})\dots(-q^{(2k+1)x_a})(-q^{(2k+1)y_1-j})\dots(-q^{(2k+1)y_b-j})$$
$$\times(-q^{(2k+1)(z_1-1)+j})\dots(-q^{(2k+1)(z_c-1)+j}).$$

The exponent of q in this term is a partition of the required form with $a+b+c$ parts, and the sign of this term is $(-1)^{a+b+c}$. The result follows as in Theorem 1.4 for exceptional numbers $((2k+1)m(m+1)-2jm)/2$; the other cases are similar. □

In the next result we collect together some further identities for later use.

THEOREM 2.3 (i) $\sum_{n=-\infty}^{\infty} q^{n^2} = \Pi_n^{\infty}(1-q^{2n})(1+q^{2n-1})^2$,

(ii) $\sum_{n=0}^{\infty} (-1)^n(2n+1)q^{n(n+1)/2} = \Pi_n^{\infty}(1-q^n)^3$,

(iii) $\sum_{n=0}^{\infty} q^{n(n+1)/2} = \Pi_n^{\infty}(1-q^{2n})(1+q^n)$

$$= \frac{(1-q^2)(1-q^4)\dots}{(1-q)(1-q^3)\dots}.$$

Note. (i) is sometimes also called Jacobi's identity.

Proof. (i) This is Theorem 2.1 with $z=1$.

(ii) In Jacobi's identity replace q by $q^{1/2}$ and z by $q^{1/2}x$, and we have, multiplying by $x/(x+1)$,

$$\frac{x}{x+1}\sum_{n=-\infty}^{\infty} q^{n(n+1)/2}x^n = \frac{x}{x+1}\Pi_n^{\infty}(1-q^n)(1+q^n x)(1+q^{n-1}x^{-1})$$
$$= \Pi_n^{\infty}(1-q^n)(1+q^n x)(1+q^n x^{-1})$$

as $1+q^{n-1}x^{-1}=(x+1)/x$ if $n=1$. Thus if $T(x)$ denotes the first expression in the above equation

$$T(x)\to\Pi_n^{\infty}(1-q^n)^3 \quad \text{as} \quad x\to -1. \tag{*}$$

Further

$$\sum_{n=-\infty}^{\infty}(-1)^n q^{n(n+1)/2}=\left\{\sum_{n=0}^{\infty}+\sum_{n=-1}^{-\infty}\right\}(-1)^n q^{n(n+1)/2}=0$$

because $(-1)^n q^{n(n+1)/2}=-(-1)^{-1-n}q^{(-1-n)(-n)/2}$. This gives

$$T(x)=\frac{x}{x+1}\sum_{n=-\infty}^{\infty}q^{n(n+1)/2}[x^n-(-1)^n]=\sum_{n=-\infty}^{\infty}q^{n(n+1)/2}\frac{x(x^n-(-1)^n)}{x+1}.$$

But

$$\frac{x(x^n-(-1)^n)}{x+1}=x(x^{n-1}+(-1)x^{n-2}+\ldots+(-1)^{n-1})\to(-1)^n n$$

as $x\to-1$, hence by uniform convergence we have

$$T(-1)=\sum_{n=-\infty}^{\infty}(-1)^n nq^{n(n+1)/2}$$

The result (ii) follows using (*) by combining the nth and $(-n-1)$th terms, where $n>0$, in this final sum.

(iii) Use the first equation in the proof of (ii) with $x=1$; note that $\Pi_n^{\infty}(1+q^n)=\Pi_n^{\infty}(1+q^{n-1})$. Now use formula for this product on p. 202.

□

By studying the tables, Ramanujan discovered some remarkable divisibility properties for the partition function. For example he noticed that $p(4)$, $p(9)$, $p(14)$, ... are each divisible by 5, and $p(5)$, $p(12)$, $p(19)$, ... are each divisible by 7; with similar results for division by 11, 25, 35, 49, 55, ... The full result, which has only recently been established in full, states:

> if $m=5^\alpha 7^\beta 11^\gamma$ and $24n\equiv 1 \pmod m$ then $5^\alpha 7^\delta 11^\gamma | p(n)$, where δ is the integer part of $(\beta+2)/2$.

Further details with some proofs are given in Andrews (1976) and Knopp (1970), see Problems 15 and 16.

We shall give a proof of this result for division by 5.

THEOREM 2.4 $p(5n+4)\equiv 0 \pmod 5$.

Proof. We begin with some notation. Suppose $f(x)$ and $g(x)$ are power series in x, we write

$$f(x)\equiv g(x) \pmod m$$

if and only if the coefficients of each power of x in f and g are congruent modulo m. Note that

$$(1-x)^{-5}\equiv(1-x^5)^{-1} \pmod 5, \qquad (*)$$

for

$$(1-x)^{-5}=1+\sum_{n=1}^{\infty}\frac{5\cdot 6\cdot \ldots (n+4)}{n!}x^n \equiv \sum_{n=0}^{\infty} x^{5n}=(1-x^5)^{-1} \pmod 5$$

as each coefficient of the first series is congruent to zero modulo 5 except for the coefficients of x^{5n}, and these are congruent to 1 (mod 5).

The result will follow if we can show that the coefficients of q^{5n} $(n=1, 2, \ldots)$ in $q\Pi_n^\infty(1-q^n)^4$ are divisible by 5, for, by Theorem 1.1,

$$\begin{aligned} q+\sum_{n=2}^{\infty} p(n-1)q^n &= q\Pi_n^\infty(1-q^n)^{-1} \\ &= \frac{q\Pi_n^\infty(1-q^n)^4}{\Pi_n^\infty(1-q^n)^5} \\ &\equiv \frac{q\Pi_n^\infty(1-q^n)^4}{\Pi_n^\infty(1-q^{5n})} \pmod 5 \end{aligned}$$

by (*). Note that only terms of the form q^{5n} occur when this denominator is expanded. We have, by Theorems 1.5 and 2.3(ii),

$$\begin{aligned} q\Pi_n^\infty(1-q^n)^4 &= q\Pi_n^\infty(1-q^n)[\Pi_n^\infty(1-q^n)]^3 \\ &= \sum_{m=-\infty}^{\infty}\sum_{n=0}^{\infty}(-1)^{m+n}(2n+1)q^t \end{aligned} \qquad (**)$$

where $t=1+m(3m-1)/2+n(n+1)/2$. Further we see that

$$2(m-1)^2+(2n+1)^2=8t-10m^2-5\equiv 8t \pmod 5,$$

and

$$2(m-1)^2\equiv 0,\ 2,\ \text{or}\ 3, \quad \text{and} \quad (2n+1)^2\equiv 0,\ 1,\ \text{or}\ 4 \pmod 5.$$

Hence if $5|t$ then $5|m-1$ and $5|2n+1$, that is the coefficient of q^{5k} in (**) is divisible by 5. Combining these results we have $p(5n-1)\equiv 0$ (mod 5) if $n\geq 1$. □

A number of further applications of Jacobi's identity are given in the problem section, including the result $p(7n+5)\equiv 0 \pmod 7$.

11.3 Estimates for $p(n)$

Formulas, both asymptotic and exact, have been developed for the partition function p, they are complex and require methods beyond the scope of this book for their derivation. One version of the asymptotic formula is

$$p(n)\sim\frac{1}{4n\sqrt{3}}e^{\pi\sqrt{(2n/3)}}$$

as $n\to\infty$. See Andrews (1976) for the exact formula, and Knopp (1970) for the asymptotic expansion.

Using elementary methods a reasonable lower bound and a good upper bound can be found for $p(n)$; we shall prove

THEOREM 3.1 *For all* $n>1$,

$$2^{[\sqrt{n}]}<p(n)<e^{\pi\sqrt{(2n/3)}}.$$

Proof. The lower bound.

Let $a_1,\ldots,a_r$ be distinct integers satisfying $1\le a_i\le[\sqrt{n}]$, and so $r\le[\sqrt{n}]$. Since

$$a_1+\ldots+a_r\le 1+2+\ldots+[\sqrt{n}]<[\sqrt{n}]^2\le n$$

we have the partition, with $r+1$ parts,

$$n=a_1+a_2+\ldots+a_r+(n-a_1-\ldots-a_r).$$

For fixed r the number of partitions of this type is given by the binomial term $\binom{[\sqrt{n}]}{r}$. Hence the total number of partitions of this type is

$$1+\binom{[\sqrt{n}]}{1}+\ldots+\binom{[\sqrt{n}]}{r}+\ldots=(1+1)^{[\sqrt{n}]}=2^{[\sqrt{n}]}$$

and the first inequality follows.

For the upper bound we require two lemmas.

LEMMA 3.2 *If* $0<q<1$ *and* $j\ge1$ *then* $\frac{1}{j}\left(\frac{q^j}{1-q^j}\right)<\frac{q}{j^2(1-q)}$.

To prove this we apply the mean value theorem to the function x^j over the range $q\le x\le1$, hence

$$\frac{1-q^j}{1-q}=jy^{j-1}$$

for some y satisfying $q<y<1$. But $jy^{j-1}>jq^{j-1}$ and so we have

$$1-q^j>jq^{j-1}(1-q)$$

and Lemma 3.2 follows. □

LEMMA 3.3 *If* $0<q<1$ *then* $-\ln q<(1-q)/q$.

To prove this we again use the mean value theorem applied this time to the function $\ln x$ over the range $1\le x\le 1/q$, so for some t satisfying $1<t<1/q$

$$-\ln q=\ln 1/q-\ln 1=\frac{1}{t}\left(\frac{1}{q}-1\right)<(1-q)/q. \qquad \square$$

We come now to the derivation of the upper bound in Theorem 3.1. If $0<q<1$ we have

$$\ln\left\{\prod_{n=1}^{\infty}(1-q^n)^{-1}\right\} = -\sum_{n=1}^{\infty}\ln(1-q^n) = \sum_{n=1}^{\infty}\sum_{j=1}^{\infty}\frac{q^{nj}}{j}$$

$$= \sum_{j=1}^{\infty}\frac{1}{j}\sum_{n=1}^{\infty}q^{nj} = \sum_{j=1}^{\infty}\frac{q^j}{j(1-q^j)}$$

$$< \frac{q}{1-q}\sum_{j=1}^{\infty}\frac{1}{j^2} \qquad \text{by Lemma 3.2}$$

$$= \frac{\pi^2 q}{6(1-q)}, \qquad (*)$$

see p. viii. Further, using Theorem 1.1, we have $p(n) \leq \prod_{j=1}^{\infty}(1-q^j)^{-1}/q^n$, and so

$$\ln p(n) \leq \ln\left\{\prod_{j=1}^{\infty}(1-q^j)^{-1}\right\} - n\ln q$$

$$< \frac{\pi^2 q}{6(1-q)} + \frac{n(1-q)}{q} \qquad \text{by } (*) \text{ and Lemma 3.3.}$$

Now this is valid for all q satisfying $0<q<1$, hence we can choose $q = \sqrt{(6n)}/(\pi+\sqrt{(6n)})$ and then the two terms on the right-hand side of the above inequality are equal. This gives, as $(1-q)/q = \pi/\sqrt{6n}$,

$$\ln p(n) < \pi\sqrt{(2n/3)}$$

and the result follows. □

11.4 Problems 11

1. Calculate $p(10)$, $p(20)$, and $p(30)$.

2. Give a proof of Theorem 1.2 without using generating functions. Use the fact that every positive integer a has the unique binary representation $a = 2^{a_1}+2^{a_2}+2^{a_3}+\ldots$ where $0 \leq a_1 < a_2 < a_3 \ldots$, and begin by considering numbers of the form $2^c b$ where b is odd.

Use generating functions in Problems 3, 4, and 5.

3. Prove that the number of partitions of n in which no part occurs more than three times equals the number of partitions of n in which only odd parts may be repeated. Generalize this result to partitions in which no part occurs more than $2^c - 1$ times.

4. Show that the number of partitions of n in which each part occurs two, three, or five times equals the number of partitions of n in which each part a satisfies: $a \equiv 2, 3, 6, 9,$ or $10 \pmod{12}$.

5. Derive the following result due to Euler: the absolute value of the difference between the number of partitions of n with an even number of parts and the number of partitions of n with an odd number of parts equals the number of partitions of n with distinct odd parts.

6. By adapting dot graphs show that the number of partitions of n with distinct odd parts equals the number of partitions of n which are identical to their conjugates.

7. Show that the number of partitions of $a-c$ into $b-1$ parts where each part is less than $c+1$ equals the number of partitions of $a-b$ into $c-1$ parts where each part is less than $b+1$. [Hint. Use dot graphs and add or remove rows and columns.]

8. Prove that the number of partitions of n in which the largest part is odd and the smallest part is larger than half the largest part equals the number of partitions of n with unique smallest part, and largest part at most twice the smallest part. To establish this split the dot graph into two pieces, one rectangular, and make a new graph by moving these blocks to new positions.

Deduce that, for $|q|<1$,

$$1+\sum_{m=0}^{\infty}\frac{q^{2m+1}}{(1-q^{m+1})\ldots(1-q^{2m+1})}=\sum_{m=0}^{\infty}\frac{q^m}{(1-q^{m+1})\ldots(1-q^{2m})}.$$

9. Given a partition $\mathbf{a}$ of n, $n=a_1+a_2+\ldots+a_k$ with $a_i\geq a_{i+1}$ for all i, let $d(\mathbf{a})$ be the largest j such that $a_j\geq j$. Check that the largest square of dots contained in the graphical representation of $\mathbf{a}$ has sides of length $d(\mathbf{a})$, this is called the *Durfee square.* The graph of $\mathbf{a}$ can be split into three disjoint pieces (i) the Durfee square, (ii) the parts of $\mathbf{a}$ which lie below this square, and (iii) a collection of dots to the right of this square containing rows of length $a_1-d(\mathbf{a})$, $a_2-d(\mathbf{a}),\ldots$. Show that the set of partitions with Durfee square having $(d(\mathbf{a}))^2=s^2$ dots is generated by

$$q^{s^2}/\Pi_n^s(1-q^n)^2.$$

Deduce that

$$\Pi_n^\infty(1-q^n)^{-1}=1+\frac{q}{(1-q)^2}+\frac{q^4}{(1-q)^2(1-q^2)^2}+\ldots.$$

10. Show that

$$q^N/\Pi_i^m(1-q^{2i})$$

is the generating function for partitions of $(n-N)/2$ into (i) parts not exceeding m, or (ii) at most m parts. Hence using Problem 6 and Durfee squares, show that

$$\Pi_n^\infty(1+q^{2n+1})=1+\frac{q}{1-q^2}+\ldots+\frac{q^{n^2}}{\Pi_i^n(1-q^{2i})}+\ldots.$$

11. Given the formula

$$\Pi_n^\infty(1-q^{5n+1})^{-1}(1-q^{5n+4})^{-1}=1+\sum_{n=1}^{\infty}\frac{q^{n^2}}{\Pi_i^n(1-q^i)}$$

(for a proof see Hardy and Wright, 1954, pp. 292–5), show that the number of partitions of n with minimum difference two is equal to the number of partitions of n in which each part has the form $5n+1$ or $5n+4$.

12. Show that $q\Pi_n^\infty(1-q^{8n})^3 = q - 3q^9 + 5q^{25} - 7q^{49} + \ldots$.

13. Prove that

$$\sum_{n=-\infty}^{\infty}(-1)^n q^{n^2} = \frac{(1-q)(1-q^2)(1-q^3)}{(1+q)(1+q^2)(1+q^3)}\cdots.$$

14. Using Jacobi's identity with q^2 replacing q, and q replacing z, show that, if $a_i = i(8i \pm 1)$,

$$p(4n) \equiv p(n) + p(n-7) + p(n-9) + \ldots + p(n-a_i) + \ldots (\mathrm{mod}\ 2).$$

15. Show that $p(7n+5) \equiv 0 \pmod 7$ by using a method similar to the proof of Theorem 2.4 and considering first $q^2\Pi_n^\infty(1-q^n)^6$.

16. Using a computer investigate those n for which $7^2|p(n)$ and make a conjecture for the general case. (See Knopp, 1970, Chapter 8.) Secondly, in the general statement of the divisibility properties of $p(n)$ (see p. 209) show that δ cannot be replaced by β. You should consider the number 243 and show that $7^3 \nmid p(243)$, note that $p(243)$ is about 1.3×10^{14}.

17. Using Problem 2(iii) in Chapter 6 show that

$$\Pi_n^\infty(1-q^{2n})^2(1+q^{2n-1})^4 = \left(\sum_{n=-\infty}^{\infty} q^{n^2}\right)^2 = 1 + 4\left(\frac{q}{1-q} - \frac{q^3}{1-q^3} + \frac{q^5}{1-q^5} - \ldots\right).$$

(Note. Generating functions can be used to prove a number of results about sums of squares, see for example Hua (1982, pp. 204–16).

18. Using Durfee squares (see Problem 9) and the fact that there are at most n^c partitions of n into c parts, give a combinatorial proof of the inequality

$$p(n) < n^{3[\sqrt{n}]}.$$

12

THE PRIME NUMBERS

The study of the primes is one of the most important branches of number theory. The only other branch of similar importance is Diophantine equations and this will be considered in the final two chapters. The literature on the prime numbers is very extensive beginning with Pythagorus and Euclid; even so, many problems remain. There are two pre-eminent results: Dirichlet's theorem and the prime number theorem (PNT). In 1837 Dirichlet proved that every arithmetic progression $\{an + b : n = 0, 1, 2, \ldots\}$, where $(a, b) = 1$, contains infinitely many primes. When one considers the state of mathematical knowledge at that time, with both analysis and group theory in their infancy, it is clear that this was a monumental achievement.

The PNT was postulated by both Legendre and Gauss on numerical evidence and it was finally established by Hadamard and (independently) de la Vallee Poussin in 1896—a no-less considerable achievement. PNT states that if $\pi(x)$ is the number of primes p which satisfy $2 \leq p \leq x$ then

$$\pi(x) \sim x/\ln x.$$

A more precise version of this result is

$$\pi(x) \sim Li(x) = \int_1^x \frac{dt}{\ln t}.$$

We shall prove both Dirichlet's theorem and a version of PNT in the next chapter.

Over two thousand years ago the Greeks began the study of the primes and two of their constructions stand out. The first gives the result, attributed to Euclid, that there are infinitely many prime numbers, see Theorem 1.8, Chapter 1. The second is the sieve of Eratosthenes and, even today, it provides an efficient method for generating primes. To enumerate all primes less than n write down a list of all positive integers between 2 and n

2, 3, ~~4~~, 5, ~~6~~, 7, ~~8~~, ~~9~~, ~~10~~, 11, ~~12~~, 13, ~~14~~, ~~15~~, ~~16~~, 17, . . . ,

take the leftmost entry not crossed out and cross out all its multiples except itself. Continue this process until all multiples of each integer less than $1 + \sqrt{n}$ have been crossed out. The result is a list of all primes less than or equal to n. In the last 75 years this construction has been greatly extended, see below.

Before discussing the material in this chapter we consider some of the many unsolved problems in prime number theory. Most of the conjectures associated with these problems are backed up with ample numerical evidence, but this must be treated with some scepticism. For example the numerical evidence (see, for instance, LeVeque, 1977, p. 268) suggests that $Li(x) > \pi(x)$ for all x, but Littlewood, in 1909, showed that this statement is false for some large x. It is now known that $Li(x) - \pi(x)$ changes sign infinitely often as x increases and the first change of sign occurs before $x = 6.69 \times 10^{370}$ (see te Riele 1986). A good discussion of this and many of the more advanced topics in prime number theory is given in Davenport (1967).

Conjectures concerning prime numbers

(i) *The twin prime conjecture.* A brief study of any table of prime numbers (see, for example, p. 337) shows that there are many pairs of primes p, q where $q = p + 2$, examples are

$$3, 5; \quad 17, 19; \quad 881, 883; \quad 1997, 1999; \quad \text{and} \quad 10^9 + 7,\ 10^9 + 9.$$

Let $\pi_2(x)$ be the number of prime pairs less than x, so for example

$$\pi_2(10^3) = 35 \quad \text{and} \quad \pi_2(10^6) = 8164.$$

The twin prime conjecture states that

$$\pi_2(x) \to \infty \quad \text{as} \quad x \to \infty.$$

In fact the tables suggest

$$\pi_2(x) = O\left(\int_2^x (\ln t)^{-2}\, dt\right).$$

Using very complicated arguments based on the idea of a sieve Chen has shown that there are infinitely many pairs of integers p, $p + 2$ where p is prime and $p + 2$ has at most two prime factors. Extensions of the twin prime conjecture have also been considered. For instance are there infinitely many triples of primes p, q, r where $q = p + 2$ and $r = p + 6$?

(ii) *The Goldbach conjecture.* It is an elementary fact that any reasonably small even positive integer, greater than 2, can be expressed as a sum of two primes. For example

$$8 = 3 + 5, \quad 80 = 37 + 43, \quad 800 = 379 + 421, \quad \text{and} \quad 8000 = 3943 + 4057.$$

It seems reasonable to conjecture that every even positive integer is the sum of two primes (the Goldbach conjecture). Using the sieve methods mentioned in (i), Chen has shown that every large even integer can be expressed as a sum $p + q$ where p is prime and q has at most two prime factors. Using different methods, Vinogradov had shown earlier that every large odd integer is the sum of three primes. For details of the sieve methods see Halberstam and Richert (1974).

(iii) *Primes in intervals.* In Section 1 we shall show that there is a prime between n and $2n$ for all positive n. This result can be extended to: there is a prime between n and $(1+\varepsilon)n$ for $\varepsilon > 0$ and $n > n(\varepsilon)$, see Problem 17, Chapter 13. But it seems very difficult to prove similar results for smaller intervals. For example, it is not known if a prime occurs between n^2 and $(n+1)^2$ for all large n.

(iv) *Extensions of Dirichlet's theorem.* Very little is known about the set D of those n for which $an+b$ is prime. For example, does D itself contain infinitely many primes? Or, at least one prime? Another extension which has been proposed replaces the linear form $an+b$ by a quadratic one. For example, it is not known if there are infinitely many primes of the form n^2+n+1 or n^2+n+p for some prime p (see p. 280).

(v) *Formulas for the primes.* We discussed this topic in Chapter 1. Another problem under this heading is: what can be proved about the difference $p_n - p_{n-1}$ where p_n is the nth prime, what is $O(p_n - p_{n-1})$ as $n \to \infty$? Some information is given in Heath-Brown (1978).

(vi) *The Riemann hypothesis.* This will be considered in Section 3. If the hypothesis is established then many estimates given in analytic number theory will be greatly improved, for example those in PNT.

In the middle of the last century Chebyshev laid the foundations for a substantial part of modern prime number theory. We discuss some of his results in Sections 1 and 2. For instance we show that if

$\lim_{x\to\infty} \pi(x) \ln x/x$ exists then it must be 1 (Theorem 1.2). We shall also consider a few sums defined over the primes, and prove

$$\sum_{p \le x} 1/p \sim \ln \ln x.$$

Brun, who was the instigator of modern sieve theory, studied the series $\sum' 1/p$ taken over all primes p where $p+2$ is also prime. He hoped to prove the twin prime conjecture by showing that this series diverges, if only slowly. In fact it converges and so the only conclusion is that the density of prime pairs is much lower than the density of the primes. For a proof of this result see Landau (1958) or Gelfond and Linnik (1965).

In this and the following chapter x, y, and z denote real variables, whilst j, k, m, n, r, and t denote integer variables.

12.1 The results of Chebyshev and Bertrand

For our first results on the primes we use some properties of binomial coefficients, and to begin with we derive a few simple inequalities concerning these coefficients. Note that (v) below is of interest in its own right; it implies that, on average, gaps between large primes cannot be too small.

LEMMA 1.1 *Let $n \geq 1$ throughout.*

(i) $2^n \leq \binom{2n}{n} < 2^{2n}$.

(ii) $\prod_{n<p\leq 2n} p \,\Big|\, \binom{2n}{n}$.

(iii) *Let $r(p)$ satisfy $p^{r(p)} \leq 2n < p^{r(n)+1}$, then*

$$\binom{2n}{n} \Big| \prod_{p\leq 2n} p^{r(p)}.$$

(iv) *If $n > 2$ and $2n/3 < p \leq n$, then $p \nmid \binom{2n}{n}$.*

(v) $\prod_{p\leq n} p < 4^n$.

Proof. (i) As $2n - k \geq 2(n-k)$ for $0 \leq k < n$ we have

$$2^n \leq \frac{2n}{n}\frac{2n-1}{n-1}\cdots\frac{n+1}{1} = \binom{2n}{n}.$$

Also as $\binom{2n}{n}$ is one of the terms in the binomial expansion of $(1+1)^{2n}$ we have

$$\binom{2n}{n} < (1+1)^{2n} = 2^{2n}.$$

(ii) This follows as each prime in the interval $[n+1, 2n]$ divides $(2n)!$ but not $n!$

(iii) The exponent of p in $n!$ is $\sum_{j=1}^{r(p)} [n/p^j]$, see Problem 14, Chapter 1, and so the exponent of p in $\binom{2n}{n}$ is

$$\sum_{j=1}^{r(p)} \{[2n/p^j] - 2[n/p^j]\} \leq \sum_{j=1}^{r(p)} 1 = r(p).$$

This last inequality holds as each term in curly brackets is either 0 or 1 [use (ii) on p. 23]. Now take the product over primes $p \leq 2n$.

(iv) If p satisfies $2n/3 < p \leq n$ then p occurs once in the prime factorization of $n!$ and twice in $(2n)!$ (as $3p > 2n$), hence as $p > 2$, $p \nmid \binom{2n}{n}$.

(v) This is proved by complete induction. Let $P(n)$ denote the proposition to be proved. Clearly $P(1)$, $P(2)$, and $P(3)$ hold, and if $m>1$ we have $P(2m-1)$ implies $P(2m)$ as

$$\prod_{p\leq 2m} p = \prod_{p\leq 2m-1} p < 4^{2m-1} < 4^{2m}.$$

So we may suppose $n=2m+1$ with $m\geq 2$. Each prime p in the interval $[m+2,\ 2m+1]$ is a factor of $\binom{2m+1}{m}$, hence, if we assume that $P(m+1)$ holds,

$$\prod_{p\leq 2m+1} p \leq \binom{2m+1}{m} \prod_{p\leq m+1} p < \binom{2m+1}{m} 4^{m+1}.$$

But $\binom{2m+1}{m}$ is one of the two central terms in the binomial expansion of $(1+1)^{2m+1}$, and so

$$\binom{2m+1}{m} < \frac{1}{2}(1+1)^{2m+1} = 4^m.$$

Thus $P(m+1)$ implies $P(2m+1)$ and the inductive proof is complete. □

We come now to a version of Chebyshev's result; note that the constants can be improved by using more precise inequalities.

THEOREM 1.2 *If $n>1$, then*

$$\frac{n}{8\ln n} < \pi(n) < \frac{6n}{\ln n}.$$

Proof. As $a|b$ implies $a\leq b$, we have by (ii) and (iii) above and the definition of $r(p)$

$$n^{\pi(2n)-\pi(n)} < \prod_{n<p\leq 2n} p \leq \binom{2n}{n} \leq \prod_{p\leq 2n} p^{r(p)} \leq (2n)^{\pi(2n)},$$

and so using (i) above

$$n^{\pi(2n)-\pi(n)} < 2^{2n}, \qquad 2^n \leq (2n)^{\pi(2n)}.$$

If we let $n=2^k$ $(k\geq 0)$, these inequalities give, considering exponents only,

$$k(\pi(2^{k+1})-\pi(2^k)) < 2^{k+1}, \qquad 2^k \leq (k+1)\pi(2^{k+1}) \tag{1}$$

as $2n=2^{k+1}$. As even numbers (except 2) are not prime we have $\pi(2^{k+1})\leq 2^k$, and hence (1) gives

$$(k+1)\pi(2^{k+1}) - k\pi(2^k) < \pi(2^{k+1}) + 2^{k+1} < 3\cdot 2^k.$$

Setting $k = 0, 1, \ldots, m$ in turn and adding we obtain [as $\pi(1) = 0$]

$$(m+1)\pi(2^{m+1}) < 3(2^0 + 2^1 + \ldots + 2^m) < 3 \cdot 2^{m+1}. \tag{2}$$

Hence by (1) and (2) we have

$$\frac{2^{m+1}}{2(m+1)} < \pi(2^{m+1}) < \frac{3 \cdot 2^{m+1}}{m+1}. \tag{3}$$

Now given n choose m to satisfy $2^{m+1} \leq n < 2^{m+2}$, $m \geq 0$ as $n > 1$. Note that for any positive k we have

$$k/2 < \ln 2^k < k$$

and so (3) gives, as π is monotonic increasing,

$$\pi(n) \leq \pi(2^{m+2}) < \frac{3 \cdot 2^{m+2}}{m+2} < \frac{6 \cdot 2^{m+1}}{\ln(2^{m+2})} < \frac{6n}{\ln n},$$

and

$$\pi(n) \geq \pi(2^{m+1}) > \frac{2^{m+1}}{2(m+1)} = \frac{2^{m+2}}{8((m+1)/2)} > \frac{2^{m+2}}{8 \ln(2^{m+1})} > \frac{n}{8 \ln n}. \qquad \square$$

This result shows that the order of magnitude of $\pi(n)$ is $n/\ln n$. The next result gives a bound on the longest gap that can occur between consecutive primes. The proof is due to Erdös.

THEOREM 1.3 (Bertrand's postulate) *If $n > 0$ then there is a prime p satisfying $n < p \leq 2n$.*

Proof. We again consider the binomial coefficient $\binom{2n}{n}$. The result clearly holds when $n \leq 3$, we shall assume that it is false for some $n > 3$ and derive a contradiction. It follows from Lemma 1.1(iv) that for this n all prime factors p of $\binom{2n}{n}$ satisfy $p \leq 2n/3$. Let $s(p)$ be the largest power of p which divides $\binom{2n}{n}$, so by Lemma 1.1(iii) we have

$$p^{s(p)} \leq 2n. \tag{$*$}$$

Hence if $s(p) > 1$ then $p \leq \surd(2n)$, and it follows that no more than $[\surd(2n)]$ primes occur in $\binom{2n}{n}$ with exponent larger than 1. Using $(*)$ and our assumption this gives

$$\binom{2n}{n} \leq (2n)^{[\surd(2n)]} \prod_{p \leq 2n/3} p.$$

Now $\binom{2n}{n} > \dfrac{4^n}{2n+1}$ $\left[\text{as } \binom{2n}{n} \text{ is the largest term in the binomial expansion of } (1+1)^{2n} \text{ which has } 2n+1 \text{ summands}\right]$. Thus we have, using Lemma 1.1(v),

$$\frac{4^n}{2n+1} < (2n)^{[\sqrt{(2n)}]} \prod_{p \le 2n/3} p < 4^{2n/3}(2n)^{\sqrt{(2n)}}.$$

But $2n+1 < (2n)^2$ and so cancelling $4^{2n/3}$ we have

$$4^{n/3} < (2n)^{2+\sqrt{(2n)}}$$

or, taking logarithms,

$$\frac{n \ln 4}{3} < (2 + \sqrt{(2n)}) \ln 2n.$$

This is clearly false for large n. In fact if $n = 750$ we have (as $1.3 < \ln 4$ and $\ln 1500 < 7.5$)

$$325 = \frac{750 \cdot 1.3}{3} < (2 + \sqrt{1500}) \ln 1500 < 41 \cdot 7.5 < 308.$$

Hence the result holds if $n \ge 750$, but the result also holds by inspection (see Table 1) if $n < 750$, for

2, 3, 5, 11, 19, 37, 73, 139, 277, 547, 751

are all prime and each is less than twice its predecessor. □

12.2 Series involving primes

Here we shall evaluate two important series defined on the primes, they will play a vital role in the next chapter. Note that if x is a real variable then $\sum_{p \le x} f(p)$ stands for the sum $f(2) + f(3) + \ldots + f(p)$ taken over all primes $p \le x$. We shall assume that $x \ge 1$ throughout.

THEOREM 2.1

$$\sum_{p \le x} \frac{\ln p}{p} = \ln x + O(1).$$

Proof. This is derived for an integer variable first by estimating $\ln(n!)$ in two distinct ways. By Problem 14, Chapter 1, we have

$$\ln(n!) = \ln \prod_{p \le n} p^{([n/p]+[n/p^2]+\dots)}$$
$$= \sum_{p \le n} [n/p] \ln p + \sum_{p \le n} ([n/p^2]+[n/p^3]+\dots) \ln p.$$

We consider these sums seperately. Clearly

$$\sum_{p \le n} [n/p] \ln p \le \sum_{p \le n} \frac{n \ln p}{p}$$

and

$$\sum_{p \le n} [n/p] \ln p > \sum_{p \le n} \left(\frac{n}{p} - 1\right) \ln p = \sum_{p \le n} \frac{n \ln p}{p} - \sum_{p \le n} \ln p$$
$$> \sum_{p \le n} \frac{n \ln p}{p} - \pi(n) \ln n$$
$$> \sum_{p \le n} \frac{n \ln p}{p} - 6n$$

by Theorem 1.2. Further

$$\sum_{p \le n} ([n/p^2]+[n/p^3]+\dots) \ln p \le n \sum_{p \le n} \left(\frac{1}{p^2} + \frac{1}{p^3} + \dots\right) \ln p$$
$$= n \sum_{p \le n} \frac{\ln p}{p(p-1)}$$
$$< n \sum_{j \le n} \frac{\ln j}{j(j-1)} = cn,$$

for some constant c, as this last series is convergent. Combining these inequalities we have

$$n \sum_{p \le n} \frac{\ln p}{p} + (c-6)n \le \ln(n!) \le n \sum_{p \le n} \frac{\ln p}{p} + cn.$$

For our second estimate we have $\ln n! = \sum_{j \le n} \ln j$ and

$$\sum_{j \le n} \ln j < \int_1^n \ln y \, dy < \sum_{j \le n+1} \ln j.$$

So, by integration,

$$n \ln n - 2n < \ln n! < n \ln n.$$

Now these two estimates give

$$n \ln n - (2+c)n < n \sum_{p \le n} \frac{\ln p}{p} < n \ln n + (6-c)n$$

or

$$\sum_{p \le n} \frac{\ln p}{p} = \ln n + O(1).$$

Finally if we replace the integer variable n by a real variable x we have, if $n \le x < n+1$,

$$\sum_{p \le x} \frac{\ln p}{p} = \sum_{p \le n} \frac{\ln p}{p} = \ln n + O(1) = \ln x + O(1)$$

as $\ln x - \ln n < 1$. □

For our second result we need the following important lemma which has many applications, see Problem 9 and the next chapter.

LEMMA 2.2 *Let $t_1, t_2, \ldots$ be a non-decreasing sequence of real numbers with limit infinity, let $z_1, z_2, \ldots$ be any sequence of real numbers, and suppose f is a real function with a continuous derivative for arguments greater than or equal to t_1. Define Z by*

$$Z(x) = \sum_{t_n \le x}^{*} z_n,$$

the sum of all z_n for those n which satisfy $t_n \le x$. Then

$$\sum_{t_n \le x}^{*} z_n f(t_n) = Z(x) f(x) - \int_{t_1}^{x} Z(y) f'(y)\, dy.$$

As an example let $t_n = p_n$, the nth prime, and let $z_n = \dfrac{\ln p_n}{p_n}$ then $Z(x) = \sum_{p \le x} \dfrac{\ln p}{p}$. We shall use this below.

Proof. As $Z(t_{n+1}) - Z(t_n) = z_{n+1}$ we have, if m is the largest subscript such that $t_m \le x$,

$$\begin{aligned}\sum_{t_n \le x}^{*} z_n f(t_n) &= Z(t_1) f(t_1) + (Z(t_2) - Z(t_1)) f(t_2) + \ldots \\ &\qquad + (Z(t_m) - Z(t_{m-1})) f(t_m) \\ &= Z(t_1)(f(t_1) - f(t_2)) + \ldots + Z(t_{m-1})(f(t_{m-1}) - f(t_m)) \\ &\qquad + Z(t_m)(f(t_m) - f(x)) + Z(x) f(x)\end{aligned}$$

as $Z(t_m) = Z(x)$. Now because Z is constant on the interval $[t_i, t_{i+1})$ we have

$$Z(t_i)(f(t_i)-f(t_{i+1})) = -\int_{t_i}^{t_{i+1}} Z(y)f'(y)\,dy.$$

Combining these equations we have

$$\sum_{t_n \le x}^{*} z_n f(t_n) = Z(x)f(x) - \left\{\sum_{i=1}^{m-1}\int_{t_i}^{t_{i+1}} + \int_{t_m}^{x}\right\} Z(y)f'(y)\,dy$$

and the proof is completed by combining these integrals. □

The method employed at the beginning of this proof is called *partial summation,* we shall use it often in the following pages.

THEOREM 2.3

$$\sum_{p \le x} \frac{1}{p} = \ln \ln x + c + O(1/\ln x)$$

for some constant c.

Proof. In Lemma 2.2 let $t_n = p_n$, the nth prime, $z_n = (\ln p_n)/p_n$ and $f(y) = 1/\ln y$, so we have if $x \ge 2$ (see example above),

$$\sum_{p \le x} \frac{1}{p} = \sum_{p \le x} \frac{\ln p}{p} \frac{1}{\ln p} = \frac{1}{\ln x} \sum_{p \le x} \frac{\ln p}{p} + \int_2^x \left(\sum_{p \le t} \frac{\ln p}{p}\right) \frac{dt}{t \ln^2 t}$$

$$= \frac{1}{\ln x}(\ln x + O(1)) + \int_2^x \frac{\ln t \, dt}{t \ln^2 t} + \int_2^x \left(\sum_{p \le t} \frac{\ln p}{p} - \ln t\right) \frac{dt}{t \ln^2 t}$$

by Theorem 2.1

$$= 1 + O\left(\frac{1}{\ln x}\right) + \ln \ln x - \ln \ln 2 + \left(\int_2^{\infty} - \int_x^{\infty}\right) \frac{O(1)\, dt}{t \ln^2 t}$$

by Theorem 2.1 again

$$= \ln \ln x + c + O(1/\ln x).$$

The constant c exists as the first integral in the line above converges and the second has order $O(1/\ln x)$. □

12.3 Riemann's zeta function

Properties of the zeta function have an important bearing on prime number theory: we shall discuss this relationship below. As none of the main results discussed will be used in other sections, only sketch proofs will be given. We assume some familiarity with complex analysis.

It is conventional to use $s = \sigma + it$ (σ, t real) as the complex variable. The function ζ given by

$$\zeta(s) = \sum_{n=1}^{\infty} n^{-s}$$

is convergent for $\sigma>1$, $-\infty<t<\infty$; it is called the *(Riemann) zeta function*. Euler was the first to notice a connection between the zeta function and the primes, his product formula is given by part (ii) of

THEOREM 3.1 (i) *If $\sigma\geq 1+\varepsilon$, for $\varepsilon>0$, then the series*

$$\sum_{n=1}^{\infty} n^{-s}$$

is absolutely and uniformly convergent.

(ii) *For $\sigma>1$,*

$$\zeta(s)=\prod_{p}(1-p^{-s})^{-1}$$

where the product is taken over all prime numbers p.

Proof. (i) For any integer j we have

$$\left|\sum_{n=j}^{\infty} n^{-s}\right| \leq \sum_{n=j}^{\infty} |n^{-s}| = \sum_{n=j}^{\infty} n^{-\sigma} \leq \int_{j-1}^{\infty} x^{-\sigma}\,\mathrm{d}x \leq \frac{(j-1)^{1-\sigma}}{\sigma-1} < \frac{j^{-\varepsilon}}{\varepsilon}.$$

The first part of the theorem follows.

(ii) Using properties of geometric series and unique factorization we have

$$\prod_{p\leq k}(1-p^{-s})^{-1} = \prod_{p\leq k}(1+p^{-s}+\ldots+p^{-ns}+\ldots) = \sum_{n\leq k} n^{-s} + R(k,s)$$

where $R(k,s)$ is a sum of the form $\sum' m^{-s}$ taken over those $m>k$ whose prime factors are less than $k+1$. By (i), $R(k,s)\to 0$ as $k\to\infty$, provided $\sigma>1$; this completes the proof. □

It is a simple matter to extend this result to show that ζ is a non-zero analytic function for all t, provided $\sigma>1$, using (i) and (ii) above. In a memoir, published in 1859, Riemann developed the theory of this function proving some remarkable results and postulating his famous hypothesis. First he was able to extend the domain of definition of ζ to the whole plane. It is a relatively easy to show that, for $\sigma>0$,

$$\zeta(s)=\frac{1}{s-1}+1-s\int_{1}^{\infty}\frac{x-[x]}{x^{s+1}}\,\mathrm{d}x,$$

using a method similar to that in the proof of Theorem 3.3 below and 'analytic continuation' (see Problem 15). For $\sigma\leq 0$, $\zeta(s)$ can be defined using the 'functional equation' $(*)$ for ζ as follows.

Let Γ, the *gamma function*, be given by (for $\sigma>0$)

$$\Gamma(s)=\int_{0}^{\infty} \mathrm{e}^{-x}x^{s-1}\,\mathrm{d}x.$$

Note that $\Gamma(s+1)=s\Gamma(s)$ (integrate by parts), and so $\Gamma(m)=(m-1)!$ for positive integers m. This relation can also be used to extend the domain of definition of Γ to the whole plane. Γ has simple poles at $s=0$, $-1, -2, \ldots$ and nowhere else.

Riemann showed that, if $\sigma<0$,

$$\zeta(s)=2^s\pi^{1-s}\Gamma(1-s)\zeta(1-s)\sin(\pi s/2) \qquad (*)$$

and, by analytic continuation, this is valid for all s in the complex plane. It follows that ζ has a simple pole at $s=1$ and is analytic at all other points of the complex plane. It has zeros at $s=-2, -4, \ldots$ and, by Theorem 3.1 and $(*)$, no more zeros outside the so-called *critical strip* $0\leq\sigma\leq 1$. The equation $(*)$ gives no information on the line $s=1/2+it$. Riemann's hypothesis states that all zeros of ζ in the critical strip lie on this line. In 1914 Hardy showed that there are infinitely many zeros on this line, but every attempt to prove that no further zeros exist has so far failed. A proof of this hypothesis would have far-reaching consequences in many areas of mathematics.

Applications

All analytic proofs of the prime number theorem (PNT) depend on properties of the ζ function. Estimates are made of either $\ln\zeta(s)$ or $\zeta'(s)/\zeta(s)$ by considering integrals over contours in the half plane $\sigma\geq 1$, but avoiding the point $s=1$. So an essential prerequisite is

THEOREM 3.2 $\zeta(1+it)\neq 0$.

Proof. We define θ by

$$\theta(\tau, t)=\zeta(1+\tau)^3\,|\zeta(1+\tau+it)|^4\,|\zeta(1+\tau+2it)|$$

for real positive τ and $t\neq 0$. Using Theorem 3.1(ii) we have

$$\theta(\tau, t)=\prod_p \psi(p)$$

where

$$\psi(p)=|1-p^{-(1+\tau)}|^{-3}\,|1-p^{-(1+\tau+it)}|^{-4}\,|1-p^{-(1+\tau+2it)}|^{-1}.$$

Hence, using the standard expansion of $-\ln(1-z)$, we have

$$\ln\psi(p)=\sum_{m=1}^{\infty} m^{-1}p^{-(1+\tau)}[3+4\cos(mt\ln p)+\cos(2mt\ln p)].$$

The term in square brackets is non-negative, and so $\ln\psi(p)>0$ and $|\theta(\tau, t)|>1$. A contradiction follows from the assumption $\zeta(1+it)=0$ by treating the three parts of $\theta(\tau, t)$ separately, see Problem 16. □

A typical analytic proof of PNT begins with the result

THEOREM 3.3 *For* $\sigma > 1$,

$$\ln \zeta(s) = s \int_2^\infty \frac{\pi(x)}{x(x^s - 1)} \, dx.$$

Proof. By Theorem 3.1 we have using partial summation

$$\begin{aligned}\ln \zeta(s) &= -\sum_p \ln(1 - p^{-s}) = -\lim_{k\to\infty} \sum_{n=2}^{k} (\pi(n) - \pi(n-1)) \ln(1 - n^{-s}) \\ &= -\lim_{k\to\infty} \left[\sum_{n=2}^{k} \pi(n)(\ln(1 - (n+1)^{-s}) - \ln(1 - n^{-s})) \right. \\ &\qquad \left. - \pi(k) \ln(1 - (k+1)^{-s}) \right].\end{aligned}$$

Now for $\sigma > 1$ the final term in the limit above tends to zero as $k \to \infty$, and so by the fundamental theorem of the calculus

$$\begin{aligned}\ln \zeta(s) &= \sum_{n=2}^{\infty} \pi(n) \int_n^{n+1} \frac{d}{dx} \ln(1 - x^{-s}) \, dx \\ &= s \sum_{n=2}^{\infty} \int_n^{n+1} \frac{\pi(x)}{x(x^s - 1)} \, dx = s \int_2^\infty \frac{\pi(x)}{x(x^s - 1)} \, dx. \qquad \square\end{aligned}$$

Now PNT can be derived by 'solving' this equation for $\pi(x)$, it uses a version of the Mellin transform, see for example Grosswald (1984). The accuracy of the estimate for $\pi(x)$ given by PNT is limited by Theorem 3.2, this estimate could be greatly improved if the Riemann hypothesis was available. Note that the proof of PNT sketched here is quite different to that given in the next chapter.

In the second application we define the function M by

$$M(x) = \sum_{n \le x} \mu(n)$$

where μ is the Möbius function (see p. 19). As $|\mu(n)| \le 1$ holds for all n, we have immediately $M(x) = O(x)$. The question arises: what is the true order of $M(x)$? In 1897 Mertens conjectured† that, for all $\varepsilon > 0$ and $x > 0$,

$$M(x) = o(x^{1/2+\varepsilon}). \qquad (*)$$

This is equivalent to the Riemann hypothesis. One part of this equivalence can be demonstrated easily. In the proof of Lemma 2.4, Chapter

† Recently it has been shown that a stronger version of this conjecture, that $M(x) < \sqrt{x}$, is false for some large but unspecified x (see Odlyzko and te Riele, 1985).

2, we showed that, if $\sigma > 1$,

$$\frac{1}{\zeta(s)} = \sum_{n=1}^{\infty} \frac{\mu(n)}{n^s}.$$

By partial summation we derive, as $M(0) = 0$,

$$\frac{1}{\zeta(s)} = \sum_{n=1}^{\infty} \frac{M(n) - M(n-1)}{n^s} = \sum_{n=1}^{\infty} M(n)\left\{\frac{1}{n^s} - \frac{1}{(n+1)^s}\right\}$$
$$= s\int_0^{\infty} M(x)x^{-s-1}\,\mathrm{d}x.$$

It is a simple matter to show that if $M(x)$ grows less rapidly than x^c then the integral above converges when $\sigma > c$. So to prove the Riemann hypothesis it is sufficient to derive (*). For the converse see Edwards (1974, pp. 261–3).

The third application concerns Farey series (see Problem 21, Chapter 1). The sequence F_n defines a dissection of the unit interval into $t(n)$ unequal subintervals if $n > 2$. Let $f_{k,n}$ be the kth element of F_n, let

$$d_{k,n} = |f_{k,n} - k/t(n)|$$

and, if $[x] = n$, let

$$D_x = \sum_{k=1}^{t(n)} d_{k,n}.$$

D_x is a measure of the unevenness of the Farey dissection of the unit interval. Rather surprisingly

$$D_x = o(x^{1/2+\varepsilon})$$

where $\varepsilon > 0$, is another statement equivalent to the Riemann hypothesis. This was proved by Franel and Landau, see Edwards (1974, pp. 263–7).

For our fourth application we return to the problem of finding the least quadratic non-residue of a prime, see Section 3, Chapter 4. In this case we require the 'extended' Riemann hypothesis which states that if χ is a character, then all the zeros of the function

$$L(s, \chi) = \sum_{n=1}^{\infty} \frac{\chi(n)}{n^s}$$

in the strip $0 \le \sigma \le 1$ lie on the line $s = 1/2 + it$. It has been shown that if this extended hypothesis holds when χ is the Legendre symbol, then $N(p)$, the least positive quadratic non-residue of a prime p, satisfies

$$N(p) = O(\ln^2 p).$$

For further details see p. 61 and Chowla (1965). This property can be used to provide an efficient method for testing for primality of an integer; see Pollard (1971) for the basic algorithm, Miller (1975) for the connection with the Riemann hypothesis, and Section 2, Chapter 1, for an introduction to this topic.

Lastly an interesting application of the extended hypothesis arose in quadratic form theory. Gauss's conjecture states that, for negative discriminants d, $h(d) \to \infty$ as $d \to -\infty$ (see p. 188). Hecke derived Gauss's conjecture on the assumption that the hypothesis is true, whilst Heilbronn proved the conjecture assuming that a zero occurred off the critical line. Thus, although essential use of the hypothesis was made in the proofs, Gauss's conjecture is independent of it. Later Siegel gave a more direct proof. A good account can be found in Goldfeld (1985).

Many further results depend on the Riemann hypothesis or one of its extended versions, see Titchmarsh (1951).

Bernoulli numbers

To end this section we introduce the Bernoulli numbers and use them to evaluate the zeta function at some integer arguments, there are many other applications.

DEFINITION The sequence $B_0, B_1, B_2, \ldots$ of rational numbers, called the *Bernoulli numbers*, is given by $B_0 = 1$ and, if $m > 0$,

$$(m+1)B_m = -\sum_{j=0}^{m-1} \binom{m+1}{j} B_j.$$

So we have

$$\begin{aligned} 2B_1 &= -1 \\ 3B_2 &= -3B_1 - 1 \\ 4B_3 &= -6B_2 - 4B_1 - 1 \end{aligned}$$

and so on. Hence

$$B_1 = -1/2,\ B_2 = 1/6,\ B_3 = 0,\ B_4 = -1/30,\ B_5 = 0,\ B_6 = 1/42, \ldots.$$

We shall see below that, for all positive integers n, $B_{2n+1} = 0$ and B_{2n} alternates in sign as n increases. First we give an equivalent definition.

LEMMA 3.4 *If we expand $t/(\mathrm{e}^t - 1)$ as a power series in t then*

$$\frac{t}{\mathrm{e}^t - 1} = \sum_{n=0}^{\infty} \frac{B_n t^n}{n!}.$$

Proof. Let $t/(e^t-1)=\sum_{n=0}^{\infty} b_n t^n/n!$, then multiplying by e^t-1 we have

$$t=\sum_{m=1}^{\infty}\frac{t^m}{m!}\sum_{n=0}^{\infty}\frac{b_n t^n}{n!}.$$

Equating coefficients we have $b_0=1$ and (if $k+1=m+n$)

$$\frac{1}{(k+1)!}\sum_{j=0}^{k}\binom{k+1}{j}b_j=0.$$

The result follows immediately from the definition. □

COROLLARY 3.5 *If $n>0$, $B_{2n+1}=0$.*

Proof. As $B_1=-1/2$ we have

$$\frac{t}{e^t-1}+\frac{t}{2}=1+\sum_{n=2}^{\infty}\frac{B_n t^n}{n!},$$

but

$$\frac{t}{e^t-1}+\frac{t}{2}=\frac{t(e^t+1)}{2(e^t-1)}$$

which is an even function of t. Hence the odd power coefficients $B_3/3!$, $B_5/5!, \ldots$ are all zero. □

In the next result we evaluate $\zeta(2t)$ in terms of B_{2t} where t is a positive integer.

THEOREM 3.6 (Euler) *If t is a positive integer,*

$$2(2t)!\,\zeta(2t)=(-1)^{t+1}(2\pi)^{2t}B_{2t}.$$

Proof. A standard result from analysis is

$$\sin x=x\prod_{n=1}^{\infty}(1-x^2/n^2\pi^2).$$

(This can be derived by using the sine formula on p. 58 and the Weierstrass M-test for products.) Taking the logarithmic derivative we have

$$\cot x=\frac{1}{x}-2\sum_{n=1}^{\infty}\frac{x}{n^2\pi^2-x^2}$$

and so, expanding each term in this sum as a geometric series,

$$\begin{aligned}x\cot x&=1-2\sum_{n=1}^{\infty}\sum_{t=1}^{\infty}(x/n\pi)^{2t}\\&=1-2\sum_{t=1}^{\infty}\zeta(2t)(x/\pi)^{2t}.\end{aligned}$$

Further, as $\cot x = i(e^{ix}+e^{-ix})/(e^{ix}-e^{-ix})$, we also have

$$\begin{aligned} x\cot x &= ix(e^{2ix}+1)/(e^{2ix}-1)\\ &= ix + 2ix/(e^{2ix}-1)\\ &= 1+\sum_{t=2}^{\infty} B_t(2ix)^t/t! \end{aligned}$$

by Lemma 3.4. Equating coefficients of x^{2t} $(t>0)$ these equations yield

$$-2\zeta(2t)/\pi^{2t} = (-1)^t 2^{2t} B_{2t}/(2t)!. \qquad \square$$

COROLLARY 3.7 (i) *If* $t>0$, $(-1)^{t+1}B_{2t}>0$.
(ii) $|B_{2t}|>2(t/\pi e)^{2t}$.

Proof. These follow immediately from Theorem 3.6 using the inequality $e^n>n^n/n!$. □

Using this theorem we see that

$$\zeta(2)=2\pi^2 B_2/2! = \pi^2/6,$$

a result we have used before. We also have $\zeta(4)=\pi^4/90$, $\zeta(6)=\pi^6/945$, and so on. No similar expression is known for $\zeta(2n+1)$. Note further that if we accept the functional equation $(*)$ (on p. 226) we can calculate $\zeta(-n)$ for $n\geq 0$. We have $\zeta(-2n)=0$ if $n>0$ (see Problem 18), and the functional equation and Theorem 3.6 give

$$\begin{aligned} \zeta(1-2n) &= 2^{1-2n}\pi^{-2n}\zeta(2n)\Gamma(2n)\sin(\pi/2-n\pi)\\ &= 2^{1-2n}\pi^{-2n}(-1)^{(n-1)}\frac{2^{2n}\pi^{2n}}{2(2n)!}B_{2n}(2n-1)!\,(-1)^n = -B_{2n}/2n. \end{aligned}$$

Some further properties of the Bernoulli numbers are given in the problem section; for more advanced topics see, for example, Ireland and Rosen (1982, Chapter 15).

12.4 Problems 12

1. Let p_n be the nth prime, find positive constants c and c' such that for $n>1$

$$cn\ln n < p_n < c'n\ln n.$$

[Hint. Use the result $\ln x = o(\sqrt{x})$.]

2. Define $\pi(x, r)$ to be the number of positive integers less than x which are not divisible by the first r primes, so we have

$$\pi(x)\leq \pi(x,r)+r.$$

Using Problem 10, Chapter 1, find a formula for $\pi(x,r)$ and so derive the inequality

$$\pi(x) < x\prod_{i=1}^{r}(1-1/p_i)+2^{r+1}.$$

Hence prove that almost all integers are composite by showing that

$$\lim_{x\to\infty} \frac{\pi(x)}{x} = 0.$$

3. Show that if $n > 4$ then $\binom{2n}{n} > \frac{2^{2n}}{2n}$, hence using arguments similar to those in the proof of Theorem 1.3 find constants e and e' so that, for large n,

$$\frac{en}{\ln n} < \pi(2n) - \pi(n) < \frac{e'n}{\ln n}.$$

4. Show that if $n > 6$ then n can be expressed as a sum of distinct primes.

5. Let k, m, and n be positive integers.
   (i) By considering the cases $k < n$ and $k \geq n$ separately, show that

$$\frac{1}{n} + \frac{1}{n+1} + \ldots + \frac{1}{n+k} = m$$

is not soluble.
   (ii) Show that the equation

$$n! = m^k$$

is only soluble if at least one of k, m, or n is 1.

6. Use Problem 1 to show that consecutive primes can be arbitrarily far apart. (See also Problem 16, Chapter 1.)

7. (i) Show that if $0 < x < 1/2$ then $x < -\ln(1-x) < 2x$.
   (ii) Prove directly that $\sum_{p \leq x} 1/p$ diverges using only (i) and Theorem 3.1.

8. Prove that

$$\sum_{n \leq x} \frac{\Lambda(n)}{n} = \ln x + O(1).$$

For a definition of the Mangoldt function Λ, see Problem 13, Chapter 2.

9. Use Lemma 2.2 to prove
   (i) Euler's formula relating $\ln x$ with $\sum_{n \leq x} 1/n$ (see p. viii),
   (ii) $\sum_{n \leq x} n^{-1/2} = 2\sqrt{x} + c + O(x^{-1/2})$,
   (iii) $\sum_{n \leq x} \ln n = x \ln x - x + O(\ln x)$,
   (iv) $\sum_{n \leq x} \ln^2 n = x \ln^2 x - 2x \ln x + 2x + O(\ln^2 x)$,
   (v) $\sum_{n \leq x} \ln^2(x/n) = O(x)$,
   (vi) $\sum_{n \leq x} \frac{\ln n}{n} = \frac{1}{2}\ln^2 x + c_1 + O\left(\frac{\ln x}{x}\right)$,
   (vii) $\sum_{n \leq x} \frac{\Lambda(n)}{n} \ln\frac{x}{n} = \sum_{n \leq x} \frac{\Lambda(n)\ln n}{n} + O(\ln x) = \frac{1}{2}\ln^2 x + O(\ln x)$,

where Λ is the Mangoldt function, see Problem 8 above.

10. (i) Show that

$$\sum_{p\le x}\left(\ln\left(1-\frac{1}{p}\right)+\frac{1}{p}\right)=O(1).$$

(ii) Deduce

$$\prod_{p\le x}\left(1-\frac{1}{p}\right)=\frac{c}{\ln x}+O\left(\frac{1}{\ln^2 x}\right).$$

It can be shown that $c=e^{-\gamma}$ where γ is Euler's constant. This result is known as Mertens' theorem, see for example Hardy and Wright (1954, p. 351).

11. (i) Show that

$$\sum_{pq\le x}\frac{1}{pq}=(\ln\ln x)^2+O(\ln\ln x).$$

[Hint. Use Theorem 2.3 and Problem 7(i).]

(ii) Use Lemma 2.2 to show that

$$\sum_{p\le x}\frac{\ln^2 p}{p}=\frac{1}{2}(\ln x)^2+O(\ln x).$$

(iii) Deduce

$$\sum_{pq\le x}\frac{\ln p\ln q}{pq}=\frac{1}{2}(\ln x)^2+O(\ln x).$$

In this problem the double sums are taken over all (unordered) pairs of prime numbers $\{p, q\}$ where $pq\le x$.

12. (i) Prove that there is a positive constant c such that

$$\phi(n)>\frac{cn}{\ln\ln n}.$$

[Hint. Use (i) of Problem 7 and consider the primes less than $\ln n$ which divide n.]

(ii) Deduce, using Problem 9, Chapter 2,

$$\sigma(n)<\frac{1}{c}n\ln\ln n.$$

13. The number of prime factors of a positive integer n. Let $\omega(n)$ denote the number of distinct prime factors of n, and let $\Omega(n)$ denote the total number of prime factors of n; so if $n=\prod_{i\le k}p_i^{\alpha_i}$, $\omega(n)=k$ and $\Omega(n)=\alpha_1+\ldots+\alpha_k$.

(i) Find three sequences $\{n_i\}$ of positive integers to satisfy (a) $\omega(n_i)=\Omega(n_i)=1$, (b) $\omega(n_i)=1$ and $\lim_{i\to\infty}\Omega(n_i)=\infty$, and (c) $\lim_{i\to\infty}\omega(n_i)=\infty$.

(ii) Using Lemma 2.1, Chapter 2, and Theorems 2.1 and 2.3 show that constants c and c' exist satisfying

$$\sum_{n\le x}\omega(n)=x\ln\ln x+cx+o(x),$$

$$\sum_{n\le x}\Omega(n)=x\ln\ln x+c'x+o(x).$$

[Hint. In the second part treat the prime and prime power terms separately.]

There are no asymptotic formulas for either of these functions [see (i)], but it can be shown that, given $\varepsilon > 0$, we have

$$(1-\varepsilon)\ln\ln n < \frac{\omega(n)}{\Omega(n)} < (1+\varepsilon)\ln\ln n \qquad (*)$$

both hold for almost all n. This is expressed by saying that $\omega(n)$, or $\Omega(n)$, has *normal order* $\ln\ln n$, see Hardy and Wright (1954, pp. 356–8).

(iii) Assuming $(*)$ show that the normal order of $\tau(n)$ is $2^{\ln\ln n} = (\ln n)^{\ln 2}$. (See comments on pp. 22–3.)

14. If $\sigma > 1$ where $s = \sigma + it$, show that $|\zeta(s)| \le \zeta(\sigma)$.

15. Let $S = \sum_{n=m+1}^{k} n^{-s}$ with $s = \sigma + it$ and $\sigma > 0$. By replacing n^{-s} with $[n-(n-1)]/n^s$ (that is, using partial summation), show that

$$S = -s\int_m^k \frac{x-[x]}{x^{s+1}}\,dx + \frac{m^{1-s}-k^{1-s}}{s-1}.$$

Deduce that, if $\sigma > 0$,

$$\zeta(s) = \frac{1}{s-1} + 1 - s\int_1^\infty \frac{x-[x]}{x^{s+1}}\,dx.$$

16. Prove that $\lim_{s\to 1^+}(s-1)\zeta(s) = 1$ and so complete the proof of Theorem 3.2.

17. Use the functional equation to calculate $\zeta(0)$.

18. Show that $\zeta(-2n) = 0$, where n is a positive integer.

19. Let ψ be given by (see Problem 8)

$$\psi(x) = \sum_{n\le x} \Lambda(n).$$

(i) Use Theorem 1.1 to show that $\psi(x) = O(x)$.

If $\sigma > 1$, where $s = \sigma + it$, prove

(ii) $\zeta'(s) = -\sum_{n=1}^{\infty} n^{-s}\ln n$,

(iii) $\zeta'(s)/\zeta(s) = -\sum_{n=1}^{\infty} n^{-s}\Lambda(n) = -s\int_1^\infty \psi(x)x^{-s-1}\,dx$.

20. Calculate the Bernoulli numbers up to B_{16}.

21. Let $S_k(m) = \sum_{n=1}^{m-1} n^k = 1^k + 2^k + \ldots + (m-1)^k$. Show that

$$\sum_{j=0}^{m-1} e^{jt} = \sum_{k=0}^{\infty} S_k(m)t^k/k!,$$

and so derive

$$(k+1)S_k(m) = \sum_{j=0}^{k}\binom{k+1}{j}B_j m^{k+1-j}.$$

Deduce that $S_3(m) = m^2(m-1)^2/4$.

22. The Bernoulli polynomials $B_n(x)$ are given by

$$B_n(x) = \sum_{j=0}^{n} \binom{n}{j} B_j x^{n-j}.$$

Show that

(i) $B_n(0) = B_n$,

(ii) $(k+1)S_k(n) = B_{k+1}(n) - B_{k+1}$,

(iii) $te^{tx}/(e^t - 1) = \sum_{m=0}^{\infty} B_m(x)t^m/m!$,

(iv) $B_n(x+1) - B_n(x) = nx^{n-1}$,

(v) $B_n(1-x) = (-1)^n B_n(x)$,

(vi) use (iv) and (v) to re-prove the property $B_{2n+3} = 0$.

23. Show that $|B_{2n+2}| > |B_{2n}|$ provided $n > 2$.

☆24. The Clausen–von Staudt theorem which determines the denominators of the Bernoulli numbers. Let $f(t) = \sum_{n=0}^{\infty} a_n t^n/n!$ and $g(t) = \sum_{n=0}^{\infty} b_n t^n/n!$. The series f is called *integral* if each a_n is an integer, and we write $f \equiv g \pmod m$ when $a_n \equiv b_n \pmod m$ for all n, and f and g are integral.

(i) Show that if f and g are integral so are f', $\int_0^t f(x)\,dx$, fg, and $f^m/m!$ provided $f(0) = 0$. [Hint. In the last part use induction and the other three parts.]

(ii) Deduce the following: if $m > 4$ and composite,

$$(e^t - 1)^{m-1} \equiv 0 \pmod m,$$

$$(e^t - 1)^3 \equiv 2\sum_{k=1}^{\infty} t^{2k+1}/(2k+1)! \pmod 4,$$

and, if p is a prime,

$$(e^t - 1)^{p-1} \equiv -\sum_{k=1}^{\infty} t^{kp-k}/(kp-k)! \pmod p.$$

In the last part use (i) and Theorems 1.13 and 2.4 of Chapter 3.

(iii) Write $t = \ln(1 + (e^t - 1))$ and so expand $t/(e^t - 1)$ as a power series in $e^t - 1$. Now using (i) and (ii) show that

$$\sum{}^* 1/p + (-1)^{n+1}B_{2n}$$

is an integer where the sum is taken over all primes p satisfying $p - 1 | 2n$. This determines the denominator of B_{2n}.

(iv) Deduce that the denominator of the Bernoulli number B_{2n+2} is divisible by 6.

25. If $k > 0$, show that $n(n^k - 1)B_k$ is an integer.

13

TWO MAJOR THEOREMS ON THE PRIMES

This chapter has a different character to the previous work; it consists entirely of the proofs of two important results—Dirichlet's theorem and the prime number theorem (PNT). A preliminary discussion of these results is given in the Introduction to Chapter 12. The individual methods used in these proofs are similar to those in the rest of this book, but difficulties may arise because of their length and complex structure. It is essential to have an overall view and not to get bogged down in details. As the first result has an easier derivation, its methods should be thoroughly understood before the second proof is tackled.

Dirichlet's theorem has wide applicability. For example we have used it in our proof of the three-square theorem in Chapter 9, and the usual derivation of Hasse's principle (see Chapter 3) also uses it. Some further applications are given in the problem section. PNT is not used extensively outside analytic number theory, but is, of course, the cornerstone of that theory. See Problems 17 and 18 for applications.

Our two results can be combined into one major theorem. Let $\pi(x, a, c)$ be the number of primes less than x which are congruent to a modulo c, then if $(a, c) = 1$

$$\pi(x, a, c) \sim \frac{1}{\phi(c)} \frac{x}{\ln x}.$$

Note that the right-hand side of this expression is independent of a. This result shows that the density of primes in an arithmetic progression is the 'expected' one given PNT. The proof uses a combination of the methods presented below; see LeVeque (1955) for an analytic derivation or Gelfond and Linnik (1965) for an elementary derivation.

The proofs in this chapter are usually described as 'elementary'. This does not mean that there are simple, but it does mean that they avoid such topics as contour integration, analytic continuation, or any topological considerations. The only concepts from real analysis are limits and simple integrals, as well as the real numbers themselves. In fact the proofs can be rewritten avoiding even these topics by using rational approximations for real numbers, finite sums in place of integrals, and ordered pairs of rationals for complex numbers. Both proofs below can be carried out in the system the mathematical logicians call 'elementary arithmetic', a subsystem of Peano arithmetic; for details see Rose (1984). A fairly straightforward, but not 'elementary', proof of PNT is given in Newman (1980); it uses contour integration and analytic continuation.

13.1 Dirichlet's theorem

Throughout this section a and c are fixed positive integers satisfying $(a, c) = 1$. Dirichlet's theorem states:

> There are infinitely many prime numbers in the arithmetic progression $\{a + cn : n = 0, 1, 2, \ldots\}$.

This will be derived by showing that a series S, which is defined on the primes in the arithmetic progression, diverges and hence must have infinitely many terms. We use characters to code the arithmetic progressions and the Mangoldt function to relate the set of primes to the set of positive integers.

We begin by listing the subsidiary results required in the main proofs. All characters used in this section are defined modulo c, where $c > 1$. The Mangoldt function Λ is given by

$$\Lambda(n) = \begin{cases} \ln p & \text{if } n = p^k \text{ for some positive integer } k \\ 0 & \text{otherwise.} \end{cases} \tag{1}$$

As $\sum_{d|p^k} \Lambda(d) = k \ln p$ and $\ln n = \sum_{i=1}^{m} k_i \ln p_i$ if $n = \prod_{i=1}^{m} p_i^{k_i}$ we have

$$\sum_{d|n} \Lambda(d) = \ln n, \tag{2}$$

and, applying the Möbius inversion formula and Lemma 1.4(ii), Chapter 2,

$$\Lambda(n) = \sum_{d|n} \mu(d) \ln \frac{n}{d} = -\sum_{d|n} \mu(d) \ln d. \tag{3}$$

A straightforward extension (Problem 8) of Theorem 2.1, both of Chapter 12, gives

$$\sum_{n \le x} \frac{\Lambda(n)}{n} = \ln x + O(1) \tag{4}$$

and, if χ_0 is the principal character modulo c,

$$\sum_{n \le x} \frac{\chi_0(n)\Lambda(n)}{n} = \ln x + O(1). \tag{5}$$

[Equation (5) follows from (4), the reader should check this by noting that c has only finitely many prime factors.] The character results we need are as follows; for the proofs see Corollary 3.5 and Problem 26, Chapter 5. Let a^* satisfy $aa^* \equiv 1 \pmod c$, then

$$\sum_{\chi} \chi(a^*)\chi(n) = \begin{cases} \phi(c) & \text{if } n \equiv a \pmod c \\ 0 & \text{otherwise.} \end{cases} \tag{6}$$

$$\left|\sum_{n=k}^{m} \chi(n)\right| < \phi(c), \text{ provided } \chi \neq \chi_0. \tag{7}$$

If $\{f_n\}$ is a decreasing sequence of non-negative reals and $\chi \neq \chi_0$,

$$\left|\sum_{n=k}^{m} \chi(n) f_n\right| < 2\phi(c) f_k. \tag{8}$$

Our last preliminary result is

$$\sum_{n \le x} n^{-1/2} = 2\sqrt{x} + c_1 + O(x^{-1/2}) \tag{9}$$

where c_1 is a constant, see Problem 9(ii), Chapter 12.

A sketch of the main proof

The basic step is the replacement of a series $S(x)$ defined on the primes in the arithmetic progression $\{a + cn\}$ less than $x + 1$, by a series $U(x)$ defined on the positive integers less than $x + 1$. Let

$$U(x) = \sum_{\chi} \chi(a^*) \sum_{n \le x} \frac{\chi(n)\Lambda(n)}{n}.$$

Combining the inner and outer sums we have by (6)

$$\begin{aligned} U(x) &= \phi(c) \sum_{\substack{n \le x \\ n \equiv a(c)}} \frac{\Lambda(n)}{n} \\ &= \phi(c) \sum_{\substack{p \le x \\ p \equiv a(c)}} \frac{\ln p}{p} + O(1) = S(x) + O(1), \end{aligned}$$

by (1), as the sum of the remaining terms in the first series is bounded. Now $U(x)$ is a finite sum of terms of the form

$$T(\chi, x) = \sum_{n \le x} \frac{\chi(n)\Lambda(n)}{n}. \tag{10}$$

By (5), $T(\chi_0, x) \to \infty$ as $x \to \infty$, where χ_0 is the principal character, and so if we can show that $T(\chi, x)$ is bounded as $x \to \infty$ when $\chi \neq \chi_0$ then $U(x) \to \infty$ as $x \to \infty$, and Dirichlet's theorem follows. To do this we first need to show that

$$L(\chi) = \sum_{n=1}^{\infty} \frac{\chi(n)}{n} \neq 0 \tag{11}$$

for all $\chi \neq \chi_0$.

In Dirichlet's original proof he used the Euler product formula (p. 225) instead of the Mangoldt function. The method used here has the advantage that it avoids the complex logarithm function. Both proofs require the result (11), see Problem 3.

We begin the main derivation by proving

THEOREM 1.1 *Let χ be a real character modulo c such that $\chi \neq \chi_0$, then $L(\chi) \neq 0$.*

We shall give two proofs. The first was given by Dirichlet in his 1837 memoir, it uses a result from quadratic form theory. The second is due to Shapiro and has the advantage that it is self-contained.

First proof. By Theorem 3.6, Chapter 5, χ can be replaced by a Kronecker symbol, that is

$$\chi(n) = (d/n)$$

for some suitably chosen integer d (congruent to 0 or 1 modulo 4), and so

$$L(\chi) = \sum_{n=1}^{\infty} \frac{(d/n)}{n}.$$

The term on the right-hand side is the sum $K(d)$ which was used in the derivation of the formula for the class number $h(d)$ of the class of quadratic forms with discriminant d. As $h(d) \geq 1$ for all d (there is always at least one class of forms with discriminant d) we have $L(\chi) \neq 0$ by Theorem 3.7, Chapter 10.

Second proof. Note first that, by (8) above, the series for $L(\chi)$ converges. Define f by

$$f(n) = \sum_{d|n} \chi(d).$$

We claim

$$\begin{aligned} f(n) &\geq 0 \qquad \text{for all } n, \text{ and} \\ f(n) &\geq 1 \qquad \text{if } n \text{ is a square.} \end{aligned} \tag{12}$$

By Theorem 1.1, Chapter 2, f is multiplicative, so we need to consider

$f(p^\alpha)$. We have

$$\begin{aligned}&\text{if } \chi(p)=1,\ f(p^\alpha)=1+1+\ldots+1\geq 1,\\ &\text{if } \chi(p)=0,\ f(p^\alpha)=1+0+\ldots+0=1,\\ &\text{if } \chi(p)=-1 \text{ and } \alpha \text{ is odd, } f(p^\alpha)=1-1+\ldots+1-1=0,\\ &\text{if } \chi(p)=-1 \text{ and } \alpha \text{ is even, } f(p^\alpha)=1-1+\ldots+1=1,\end{aligned}$$

and (12) follows.

Consider the series

$$F(x)=\sum_{n\leq x}\frac{f(n)}{\sqrt{n}}.$$

By (12) each term in this series is non-negative and

$$F(x)\geq\sum_{n^2\leq x}\frac{1}{(n^2)^{1/2}}=\sum_{n\leq\sqrt{x}}\frac{1}{n}\to\infty \tag{13}$$

as $x\to\infty$. On the other hand we show that

$$F(x)=2L(\chi)\sqrt{x}+O(1). \tag{14}$$

The theorem follows because (14) contradicts (13) if $L(\chi)=0$. We derive (14) by splitting the series for $F(x)$ into two parts and estimating each part separately. We have by DSI (see p. 23)

$$F(x)=\sum_{n\leq x}n^{-1/2}\sum_{d\mid n}\chi(d)=\sum_{rs\leq x}\frac{\chi(r)}{\sqrt{(rs)}}.$$

Split the set of points in the r, s plane with positive integer coordinates under the curve $rs=x$ into two subsets, those above the horizontal line $s=\sqrt{x}$ and those below the line; we have (upper set first)

$$\begin{aligned}F(x)&=\left\{\sum_{s\leq\sqrt{x}}\sum_{r=[\sqrt{x}]+1}^{x/s}+\sum_{r\leq\sqrt{x}}\sum_{s\leq x/r}\right\}\frac{\chi(r)}{\sqrt{(rs)}}\\ &=\sum_{s\leq\sqrt{x}}s^{-1/2}\sum_{r=[\sqrt{x}]+1}^{x/s}\frac{\chi(r)}{\sqrt{r}}+\sum_{r\leq\sqrt{x}}\frac{\chi(r)}{\sqrt{r}}\sum_{s\leq x/r}s^{-1/2}=S_1+S_2.\end{aligned}$$

Now by (8)

$$\begin{aligned}S_1&=\sum_{s\leq\sqrt{x}}s^{-1/2}O(x^{-1/4})\\ &=O(x^{1/4})O(x^{-1/4})=O(1)\end{aligned}$$

by (9). Secondly using (9) again we have

$$S_2 = \sum_{r \le \sqrt{x}} \frac{\chi(r)}{\sqrt{r}} \{2\sqrt{(x/r)} + c_1 + O(\sqrt{(r/x)})\}$$

$$= 2\sqrt{x} \sum_{r \le \sqrt{x}} \frac{\chi(r)}{r} + c_1 O(1) + \frac{O(1)}{\sqrt{x}}$$

$$= 2\sqrt{x}\left\{L(\chi) - \sum_{r=[\sqrt{x}]+1}^{\infty} \frac{\chi(r)}{r}\right\} + O(1) = 2L(\chi)\sqrt{x} + O(1)$$

by (8) again. Combining this with the estimate for S_1 the results follows. □

Our next two results enable us to show that $L(\chi) \neq 0$ for all $\chi \neq \chi_0$. Note we write $\sum_{t>k} f(t)$ for the series $\sum_{t=k+1}^{\infty} f(t)$.

LEMMA 1.2 *Let χ be a non-principal character.*

(i) $L(\chi)T(\chi, x) = O(1)$,

(ii) $T(\chi, x) = -\ln x + L(\chi)A(x) + O(1)$ *for some term* $A(x)$ *given below. See* (10) *for the definition of* T.

Proof. (i) Consider the series $V(x) = \sum_{n \le x} \frac{\chi(n) \ln n}{n}$. We have

$$V(x) = \sum_{n \le x} \frac{\chi(n)}{n} \sum_{d|n} \Lambda(d) \qquad \text{by (2)}$$

$$= \sum_{m \le x} \frac{\chi(m)\Lambda(m)}{m} \sum_{k \le x/m} \frac{\chi(k)}{k} \qquad \text{as } \chi \text{ is multiplicative and using DSI,}$$

$$= L(\chi)T(\chi, x) - R_1(x) \qquad \text{using (10) and (11),}$$

where

$$R_1(x) = \sum_{m \le x} \frac{\chi(m)\Lambda(m)}{m} \sum_{k > x/m} \frac{\chi(k)}{k}.$$

Now, as $|\chi(m)| \le 1$,

$$|R_1(x)| \le \sum_{m \le x} \frac{\Lambda(m)}{m} \left| \sum_{k > x/m} \frac{\chi(k)}{k} \right| = \sum_{m \le x} \frac{\Lambda(m)}{m} O(m/x) \qquad \text{by (8)}$$

$$= O\left(\frac{1}{x} \sum_{m \le x} \Lambda(m)\right) = O(1)$$

by (5). But $V(x)$ is bounded for all x as the series $\sum_{n=1}^{\infty} \frac{\chi(n) \ln n}{n}$ is convergent by (8), and so (i) follows.

(ii) This is derived in a similar manner. Consider the series $W(x)=\sum_{n\le x}\frac{\chi(n)}{n}\sum_{d|n}\mu(d)\ln\frac{x}{d}$. We have

$$\begin{aligned}W(x)&=\sum_{n\le x}\frac{\chi(n)}{n}\Big\{\ln x\sum_{d|n}\mu(d)-\sum_{d|n}\mu(d)\ln d\Big\}\\&=\ln x+T(\chi,x)\end{aligned}$$

by Lemma 1.4, Chapter 2, and (3). Rewriting the sum defining $W(x)$ we have by DSI, as χ is multiplicative,

$$\begin{aligned}W(x)&=\sum_{m\le x}\frac{\chi(m)\mu(m)}{m}\ln\frac{x}{m}\sum_{k\le x/m}\frac{\chi(k)}{k}\\&=L(\chi)\sum_{m\le x}\frac{\chi(m)\mu(m)}{m}\ln\frac{x}{m}-R_2(x)\end{aligned}\tag{15}$$

where

$$R_2(x)=\sum_{m\le x}\frac{\chi(m)\mu(m)}{m}\ln\frac{x}{m}\sum_{k>x/m}\frac{\chi(k)}{k}.$$

Now as $|\chi(m)\mu(m)|\le 1$ we have

$$\begin{aligned}|R_2(x)|&\le\sum_{m\le x}\frac{1}{m}\ln\frac{x}{m}\left|\sum_{k>x/m}\frac{\chi(k)}{k}\right|\\&\le\sum_{m\le x}\frac{1}{m}\ln\frac{x}{m}O(m/x)\qquad\text{by (8)}\\&=O\Big(\frac{1}{x}\sum_{m\le x}\ln\frac{x}{m}\Big)=O(1).\end{aligned}$$

This last inequality is given by

$$\sum_{m\le x}\ln\frac{x}{m}\le\ln\frac{x^x}{x!}\le\ln \mathrm{e}^x=x.$$

Using (15), equation (ii) follows if we let $A(x)=\sum_{m\le x}\frac{\chi(m)\mu(m)}{m}\ln\frac{x}{m}$. □.

Now define $t(\chi)$ by

$$t(\chi)=\begin{cases}-1 & \text{if } L(\chi)=0\\ 0 & \text{if } L(\chi)\ne 0.\end{cases}$$

Combining the results in Lemma 1.2 we have

$$T(\chi,x)=t(\chi)\ln x+O(1)$$

provided $\chi \neq \chi_0$. Using (5) for the case $\chi = \chi_0$ this gives

$$\ln x + \sum_{\substack{\chi \\ \chi \neq \chi_0}} t(\chi) \ln x + O(1) = \sum_{\chi} T(\chi, x)$$

$$= \sum_{n \leq x} \frac{\Lambda(n)}{n} \sum_{\chi} \chi(n) \geq 0 \qquad (16)$$

where the last inequality is given by Theorem 3.4, Chapter 5.

THEOREM 1.3 *Let* $\chi \neq \chi_0$.
(i) $L(\chi) \neq 0$.
(ii) $T(\chi, x) = O(1)$.

Proof. (i) Suppose first $L(\chi) = 0$ for a non-real character χ. If $\bar{\chi}$ is its complex conjugate then we also have $L(\bar{\chi}) = 0$, and so

$$t(\chi) = t(\bar{\chi}) = -1 \quad \text{and} \quad \sum_{\substack{\chi \\ \chi \neq \chi_0}} t(\chi) \leq -2.$$

This contradicts (16) for large x, hence $L(\chi) \neq 0$ for non-real characters χ (for a second proof of this result see Problem 4). Using Theorem 1.1 this gives (i). (ii) follows by Lemma 1.2(i). □

The main result of this section is

THEOREM 1.4 *If* $c > 1$ *and* $(a, c) = 1$ *then*

$$\sum_{\substack{p \leq x \\ p \equiv a(c)}} \frac{\ln p}{p} = \frac{\ln x}{\phi(c)} + O(1).$$

The reader should compare this result with Theorem 2.1, Chapter 12, and also note that the right-hand side of the above equation is independent of a.

Proof. We show first

$$\sum_{\substack{n \leq x \\ n \equiv a(c)}} \frac{\Lambda(n)}{n} = \sum_{\substack{p \leq x \\ p \equiv a(c)}} \frac{\ln p}{p} + O(1). \qquad (17)$$

By (1) the left-hand side is bounded above by

$$\sum_{\substack{p \leq x \\ p \equiv a(c)}} \frac{\ln p}{p} + \sum_{\substack{p^2 \leq x \\ p^2 \equiv a(c)}} \frac{\ln p}{p^2} + \ldots$$

$$< \sum_{\substack{p \leq x \\ p \equiv a(c)}} \frac{\ln p}{p} + \sum_{p \leq x} \frac{\ln p}{p^2}\left(1 + \frac{1}{p} + \ldots\right)$$

$$= \sum_{\substack{p \leq x \\ p \equiv a(c)}} \frac{\ln p}{p} + \sum_{p \leq x} \frac{\ln p}{p(p-1)} = \sum_{\substack{p \leq x \\ p \equiv a(c)}} \frac{\ln p}{p} + O(1)$$

as the series $\sum_{n \le x} \frac{\ln n}{n(n-1)}$ is convergent. (17) follows because the sum on its left-hand side is larger than the sum on its right-hand side.

Now if $aa^* \equiv 1 \pmod{c}$ we have, as $(a^*, c) = 1$,

$$\sum_{\chi} \chi(a^*)T(\chi, x) = \ln x + O(1)$$

by (5) and Theorem 1.3(ii). But by (6)

$$\sum_{\chi} \chi(a^*)T(\chi, x) = \phi(c) \sum_{\substack{n \le x \\ n \equiv a(c)}} \frac{\Lambda(n)}{n} + O(1)$$

$$= \phi(c) \sum_{\substack{p \le x \\ p \equiv a(c)}} \frac{\ln p}{p} + O(1)$$

by (17). Our result, and hence Dirichlet's theorem, follows. □

13.2 PNT: Preliminaries and Selberg's theorem

Before proceeding the reader should refer to Section 2, Chapter 2, for the basic definitions and the notions DSI and partial summation.

There are a number of propositions which are equivalent to PNT in the sense that each can be deduced from any of the remaining ones using elementary methods similar to those used in the first two sections of Chapter 12. We list the most important in the theorem below. Note that $f(x) \sim g(x)$ stands for

$$\lim_{x \to \infty} \frac{f(x)}{g(x)} = 1.$$

DEFINITION For $x \ge 1$ let the functions θ and ψ be given by

$$\theta(x) = \sum_{p \le x} \ln p$$

$$\psi(x) = \sum_{p^k \le x} \ln p$$

where the second sum is defined over all primes p and all k such that $p^k \le x$. We have immediately

$$\psi(x) = \sum_{n \le x} \Lambda(n)$$

and

$$\psi(x) = \theta(x) + \theta(x^{1/2}) + \ldots + \theta(x^{1/k})$$

where $k = [\ln x / \ln 2]$ (for if $t > \ln x / \ln 2$ then $x^{1/t} < 2$, but $\theta(y) = 0$ if $y < 2$).

THEOREM 2.1 *The following propositions are equivalent in the sense above.* c_1 *and* c_2 *are constants.*

(i) $\pi(x) \sim x/\ln x$.
(ii) $\theta(x) \sim x$.
(iii) $\psi(x) \sim x$.
(iv) $\sum_{n \le x} \Lambda(n)/n = \ln x + c_1 + o(1)$.
(v) $\sum_{p \le x} \ln p/p = \ln x + c_2 + o(1)$.
(vi) $M(x) = \sum_{n \le x} \mu(n) = o(x)$.
(vii) $\sum_{n \le x} \mu(n)/n = o(1)$.
(viii) $p_n \sim n \ln n$, p_n the nth prime.

Proof. We show that (ii) is equivalent to (iii), and (ii) implies (i). The remaining equivalences will be considered in the problem section. The first equivalence follows from

$$0 \le \psi(x) - \theta(x) \le 6\sqrt{x}\,\frac{\ln x}{\ln 2}. \qquad (*)$$

To prove this we have (see above)

$$0 \le \psi(x) - \theta(x) \le \sum_{2 \le j \le k} \theta(x^{1/j})$$

where $k = [\ln x/\ln 2]$. Now clearly $\theta(x) < \pi(x)\ln x < 6x$, by Theorem 1.1, Chapter 12, and so

$$\psi(x) - \theta(x) < \sum_{2 \le j \le k} 6x^{1/j} < 6kx^{1/2} \le 6\sqrt{x}\,\frac{\ln x}{\ln 2}.$$

The inequality $(*)$ shows that (ii) is equivalent to (iii) as $(\sqrt{x})\ln x = o(x)$.

To show that (ii) implies (i) we use Lemma 2.2, Chapter 12, and set $t_n = n$, $z_n = \ln n$ if n is a prime and $z_n = 0$ otherwise, and $f(t) = (\ln t)^{-1}$. This gives $Z(x) = \theta(x)$ and

$$\pi(x) = \sum_{n \le x} z_n f(t_n) = \frac{\theta(x)}{\ln x} + \int_2^x \frac{\theta(t)\,dt}{t \ln^2 t};$$

hence we need to show that

$$S(x) = \frac{\ln x}{x} \int_2^x \frac{\theta(t)\,dt}{t \ln^2 t} \to 0$$

as $x \to \infty$. We have

$$\int_2^x \frac{dt}{\ln^2 t} = \left(\int_2^{\sqrt{x}} + \int_{\sqrt{x}}^x\right) \frac{dt}{\ln^2 t} \le \frac{-2 + \sqrt{x}}{\ln^2 2} + \frac{x - \sqrt{x}}{\ln^2 \sqrt{x}} = O\left(\frac{x}{\ln^2 x}\right),$$

and this gives, using (ii),

$$S(x) = O\left(\frac{\ln x}{x}\int_2^x \frac{dt}{\ln^2 t}\right) = O\left(\frac{1}{\ln x}\right) \to 0$$

as $x \to \infty$. □

In Problem 10 we show that (i) implies (ii). From now on we shall take (iii) as our standard statement of PNT.

Selberg's symmetry formula

Our proof of PNT is based on this celebrated formula which first appeared in print in 1948. The derivation begins with

LEMMA 2.2 (i) $\sum_{n \le x} \frac{\mu(n)}{n} \ln \frac{x}{n} = O(1)$

(ii) $\sum_{n \le x} \frac{\mu(n)}{n} \ln^2 \frac{x}{n} = 2 \ln x + O(1).$

Proof. Note first that if f and g are number-theoretic functions and

$$f(x) = \sum_{n \le x} g(x/n)$$

then (∗)

$$g(x) = \sum_{n \le x} \mu(n) f(x/n).$$

For we have

$$\sum_{n \le x} \mu(n) f(x/n) = \sum_{n \le x} \mu(n) \sum_{m \le x/n} g(x/mn)$$

$$= \sum_{mn \le x} \mu(n) g(x/mn) = \sum_{r \le x} g(x/r) \sum_{n \mid r} \mu(n) = g(x)$$

by DSI and Lemma 1.4, Chapter 2.

(i) If $f(x) = \sum_{n \le x} x/n$ we have by the Euler formula (see Problem 9, Chapter 12)

$$f(x) = x \ln x + \gamma x + O(1)$$

and so, by (∗) with $g(x) = x$,

$$x = \sum_{n \le x} \mu(n) \left[\frac{x}{n} \ln \frac{x}{n} + \frac{\gamma x}{n} + O(1)\right]$$

$$= x \sum_{n \le x} \frac{\mu(n)}{n} \ln \frac{x}{n} + x \sum_{n \le x} \frac{\mu(n)}{n} + O(x).$$

Now (i) follows by Problem 2, Chapter 2.

(ii) We use a similar argument with $g(x) = x \ln x$ in $(*)$. Then

$$f(x) = \sum_{n \le x} \frac{x}{n} \ln \frac{x}{n} = x \ln x \sum_{n \le x} \frac{1}{n} - x \sum_{n \le x} \frac{\ln n}{n}$$

$$= x \ln x(\ln x + \gamma + O(1/x)) - x\left(\frac{1}{2}\ln^2 x + c + O\left(\frac{\ln x}{x}\right)\right)$$

using Euler's formula and Problem 9(vi), Chapter 12. Hence

$$\sum_{n \le x} \frac{x}{n} \ln \frac{x}{n} = \frac{x}{2} \ln^2 x + \gamma x \ln x - cx + O(\ln x)$$

and $(*)$ gives

$$x \ln x = \sum_{n \le x} \mu(n)\left(\frac{x}{2n} \ln^2 \frac{x}{n} + \frac{\gamma x}{n} \ln \frac{x}{n} - \frac{cx}{n} + O\left(\ln \frac{x}{n}\right)\right)$$

$$= \frac{x}{2} \sum_{n \le x} \frac{\mu(n)}{n} \ln^2 \frac{x}{n} + \gamma x \sum_{n \le x} \frac{\mu(n)}{n} \ln \frac{x}{n} - cx \sum_{n \le x} \frac{\mu(n)}{n} + O\left(\sum_{n \le x} \ln \frac{x}{n}\right).$$

Now by (i), Problem 2, Chapter 2, and Problem 9(iii), Chapter 12, the second, third, and fourth terms in this expression are $O(x)$, and so equation (ii) follows. □

THEOREM 2.3 *Selberg's symmetry formula*

$$\sum_{n \le x} \Lambda(n) \ln n + \sum_{rs \le x} \Lambda(r)\Lambda(s) = 2x \ln x + O(x).$$

Proof. Note first that, by Problem 15, Chapter 2, we have $(1 * \Lambda) * \Lambda = 1 * (\Lambda * \Lambda)$, where 1 is the unit function $1(x) = 1$, and so

$$\sum_{d|n} \Lambda(n/d) \ln d = \sum_{d|n} \left(\sum_{r|d} \Lambda(r)\right)\Lambda(n/d) = \sum_{d|n} \sum_{r|d} \Lambda(r)\Lambda(d/r).$$

Now by (2) on p. 237 and as $\ln n = \ln d + \ln n/d$, we have

$$\ln^2 n = \sum_{d|n} \Lambda(d)(\ln d + \ln n/d)$$

$$= \sum_{d|n} \Lambda(d) \ln d + \sum_{d|n} \Lambda(n/d) \ln d \qquad \text{by (2) on p. 17}$$

$$= \sum_{d|n} \left(\Lambda(d) \ln d + \sum_{r|d} \Lambda(r)\Lambda(d/r)\right)$$

by above. The Möbius inversion formula (Theorem 1.5, Chapter 2) gives

$$\Lambda(n) \ln n + \sum_{d|n} \Lambda(d)\Lambda(n/d) = \sum_{d|n} \mu(d) \ln^2 (n/d)$$

and, summing over $n \leq x$ and using DSI,

$$\sum_{n \leq x} \Lambda(n) \ln n + \sum_{rs \leq x} \Lambda(r)\Lambda(s) = \sum_{n \leq x} \sum_{d|n} \mu(d) \ln^2 (n/d). \qquad (*)$$

The right-hand side of this equation equals, by DSI again and with a change of bound variables,

$$\sum_{rs \leq x} \mu(r) \ln^2 s = \sum_{r \leq x} \mu(r) \sum_{s \leq x/r} \ln^2 s$$

$$= \sum_{d < x} \mu(d) \left\{ \frac{x}{d} \ln^2 \frac{x}{d} - \frac{2x}{d} \ln \frac{x}{d} + \frac{2x}{d} + O\left(\ln^2 \frac{x}{d} \right) \right\}$$

by Problem 9(iv), Chapter 12,

$$= x \sum_{d \leq x} \frac{\mu(d)}{d} \ln^2 \frac{x}{d} - 2x \sum_{d \leq x} \frac{\mu(d)}{d} \ln \frac{x}{d} + 2x \sum_{d \leq x} \frac{\mu(d)}{d} + O\left(\sum_{d \leq x} \ln^2 \frac{x}{d} \right).$$

By Lemma 2.2(i), Problem 2, Chapter 2, and Problem 9(v), Chapter 12, the second, third, and fourth terms are of order x, and by Lemma 2.2(ii)

$$\sum_{d \leq x} \frac{\mu(d)}{d} \ln^2 \frac{x}{d} = 2x \ln x + O(x).$$

The result follows using $(*)$. □

For our first application we give a second proof of the result below, the first proof was given in Problem 19, Chapter 12.

Corollary 2.4 $\psi(x) = O(x)$.

Proof. Replacing x by $x/2$ in Theorem 2.3 and subtracting we have

$$\sum_{x/2 < n \leq x} \Lambda(n) \ln n + \sum_{x/2 < rs \leq x} \Lambda(r)\Lambda(s) = x \ln x + O(x),$$

so, as Λ never takes negative values,

$$x \ln x + O(x) > \sum_{x/2 < n \leq x} \Lambda(n) \ln n \geq (\psi(x) - \psi(x/2)) \ln(x/2).$$

This gives, if $x > 4$,

$$\psi(x) - \psi(x/2) < 2x + O(x/\ln x).$$

Substituting $x/2^k$ for x, $k = 0, 1, \ldots, [\ln x/\ln 2] - 1$, and adding we obtain

$$\psi(x) < 4x + O(x) = O(x)$$

as $\sum_{n=1}^{\infty} 2^{-n} = 2.$ □

A further application is given in Problem 14.

Selberg's formula has many equivalent formulations. We shall give two which will be required in our proof of PNT; two further particularly simple formulations are given in Problem 15.

COROLLARY 2.5

$$\psi(x)\ln x + \sum_{rs \le x} \Lambda(r)\Lambda(s) = 2x\ln x + O(x).$$

Proof. By Lemma 2.2, Chapter 12, with $t_n = n$, $z_n = \Lambda(n)$ and $f(t) = \ln t$, we have

$$\sum_{n \le x} \Lambda(n)\ln n = \psi(x)\ln x - \int_1^x \frac{\psi(t)}{t}\,dt = \psi(x)\ln x + O(x) \qquad (*)$$

by Corollary 2.4, now use Theorem 2.3. □

THEOREM 2.6(i) $\sum_{rs \le x} \Lambda(r)\Lambda(s)\ln s = \frac{\ln x}{2} \sum_{rs \le x} \Lambda(r)\Lambda(s) + O(x\ln x)$

(ii) $\psi(x)\ln^2 x = 2\sum_{rs \le x} \Lambda(r)\Lambda(s)\psi(x/rs) + O(x\ln x).$

Proof. (i) We have, by $(*)$ in the proof of Corollary 2.5,

$$\begin{aligned}\sum_{rs \le x} \Lambda(r)\Lambda(s)\ln s &= \sum_{r \le x} \Lambda(r) \sum_{s \le x/r} \Lambda(s)\ln s \\ &= \sum_{r \le x} \Lambda(r)[\psi(x/r)\ln(x/r) + O(x/r)] \\ &= \ln x \sum_{r \le x} \Lambda(r)\psi(x/r) - \sum_{r \le x} \Lambda(r)\psi(x/r)\ln r + O(x\ln x) \\ &= \ln x \sum_{r \le x} \Lambda(r)\psi(x/r) - \sum_{rs \le x} \Lambda(r)\Lambda(s)\ln r + O(x\ln x)\end{aligned}$$

as $\psi(x/r) = \sum_{t | x/r} \Lambda(t)$ and using DSI. This gives (i).

(ii) Selberg's formula (Theorem 2.3), multiplied by the factor $\Lambda(m)$ and with x/m for x, gives

$$\Lambda(m) \sum_{n \leq x/m} \Lambda(n) \ln n + \Lambda(m) \sum_{rs \leq x/m} \Lambda(r)\Lambda(s) = 2\Lambda(m)\frac{x}{m}\ln\frac{x}{m} + O\left(\frac{x}{m}\Lambda(m)\right)$$

and, summing over $m \leq x$ and using Problem 8, Chapter 12,

$$\sum_{mn \leq x} \Lambda(m)\Lambda(n) \ln n + \sum_{rst \leq x} \Lambda(r)\Lambda(s)\Lambda(t) = 2x \sum_{n \leq x} \frac{\Lambda(n)}{n} \ln\frac{x}{n} + O(x \ln x)$$
$$= x \ln^2 x + O(x \ln x)$$

by Problem 9(vii), Chapter 12. Thus using (i) we have

$$\frac{1}{2} \ln x \sum_{rs \leq x} \Lambda(r)\Lambda(s) + \sum_{rst \leq x} \Lambda(r)\Lambda(s)\Lambda(t) = x \ln^2 x + O(x \ln x).$$

We also have, multiplying Corollary 2.5 by $\frac{1}{2}\ln x$,

$$\frac{1}{2}\psi(x)\ln^2 x + \frac{1}{2}\ln x \sum_{rs \leq x} \Lambda(r)\Lambda(s) = x \ln^2 x + O(x \ln x).$$

Proposition (ii) follows by taking the difference of these two equations and noting that

$$\sum_{rst \leq x} \Lambda(r)\Lambda(s)\Lambda(t) = \sum_{rs \leq x} \Lambda(r)\Lambda(s) \sum_{t \leq x/rs} \Lambda(t) = \sum_{rs \leq x} \Lambda(r)\Lambda(s)\psi(x/rs). \quad \square$$

Our final preliminary result is a type of continuity condition for ψ.

THEOREM 2.7 *If x and y are positive and $0 < c < x/y < c'$, where c and c' are constants, then*

$$|\psi(x) - \psi(y)| \leq 2|x - y| + O(x/\ln x)$$

where the error term depends on c and c' only.

Proof. Assume $x > y$, then by Corollary 2.5, we have

$$\psi(x)\ln x - \psi(y)\ln y + \sum_{n \leq x} \Lambda(n)(\psi(x/n) - \psi(y/n))$$
$$= 2x \ln x - 2y \ln y + O(x)$$

or, as the sum in this expression is non-negative,

$$(\psi(x) - \psi(y))\ln x \leq 2(x - y)\ln x + (2y - \psi(y))\ln(x/y) + O(x).$$

But, by Corollary 2.4, we have

$$(2y - \psi(y)) \ln(x/y) - O(y) \ln(x/y) - O(x)$$

as x/y lies between c and c'. The result follows. □

13.3 PNT: The main proof

There are a number of derivations of PNT, none of which are direct. The proof given below is due to Selberg with simplifications by Shapiro; it is completely elementary in character; in particular it avoids the use of upper and lower limits which are employed in some non-analytic proofs. Further elementary proofs have been given by, amongst others, Erdös (for a good exposition see Shapiro, 1983) and Gelfond (see Gelfond and Linnik, 1965).

Let

$$\psi(x) - x = R(x). \tag{1}$$

By Theorem 2.1 the prime number theorem is equivalent to

THEOREM 3.1 *Given $\varepsilon > 0$, we can find x_0 such that, if $x > x_0$,*

$$|R(x)| < \varepsilon x. \tag{2}$$

We shall prove this theorem in three stages. In the first we derive a new formulation of Selberg's symmetry formula

$$|R(x)| \ln^2 x \le 2 \sum_{n \le x} |R(x/n)| \ln n + O(x \ln x). \tag{3}$$

This version has the advantage that the primes are not involved in the 'weighting' of the sum in (3); that is the logarithm function has replaced Λ on the right-hand side. Secondly we show that (2) holds when x lies in a subset of the real line consisting of a sequence of intervals I_t defined below. In the final state we prove, by induction on n,

$$|R(x)| < c_n x \tag{4}$$

where c_0 is given by Corollary 2.4 and $\lim_{n \to \infty} c_n = 0$. For the induction step we apply (4) to (3) to obtain an estimate for $|R(x)|$. Then we note that this estimate can be replaced by (2) when x lies in one of the intervals I_t. From this we can construct c_{n+1}, derive (4) with c_{n+1} replacing c_n, and show that $c_{n+1} = c_n - b^* c_n^3$ for some positive constant b^*. As noted above this proves the theorem.

The first stage

By manipulating the formulas derived in Section 2, we shall prove the following version of Selberg's formula.

THEOREM 3.2

$$|R(x)| \ln^2 x \le 2 \sum_{n \le x} |R(x/n)| \ln n + O(x \ln x).$$

Proof. Theorem 2.6(i) can be written in the form

$$2 \sum_{n \le x} \Lambda(n)\psi(x/n) \ln n = \ln x \sum_{rs \le x} \Lambda(r)\Lambda(s) + O(x \ln x);$$

combining this with Corollary 2.5 (multiplied by the factor $\ln x$) we have

$$\psi(x) \ln^2 x + 2 \sum_{n \le x} \Lambda(n)\psi(x/n) \ln n = 2x \ln^2 x + O(x \ln x).$$

Replace $\psi(x)$ by $R(x) + x$ and we obtain by Problem 9(vii), Chapter 12,

$$R(x) \ln^2 x + 2 \sum_{n \le x} \Lambda(n) R(x/n) \ln n = O(x \ln x)$$

or

$$|R(x)| \ln^2 x \le 2 \sum_{n \le x} \Lambda(n)\, |R(x/n)| \ln n + O(x \ln x). \tag{5}$$

Secondly by replacing $\psi(x)$ with $R(x) + x$ in Theorem 2.6(ii) and writing as an inequality we have

$$|R(x)| \ln^2 x \le 2 \sum_{rs \le x} \Lambda(r)\Lambda(s)\, |R(x/rs)| + O(x \ln x) \tag{6}$$

[note that, by Problems 8 and 9(vii), Chapter 12, $\sum_{rs \le x} \Lambda(r)\Lambda(s)/rs = \frac{1}{2} \ln^2 x + O(\ln x)$]. Now (5) and (6) give

$$\begin{aligned} |R(x)| \ln^2 x &\le \sum_{n \le x} \Lambda(n)\, |R(x/n)| \ln n + \sum_{rs \le x} \Lambda(r)\Lambda(s)\, |R(x/rs)| + O(x \ln x) \\ &= \sum_{n \le x} \left\{ \Lambda(n) \ln n + \sum_{rs = n} \Lambda(r)\Lambda(s) \right\} |R(x/n)| + O(x \ln x). \end{aligned} \tag{7}$$

Let $T(x) = \sum_{n \le x} \left\{ \Lambda(n) \ln n + \sum_{rs = n} \Lambda(r)\Lambda(s) \right\}$; we have, by Selberg's formula (Theorem 2.3),

$$T(x) = 2x \ln x + O(x). \tag{8}$$

Now (7) can be written in the form

$$\begin{aligned} |R(x)| \ln^2 x &\le \sum_{n \le x} (T(n) - T(n-1))\, |R(x/n)| + O(x \ln x) \\ &= T([x]) R(x/[x]) \\ &\quad + \sum_{n \le x-1} T(n) \left(\left| R\left(\frac{x}{n}\right) \right| - \left| R\left(\frac{x}{n+1}\right) \right| \right) + O(x \ln x) \end{aligned}$$

$$= 2[x]R(x/[x]) \ln [x] + O(x)$$
$$+ \sum_{n \le x-1} 2n\left(\left|R\left(\frac{x}{n}\right)\right| - \left|R\left(\frac{x}{n+1}\right)\right|\right) \ln n$$
$$+ O\left(\sum_{n \le x-1} n \left|\left|R\left(\frac{x}{n}\right)\right| - \left|R\left(\frac{x}{n+1}\right)\right|\right|\right) + O(x \ln x)$$
$$= S_1 + S_2 + O(S_3) + O(x \ln x) \qquad (9)$$

by (8). Clearly $S_1 = O(x \ln x)$. Secondly by Corollary 2.4 and partial summation

$$S_2 = 2\sum_{n \le x} |R(x/n)|\,(n \ln n - (n-1)\ln(n-1)) + O(x \ln x)$$
$$= 2\sum_{n \le x} |R(x/n)|\,(\ln n + O(1)) + O(x \ln x)$$
$$= 2\sum_{n \le x} |R(x/n)| \ln n + O(x \ln x).$$

Thirdly, as $||a| - |b|| \le |a - b|$,

$$S_3 \le \sum_{n \le x} n\,|R(x/n) - R(x/(n+1))|$$
$$\le \sum_{n \le x} n\left(\psi\left(\frac{x}{n}\right) - \psi\left(\frac{x}{n+1}\right)\right) + \sum_{n \le x} n\left(\frac{x}{n} - \frac{x}{n+1}\right)$$
$$\le \sum_{n \le x} \psi(x/n) + O(x \ln x) = O(x \ln x)$$

by Corollary 2.4. The result follows by substituting the estimates of S_1, S_2, and S_3 in (9). □

The second stage

We show first that the inequality (2) holds at certain points n on the real line. Secondly, using Theorem 2.7, we extend this to show that (2) also holds in the intervals I_n containing the points n.

LEMMA 3.3 *Given $\varepsilon > 0$, we can find k_0 and x_0, both dependent on ε, such that, if $k > k_0$ and $x > x_0$, an integer n exists satisfying*

$$x < n \le kx \quad \text{and} \quad |R(n)| < \varepsilon n.$$

Further we can take k_0 in the form $e^{c/\varepsilon}$ for some absolute constant c.

Proof. We consider two cases: (i) R changes sign in the interval $(x, kx]$, and (ii) R has a constant sign in this interval.

(i) In this case there is an n, $x < n \le kx$, such that $R(n)R(n+1) \le 0$. So

$$|R(n)| \le |R(n+1) - R(n)| = |\psi(n+1) - \psi(n) - 1| < \ln n$$

by (1) on p. 251 and the definition of ψ. This gives

$$|R(n)/n| < \ln n/n < \ln x/x < \varepsilon,$$

where this last inequality holds if $x > e^{2/\varepsilon}$.

(ii) Suppose $R(n) \geq 0$ for $x < n \leq kx$. [The argument is similar if $R(n) \leq 0$.] We note first that

$$\sum_{n \leq x} \frac{R(n)}{n(n+1)} < c' \tag{10}$$

for some absolute constant c'. For we have, by (4) on p. 237,

$$\begin{aligned}\ln x + O(1) &= \sum_{n \leq x} \frac{\Lambda(n)}{n} = \sum_{n \leq x} \frac{\psi(n) - \psi(n-1)}{n} \\ &= \sum_{n \leq x} \left\{\frac{1}{n} + \frac{R(n) - R(n-1)}{n}\right\} \\ &= \ln x + O(1) + R([x])/[x] + \sum_{n \leq x-1} R(n)/n(n+1),\end{aligned}$$

by partial summation, and (10) follows by Corollary 2.4. Using this result we have

$$\left(\min_{x < n \leq kx} R(n)/n\right)\left(\sum_{x < n \leq kx} \frac{1}{n+1}\right) < c',$$

or, as the term in the second bracket is bounded by $\ln k + 1$,

$$\min_{x < n \leq kx} R(n)/n < c'/(\ln k + 1) < \varepsilon,$$

where this last inequality holds if $k > e^{c'/\varepsilon}$. Hence if we take $x_0 = e^{2/\varepsilon}$ and $k_0 = e^{c'/\varepsilon}$ the result follows. □

As noted above we can extend this to

THEOREM 3.4 *Given $\varepsilon > 0$, we can find x_1, depending on ε, and absolute constants c^* and b, $0 < b < 1$, such that, if $k > e^{c^*/\varepsilon}$ and $x > x_1$, there is an interval $I = [s, (1 + b\varepsilon)s]$ contained in the interval $(x, kx]$ with the property: if $m \in I$ then $|R(m)| < \varepsilon m$.*

Proof. By Lemma 3.3 we can find k_0 and x_0 such that, if $k > 2k_0$ and $x > x_0$, an integer n exists satisfying

$$x < n \leq kx \quad \text{and} \quad |R(n)| < \varepsilon n/3 \tag{11}$$

and $k > e^{c^*/\varepsilon}$ for some absolute constant c^*.

Now given z, where $0 < z < 1$, suppose m satisfies $n \leq m \leq (1 + z)n$

[that is $((m/n)-1)\leq z$]. Then

$$\left|\frac{R(m)}{m}-\frac{R(n)}{n}\right|\leq\left|\frac{R(m)}{m}\right|\left|1-\frac{m}{n}\right|+\left|\frac{R(m)-R(n)}{n}\right|$$
$$\leq z\,|R(m)/m|+|(\psi(m)-\psi(n))-(m-n)|/n$$
$$\leq z(|R(m)/m|+1)+|\psi(m)-\psi(n)|/n.$$

Using Theorem 2.7 with $c=1$ and $c'=2>1+z$, this gives

$$\left|\frac{R(m)}{m}-\frac{R(n)}{n}\right|\leq z(|R(m)/m|+1)+2(m-n)/n+O(1/\ln n)$$
$$\leq z(|R(m)/m|+3)+O(1/\ln n).$$

Further by Corollary 2.4 there is an absolute constant c'' with the property that $|R(x)/x|<c''$ for all x, hence the inequality above gives

$$\left|\frac{R(m)}{m}\right|\leq\left|\frac{R(n)}{n}\right|+z(c''+3)+O(1/\ln x)$$
$$\leq\varepsilon/3+z(c''+3)+O(1/\ln x)$$

by (11). Now choose x' so that, if $x>x'$, the last term above is bounded by $\varepsilon/3$ and choose z so that

$$z(c''+3)=\varepsilon/3,$$

that is, let

$$z=\varepsilon/3(c''+3).$$

The result follows with $x_1=\max(x_0,x')$, $s=n$, and $b=1/3(c''+3)$ for then we have, by (11) with $x>x_1$,

$$x<s\leq m\leq(1+b\varepsilon)s\leq(1+b\varepsilon)(k/2)\leq kx. \qquad\square$$

The final stage

Suppose we have shown

$$|R(x)|<c_m x \tag{12}$$

for $x>x_m$. Using Theorems 3.2 and 3.4 we shall obtain an improved version of this inequality with c_{m+1} replacing c_m where $c_{m+1}<c_m$. (This is the inductive step in the main proof.) We shall show below that (12) implies

$$|R(x)|\ln^2 x\leq 2c_m x\sum_{\substack{n\\ x/n>x_m}}\frac{\ln n}{n}+O(x\ln x).$$

With the help of Theorem 3.4 we can improve this estimate by subtracting from the right-hand side

$$2(c_m - \varepsilon)x\sum_n{}^* \frac{\ln n}{n} \tag{13}$$

where the sum is taken over those n for which x/n belongs to one of the intervals I_t. [In these intervals we have the improved estimate for $|R(x)|$ given by (2).] This will allow us to define c_{m+1} and so complete the proof. Thus our first task is the definition of the intervals I_t and the estimation of (13).

Let $r > 1$ be a positive constant larger than c_m to be chosen later, and let

$$\varepsilon = c_m/r, \qquad k = e^{c^*/\varepsilon} = e^{c^* r/c_m} \tag{14}$$

where c^* is defined in Theorem 3.4. Note that $0 < \varepsilon < 1$ and ε can be used for all smaller c_{m+k}. Let the intervals J_t be given by

$$z \in J_t \text{ if and only if } k^t < z \leq k^{t+1}.$$

Note that these intervals are disjoint. If J_t is to satisfy the conditions of Theorem 3.4 we must have

$$k^t > x_1 \quad \text{and} \quad k^{t+1} < \sqrt{x} < x \tag{15}$$

as we are working on the interval $(1, x]$. [We use the bound $\sqrt{x}$, rather than x, so that the fraction x/s_t in (22) below is not too small.] By (15) if we let

$$u = [\ln x_1/\ln k] + 1 \quad \text{and} \quad v = [\ln x/2\ln k] - 1, \tag{16}$$

then we can apply Theorem 3.4 to J_t when $u \leq t \leq v$. Further we have

$$\begin{aligned} v - u &\geq \frac{\ln x}{2\ln k} - \frac{\ln x_1}{\ln k} - 2 \\ &> \frac{\ln x}{3\ln k} \qquad \text{for large enough } x, \\ &= \frac{c_m \ln x}{3rc^*} \qquad \text{by (14).} \end{aligned} \tag{17}$$

We can now describe the intervals I_t. By Theorem 3.4 there is a constant b and an integer $s_t \in J_t$ which defines I_t by

$$I_t \subset J_t, \quad \text{and} \quad z \in I_t \text{ if and only if } s_t \leq z \leq (1 + b\varepsilon)s_t. \tag{18}$$

We see at once

$$x/n \in I_t \text{ if and only if } \frac{x}{(1+b\varepsilon)s_t} \leq n \leq \frac{x}{s_t}. \tag{19}$$

As we pointed out above (13) we require a lower bound for

$$U = \sum_n{}^* \frac{\ln n}{n} \tag{20}$$

where the sum is taken over those n for which $x/n \in I_t$ and t satisfies $u \le t \le v$. By (19), as $1 + b\varepsilon < 2$, there are at least

$$\frac{x}{2s_t}\left(1 - \frac{1}{1+b\varepsilon}\right) = \frac{b\varepsilon x}{2(1+b\varepsilon)s_t} > \frac{b\varepsilon x}{4s_t} \tag{21}$$

points of the form x/n belonging to I_t. [The fraction 1/2 is introduced to avoid using integer parts.] Hence by (21) we have

$$U = \sum_{t=u}^{v} \sum_{\substack{n \\ x/n \in I_t}} \frac{\ln n}{n} > \sum_{t=u}^{v} \frac{b\varepsilon x}{4s_t} \frac{\ln(x/s_t)}{x/s_t}$$

$$= \frac{b\varepsilon}{4} \sum_{t=u}^{v} \ln(x/s_t) \ge \frac{b\varepsilon}{4}(v-u)\frac{\ln x}{2}$$

as (15) implies $x/s_t > \sqrt{x}$. Now (14) and (17) give

$$U > \frac{b\varepsilon}{4} \frac{c_m}{3rc^*} \frac{\ln^2 x}{2} = \frac{bc_m^2 \ln^2 x}{24r^2c^*}. \tag{22}$$

Our second task is to combine this estimate with our final version of Selberg's formula (Theorem 3.2). We have

$$|R(x)| \ln^2 x \le 2\left\{ \sum_{n \le x/x_m} + \sum_{x/x_m < n \le x} \right\} |R(x/n)| \ln n + O(x \ln x)$$

$$\le 2c_m \sum_{n \le x/x_m} \frac{x}{n} \ln n + \sum_{x/x_m < n \le x} O\left(\frac{x}{n}\right) \ln n + O(x \ln x)$$

by (12) and Corollary 2.4. The second sum in this expression equals

$$O\left(x \sum_{x/x_m < n \le x} \frac{\ln n}{n}\right) = O(x[\ln^2 x - \ln^2(x/x_m)]) = O(x \ln x)$$

by Problem 9(vi), Chapter 12, as the term in square brackets has $O(\ln x)$. Combining these equations we derive

$$|R(x)| \ln^2 x \le 2c_m x \sum_{n \le x/x_m} \frac{\ln n}{n} + O(x \ln x).$$

We now improve this estimate. When x/n lies in one of the intervals I_t we have $|R(x/n)| < \varepsilon x/n$ (Theorems 3.4) and so for these values of x/n we can replace $c_m x/n$ by $\varepsilon x/n$ as an estimate for $|R(x/n)|$. We do this by subtracting the term

$$2(c_m - \varepsilon)\sum_n{}^* \frac{x}{n} \ln n = 2(c_m - \varepsilon)xU$$

[see (20)] from the right-hand side of the above inequality. Hence by (22)

$$|R(x)| \ln^2 x \le 2c_m x \sum_{n \le x/x_m} \frac{\ln n}{n} - 2(c_m - \varepsilon)xU + O(x \ln x)$$

$$\le 2c_m x \sum_{n \le x/x_m} \frac{\ln n}{n} - \frac{(c_m - \varepsilon)xbc_m^2 \ln^2 x}{12r^2c^*} + O(x \ln x).$$

But by Problem 9(vi), Chapter 12, the first term

$$2c_m x \sum_{n \le x/x_m} \frac{\ln n}{n} = 2c_m x\left(\frac{1}{2}\ln^2(x/x_m) + O(1)\right) = c_m x \ln^2 x + O(x \ln x),$$

and so we have

$$|R(x)| \ln^2 x \le x \ln^2 x\left\{c_m - \frac{(c_m - \varepsilon)bc_m^2}{12r^2c^*}\right\} + O(x \ln x).$$

Decreasing the factor 1/12 allows us to absorb the error term, thus for sufficiently large x we have, as $\varepsilon = c_m/r$,

$$|R(x)| < \left(c_m - \frac{(c_m - \varepsilon)bc_m^2}{13r^2c^*}\right)x$$

$$= \left(c_m - \frac{(r-1)b}{13r^3c^*}c_m^3\right)x = (c_m - b^*c_m^3)x \tag{23}$$

where $b^* = (r-1)b/13r^3c^*$.

Finally we can choose r and define c_{m+1}. By Corollary 2.4 there is a constant c_0 with the property $|R(x)| < c_0 x$ for all x. Let r satisfy

$$r > c_0 \quad \text{and} \quad 1 > b^*c_0^2,$$

that is

$$r > c_0\sqrt{(1 + b/13c^*)}, \tag{24}$$

and let c_{m+1} be given by

$$c_{m+1} = c_m - b^*c_m^3 \tag{25}$$

As $b^* > 0$, (24) gives by induction

$$c_m > 0 \quad \text{and} \quad c_{m+1} < c_m.$$

Thus the sequence $\{c_n\}$ has a non-negative limit c and by (25)

$$c = c - b^*c^3,$$

that is $c = 0$. By repeated applications of (23), it follows that we can derive the inequality

$$|R(x)| < \varepsilon x$$

for all $\varepsilon > 0$ provided x is sufficiently large; this completes the proof of Theorem 3.1 and PNT. □

13.4 Problems 13

1. Let a and c be positive integers with $(a, c)=1$ and a fixed. Show that Dirichlet's theorem is equivalent to the proposition: for all c there is a n such that $a+cn$ is prime.

☆2. An algebraic proof of a special case of Dirichlet's theorem. Let $\mathrm{e}(x)=\mathrm{e}^{2\pi i x}$ and let F_n be given by

$$F_n(x)=\prod_{\substack{a\\(a,n)=1}}(x-\mathrm{e}(a/n))$$

where the product is taken over a reduced system of residues modulo n. Note that

$$x^n-1=\prod_{m|n}F_m(x).$$

Further let G_n satisfy

$$x^n-1=F_n(x)G_n(x). \qquad (*)$$

You are given the results: (i) Both F_n and G_n have integer coefficients. (ii) If r is an integer not equal to 1 or -1, then $F_n(r)G_n(r)\neq 0$ and each common factor of $F_n(r)$ and $G_n(r)$ is a divisor of n.

By letting $x=ay$ in $(*)$, show that a prime p exists so that for each proper divisor m of a

$$x^m\not\equiv 1 \pmod p \quad \text{but} \quad x^a\equiv 1 \pmod p.$$

Deduce Dirichlet's theorem for arithmetic progressions of the form $\{an+1\}$ by showing that $a|p-1$.

☆3. Assume that $L(\chi)\neq 0$ for all $\chi\neq\chi_0$. By considering the series $L(s,\chi)=\sum_{n=1}^{\infty}\frac{\chi(n)}{n^s}$, re-prove Dirichlet's theorem using the Euler product (Theorem 3.1, Chapter 12) and the complex logarithm function.

☆4. With $L(s,\chi)$ as above, prove that

$$L^3(s,\chi_0)\,|L(s,\chi)|^4\,|L(s,\chi^2)|^2\geq 1,$$

using the following properties, where $r,\theta\in\mathbb{R}$.

(i) The geometric mean of three positive reals is at most equal to the arithmetic mean.
(ii) If $z=re^{i\theta}$, $|1-z|^2=1-2r\cos\theta+r^2$.
(iii) $2\cos\theta+\cos 2\theta\geq -3/2$.
(iv) $(s-1)L(s,\chi_0)$ is bounded as $s\to 1^+$.

Now re-prove the result $L(\chi)\neq 0$ where χ is a non-real character, that is when $\chi^2\neq\chi_0$.

☆5. Show that if $(a,c)=1$ and $c>1$ then for some constant b

(i) $\displaystyle\sum_{\substack{p\leq x\\ p\equiv a(c)}}\frac{1}{p}=\frac{1}{\phi(c)}\ln\ln x+b+O(1/\ln x).$

(ii) $\sum_{\substack{n \le x \\ n \equiv a(c)}} \frac{\mu(n)}{n} = O(1)$.

In (i) use Lemma 2.2, Chapter 12, and in (ii) use Lemma 1.2 and Theorem 1.3.

The next four problems involve applications of Dirichlet's theorem.

6. Prove that for all integers m there are infinitely many primes p such that

$$|p - q| > m$$

if q is also prime.
[Hint. Show first that if p_1 is an odd prime then $(p_1!, (p_1 - 1)! - 1) = 1$.]

7. For $i = 1, \ldots, k$, let $\varepsilon_i = \pm 1$ and $a_i \in \mathbb{Z}$ where each a_i is not square and $(a_i, a_j) = 1$ if $i \neq j$. Show that there are infinitely many primes p which satisfy

$$(a_i/p) = \varepsilon_i \quad \text{for} \quad i = 1, \ldots, k.$$

[Hint. Use Theorem 3.3. Chapter 4.]

8. Let f be a polynomial of degree $n \ge 1$ with integer coefficients and suppose, for each prime p, an integer m and a prime q exist so that $f(p) = q^m$. Prove that $f(x)$ is the polynomial x^n, that is $q = p$ and $m = n$.
[Hint. Show that $q^{m+1} \mid f(p + tq^{m+1}) - f(p)$ and consider primes $q \neq p$.]

9. Let F_n be the polynomial defined in Problem 2 above. Show that F_n is irreducible over the integers, that is it cannot be expressed as a product of two polynomials with integer coefficients both having positive degree.
[Hint. Use Fermat's theorem (Corollary 1.13, Chapter 3) to show that if g is a factor of F_n and $g(\zeta) = 0$ then $g(\zeta^m) = 0$ if $(m, n) = 1$ where ζ is an nth root of unity.]

The next four problems are concerned with the equivalences in Theorem 2.1.

10. Prove

$$\theta(x) = \pi(x) \ln x - \int_2^x (\pi(t)/t) \, dt$$

and hence show that (i) implies (ii).

11. By reanalysing the proofs of Theorem 2.1 and Problems 1 and 8, Chapter 12, show that (i) implies (iv), (v), and (viii).

☆12. Use the Möbius inversion formula and Problem 13, Chapter 2, to show that

$$\mu(n) \ln n = -\sum_{d \mid n} \Lambda(d) \mu(n/d),$$

and so prove

$$\sum_{n \le x} \mu(n) \ln n = \sum_{n \le x} \sum_{d \mid n} (1 - \Lambda(d)) \mu(n/d) \; - 1.$$

Further show that the left-hand side of this equation equals $M(x) \ln x + O(x)$ (use partial summation), and the right-hand side equals

$$\sum_{n\leq x} ([x/n]-\psi(x/n))\mu(n).$$

Finally use (iii) for large x and Theorem 1.1, Chapter 12, for small x to derive (vi).

13. Let $S(x)=\sum_{n\leq x} x\mu(n)/n$. Using DSI and partial summation show that, if $x\geq 1$,

(i) $S(x)=\sum_{n\leq x} \mu(n)(x/n-[x/n]) + 1$

and, for large fixed t,

(ii) $S(x)=1+\sum_{n\leq x/t} (x/n-[x/n])\mu(n)+\int_1^t M(x/y)\,dy-\sum_{2\leq n\leq t} M(x/n)$
$-(t-[t])M(x/t)$.

Hence prove that (vi) implies (vii).

☆14. By replacing x and x/m and summing over $m\leq x$ in Corollary 2.5, show that

$$\sum_{n\leq x} \psi(x/n)=x\ln x+O(x).$$

Hence show that Problem 8, Chapter 12, can be solved using Selberg's formula.

☆15. Derive the following reformulations of Selberg's formula.

(i) $\theta(x)\ln x+\sum_{p\leq x} \theta(x/p)\ln p=2x\ln x+O(x)$

(ii) $\sum_{p\leq x} \ln^2 p+\sum_{pq\leq x} \ln p\ln q=2x\ln x+O(x)$

where p and q denote prime numbers. [Hint. Show that the corresponding terms agree with those in Theorem 2.3.]

16. Show that, if $M(x)=\sum_{n\leq x} \mu(n)$,

$$M(x)\ln x+\sum_{n\leq x} \Lambda(n)M(x/n)=O(x).$$

17. Use PNT to show that, given $\varepsilon>0$, there is an x_0 such that, if $x>x_0$, a prime number p exists satisfying

$$x<p<(1+\varepsilon)x.$$

☆18. Let $\varepsilon>0$. Show that for infinitely many n we have

(i) $\tau(n)>2^{(1-\varepsilon)\ln n/\ln\ln n}$,
(ii) $\phi(n)<(1+\varepsilon)cn/\ln\ln n$.

[Hint. Consider n equal to the product of all primes less than x, and in (ii) use Problem 10, Chapter 12, where c is defined.]

☆☆ 19. Reprove PNT by showing directly that

$$M(x) = o(x).$$

You should prove the following propositions in turn using the methods of Section 3. Note that the analogue of Theorem 2.7 is trivial in this case. Note also that this is probably the most difficult problem in the book!

(i) Use Problem 16 to show that

$$M(x)\ln^2 x + \sum_{n\le x} \Lambda(n)M(x/n)\ln n - \sum_{mn\le x} \Lambda(m)\Lambda(n)M(x/mn) = O(x\ln x).$$

(ii) Deduce using Selberg's symmetry formula

$$M(x)\ln^2 x + 2\sum_{n\le x} \Lambda(n)M(x/n)\ln n = O(x\ln x)$$

and

$$M(x)\ln^2 x - 2\sum_{mn\le x} \Lambda(m)\Lambda(n)M(x/mn) = O(x\ln x).$$

(iii) Combining these show that

$$|M(x)|\ln^2 x \le 2\sum_{n\le x} |M(x/n)|\ln n + O(x\ln x).$$

(iv) Prove that

$$\sum_{n\le x} \frac{M(n)}{n(n+1)} = O(1).$$

(v) Use (iv) to define a set of intervals I_t' (as in Theorem 3.4) such that if $z \in I_t'$ then $|M(z)| < \varepsilon z$.

(vi) Complete the proof.

14

DIOPHANTINE EQUATIONS

Diophantus worked in Alexandria in the middle of the third century AD. His six or seven surviving books show that he had a highly developed understanding of many number theoretical problems. These mainly involved the solution of equations in positive integers or rational numbers. He was one of the first to introduce symbolic notation into mathematics, for example he used a symbol similar to our 'x' for the indeterminate variable in an equation. One of the problems he considered was the representation of integers as sums of squares. It seems clear that he was aware of the identity

$$(x^2+y^2)(z^2+t^2)=(xz\pm yt)^2+(xt\mp yz)^2,$$

although we cannot be certain as many of his books are lost. He also gave necessary conditions for representation by two and three squares and it seems most probable that he knew that every integer can be expressed as a sum of four squares, but he had no formal proof of this. For a full account of the work of Diophantus see Heath (1910). It is of interest to note that the seventeenth-century number theorists, Bachet, Fermat, and others, studied the surviving books of Diophantus in great detail.

Today we use the term 'Diophantine equation' when investigating the solubility of algebraic equations in a particular number system, such as the integers $\mathbb{Z}$ or the rational numbers $\mathbb{Q}$. Although some methods recur, for example the method of descent, a large array of distinct ideas and arguments are used; this is particularly true when integer solutions are sought. When solutions in a field are required, a more organized theory exists based on ideas from algebraic geometry; we shall discuss this in the next chapter.

We have considered a number of Diophantine problems previously in this book. These include:

(i) linear Diophantine equations, Chapter 1;
(ii) Chevalley's theorem, Chapter 3;
(iii) sums of two and four squares and equations of the form $x^2+ky^2=p$, Chapter 6;
(iv) Pell's equation, Chapter 7;
(v) Thue–Siegel–Roth theorem, Chapter 8;
(vi) sums of three squares, Chapter 9;
(vii) general binary quadratic equations, Chapters 9 and 10;
(viii) the Goldbach conjecture, Chapter 12.

Below we consider four more Diophantine problems to illustrate the diversity of the methods used. These are mainly questions concerning solubility in integers and are as follows.

(ix) Legendre's theorem. This gives necessary and sufficient conditions for the solution in integers of the equation

$$ax^2 + by^2 + cz^2 = 0.$$

(x) Fermat's conjecture concerning the non-solubility of

$$x^n + y^n = z^n$$

if $n > 2$. We prove this when $n = 3$ and 4, and so illustrate some of the ideas used for larger n.

(xi) Skolem's method. Using p-adic ideas, Skolem invented a process for solving a wide range of Diophantine equations. We shall apply his method to the equation

$$x^4 - 2y^4 = 1.$$

(xii) Mordell's equation

$$y^2 = x^3 + k.$$

We shall discuss a number of methods (mainly due to Mordell) for solving this equation both in integers and in rational numbers.

Note that one or two of these results rely on theorems not proved in this book. They mainly involve properties of algebraic number fields; detailed references will be given.

In some cases simple congruence considerations can be used to show that an equation has no solutions. For example

$$4x^2 + 6y^3 = 3$$

clearly has no solutions in integers as the left-hand side is always even. The following propositions may seem obvious but can be useful.

(i) The equation $f(x_1, \ldots, x_n) = 0$ is soluble in integers or rational numbers only if it is soluble in the real field $\mathbb{R}$.

(ii) If f is an inhomogeneous polynomial, then the equation $f(x_1, \ldots, x_n) = 0$ is soluble in integers only if the congruence $f(x_1, \ldots, x_n) \equiv 0 \pmod{m}$ is soluble for all integers m.

(iii) If f is a homogeneous polynomial, then the equation $f(x_1, \ldots, x_n) = 0$ is soluble in rational numbers not all zero if and only if it has an integer solution $\{x_1, \ldots, x_n\}$ with at least one non-zero entry and $(x_1, \ldots, x_n) = 1$.

Note that we follow the usual convention and use x, y, or z (sometimes with suffices etc.) for indeterminate variables to be found. An excellent survey of the whole area of Diophantine equations is given by Mordell (1969); only the most advanced or extensively numerical material is omitted and it has a good bibliography.

14.1 Legendre's theorem

For our first Diophantine problem we consider Legendre's equation

$$ax^2 + by^2 + cz^2 = 0, \tag{1}$$

where $abc \neq 0$. A solution $\{x_0, y_0, z_0\}$ to (1) is called *non-trivial* if at least one of x_0, y_0, and z_0 is non-zero *and* $(x_0, y_0, z_0) = 1$; from now on we shall only consider non-trivial solutions. We can further assume that a, b, and c satisfy

$$a > 0, \quad b < 0, \quad \text{and} \quad c < 0, \tag{2}$$

$$a, b, \text{ and } c \text{ are square-free}, \tag{3}$$

$$(a, b) = (b, c) = (c, a) = 1. \tag{4}$$

First if a, b, and c all have the same sign then (1) only has the trivial solution $\{0, 0, 0\}$; so, renaming coefficients and multiplying by -1 where necessary, we can assume (2). Secondly if $a = a_1 r^2 (r > 1)$, (1) is soluble if and only if $a_1 x^2 + by^2 + cz^2 = 0$ is also soluble. Finally note we can clearly assume that $(a, b, c) = 1$. Suppose $p \mid (a, b)$. Then $a = a'p$, $b = b'p$, and it follows that $p \mid cz^2$ and so $p \mid z$ [as $(a, b, c) = 1$]. Hence we may replace (1) by the equation $a'x^2 + b'y^2 + cpz^2 = 0$. Now repeat this process for the remaining prime factors of (a, b), (b, c), and (c, a). Thus there is no loss of generality in assuming (2), (3), and (4).

Our proof of Legendre's theorem is due to Mordell; it requires two lemmas.

LEMMA 1.1 *Let n be a positive integer. Suppose α, β, and γ are positive irrational numbers whose product $\alpha\beta\gamma = n$. Then for every triple a_1, a_2, a_3 of integers, the congruence*

$$a_1 x + a_2 y + a_3 z \equiv 0 \pmod{n}$$

has a solution $\{x_0, y_0, z_0\}$ which satisfies $\{x_0, y_0, z_0\} \neq \{0, 0, 0\}$ and

$$|x_0| < \alpha, \quad |y_0| < \beta, \quad \text{and} \quad |z_0| < \gamma.$$

Proof. Consider the set S of triples $\{x, y, z\}$ of integers where

$$0 \leq x \leq [\alpha], \quad 0 \leq y \leq [\beta], \quad \text{and} \quad 0 \leq z \leq [\gamma].$$

This set contains $(1 + [\alpha])(1 + [\beta])(1 + [\gamma]) > \alpha\beta\gamma = n$ elements. But there are at most n residue classes modulo n, and so triples $\{x_1, y_1, z_1\}$ and $\{x_2, y_2, z_2\}$ occur in S satisfying

$$a_1 x_1 + a_2 y_1 + a_3 z_1 \equiv a_1 x_2 + a_2 y_2 + a_3 z_2 \pmod{n}.$$

The result follows if we take $x_0 = x_2 - x_1$, $y_0 = y_2 - y_1$, and $z_0 = z_2 - z_1$. □

LEMMA 1.2 *Let* $(m, n) = 1$. *Suppose the form* $ax^2 + by^2 + cz^2$ *can be expressed as a product of linear factors both modulo* m *and modulo* n, *then it can be expressed as a product of linear factors modulo* mn.

Proof. Let the conditions be expressed by

$$ax^2 + by^2 + cz^2 \equiv (a_1x + a_2y + a_3z)(a_4x + a_5y + a_6z) \pmod{m}$$
$$ax^2 + by^2 + cz^2 \equiv (a_1'x + a_2'y + a_3'z)(a_4'x + a_5'y + a_6'z) \pmod{n}.$$

By the Chinese remainder theorem (1.8, Chapter 3) we can find integers a_i^* satisfying $a_i^* \equiv a_i \pmod{m}$ and $a_i^* \equiv a_i' \pmod{n}$, for $i = 1, \ldots, 6$. Combining these congruences we have

$$ax^2 + by^2 + cz^2 \equiv (a_1^*x + a_2^*y + a_3^*z)(a_4^*x + a_5^*y + a_6^*z) \pmod{mn}. \quad \square$$

Let aRn stand for: there is an integer x satisfying $x^2 \equiv a \pmod{n}$, that is a is a quadratic residue modulo n; see Problem 14, Chapter 4.

THEOREM 1.3 *Legendre's theorem.*
Suppose a, b, *and* c *satisfy conditions* (2), (3), *and* (4) *on p.* 265. *The equation*

$$ax^2 + by^2 + cz^2 = 0 \tag{5}$$

has a (non-trivial) solution if and only if

$$-abRc, \quad -bcRa, \text{ and } \quad -caRb \tag{6}$$

all hold.

Proof. We show first that the conditions (6) are necessary. Let $\{x_0, y_0, z_0\}$ be a solution of (5); it follows that $(c, x_0) = 1$. For if $p \mid (c, x_0)$, then $p \mid by_0^2$ but $p \nmid b$ [by (4)] and so $p \mid y_0$. Consequently we have $p^2 \mid ax_0^2 + by_0^2$ and $p^2 \mid cz_0^2$. But c is square-free and so $p \mid z_0$. This contradicts our assumption $(x_0, y_0, z_0) = 1$.

As $(c, x_0) = 1$ we can find x_0' satisfying $x_0x_0' \equiv 1 \pmod{c}$. Also, clearly

$$ax_0^2 + by_0^2 \equiv 0 \pmod{c},$$

and so, multiplying by $bx_0'^2$,

$$b^2x_0'^2y_0^2 \equiv -ab(x_0x_0')^2 \equiv -ab \pmod{c}.$$

Thus $-abRc$ holds. The remaining conditions can be derived similarly.

The proof of the reverse implication begins with three special cases. Case (i) $b = c = -1$. In this case (6) gives $-1Ra$ and so, by Problem 2,

Chapter 6, integers r and s exist satisfying $r^2 + s^2 = a$. Hence in this case Legendre's equation has the solution $x = 1$, $y = r$, and $z = s$. Case (ii) $a = 1$, $b = -1$. Here the equation has the solution $x = y = 1$, $z = 0$. Case (iii) $a = 1$, $c = -1$. This is similar to (ii).

In the general case we have $-abRc$, that is an integer t can be found to satisfy

$$t^2 \equiv -ab \pmod{c}. \tag{7}$$

Also [by (4)] a^* exists satisfying $aa^* \equiv 1 \pmod{c}$. Thus working modulo c we have

$$\begin{aligned} ax^2 + by^2 + cz^2 &\equiv aa^*(ax^2 + by^2) \equiv a^*(a^2x^2 + aby^2) \\ &\equiv a^*(a^2x^2 - t^2y^2) \qquad \text{by (7)} \\ &\equiv a^*(ax - ty)(ax + ty) \\ &\equiv (x - a^*ty)(ax + ty) \pmod{c}. \end{aligned}$$

Using the remaining conditions (6) we see that $ax^2 + by^2 + cz^2$ can also be expressed as a product of linear factors modulo b and modulo a and so, by Lemma 1.2, integers $a_1, \ldots, a_6$ can be found to satisfy

$$ax^2 + by^2 + cz^2 \equiv (a_1x + a_2y + a_3z)(a_4x + a_5y + a_6z) \pmod{abc} \tag{8}$$

Note that this holds for all x, y, and z.

For the next part of the proof consider the congruence

$$a_1x + a_2y + a_3z \equiv 0 \pmod{abc}. \tag{9}$$

As we have dealt with the special cases (i), (ii), and (iii) above, and as a, b, and c satisfy (3) and (4), we may assume that $\sqrt{(bc)}$, $\sqrt{(-ca)}$, and $\sqrt{(-ab)}$ are irrational. Applying Lemma 1.1 to (9), with $\alpha = \sqrt{(bc)}$, $\beta = \sqrt{(-ca)}$, and $\gamma = \sqrt{(-ab)}$, integers x_1, y_1, and z_1 can be found to satisfy $\{x_1, y_1, z_1\} \neq \{0, 0, 0\}$,

$$a_1x_1 + a_2y_1 + a_3z_1 \equiv 0 \pmod{abc},$$

and

$$|x_1| < \sqrt{(bc)}, \ |y_1| < \sqrt{(-ca)} \quad \text{and} \quad |z_1| < \sqrt{(-ab)} \tag{10}$$

Now combining (8) and (10) we have

$$ax_1^2 + by_1^2 + cz_1^2 \equiv 0 \pmod{abc}.$$

But, as b and c are negative, (10) also gives

$$ax_1^2 + by_1^2 + cz_1^2 \leq ax_1^2 < abc$$

and, as a is positive,

$$ax_1^2 + by_1^2 + cz_1^2 \geq by_1^2 + cz_1^2 > b(-ac) + c(-ab) = -2abc.$$

These three relations imply that

$$ax_1^2 + by_1^2 + cz_1^2 = 0 \quad \text{or} \quad -abc. \tag{11}$$

If the first case holds the result follows, so we may assume that the second case holds. Let

$$x_2 = x_1z_1 - by_1,\ y_2 = y_1z_1 + ax_1,\ z_2 = z_1^2 + ab.$$

This gives

$$\begin{aligned} ax_2^2 + by_2^2 + cz_2^2 &= a(x_1z_1 - by_1)^2 + b(y_1z_1 + ax_1)^2 + c(z_1^2 + ab)^2 \\ &= (ax_1^2 + by_1^2 + cz_1^2)z_1^2 - 2abx_1y_1z_1 + 2abx_1y_1z_1 \\ &\quad + ab(by_1^2 + ax_1^2 + cz_1^2) + abcz_1^2 + a^2b^2c \\ &= (-abc)z_1^2 + ab(-abc) + abcz_1^2 + a^2b^2c = 0 \end{aligned}$$

using our assumption. This is a non-trivial solution. For if $z_1^2 + ab = 0$ then $a = 1$ and $b = -1$ as a and b are coprime and square-free, but this case has been dealt with previously [see (ii) above]. Thus non-trivial solutions have been found in all cases and the proof is complete. □

This proof can be extended to show that, provided the conditions are satisfied, (5) has a solution $\{x_1, y_1, z_1\}$ which in all cases satisfies

$$0 \le x_1 \le \sqrt{(bc)},\ 0 \le y_1 \le \sqrt{(-ca)},\ \text{and } 0 \le z_1 \le \sqrt{(-ab)},$$

see Mordell (1969, p. 47). A different proof of Legendre's theorem is given by Ireland and Rosen (1982, p. 273).

The conditions in Theorem 1.3 can be expressed in a more general way which will show the connection with the Hasse Principle (see Section 3, Chapter 3). Note that, as equation (5) is homogeneous, it has integer solutions if and only if it has solutions in the rational field $\mathbb{Q}$.

THEOREM 1.4 *Let $f(x, y, z) = ax^2 + by^2 + cz^2$. The equation*

$$f(x, y, z) = 0 \tag{12}$$

has a (non-trivial) solution in the rational field $\mathbb{Q}$ if and only if it has a (non-trivial) solution (i) *in the real field $\mathbb{R}$ and* (ii) *in the p-adic fields $\mathbb{Q}_p$ for each prime p.*

A definition of the p-adic fields is given in Section 3 of Chapter 3. Condition (ii) is equivalent to: the congruence $f(x, y, z) \equiv 0 \pmod{p^m}$ has a non-trivial solution (that is, a solution in which at least one of x, y, or z are not divisible by p) for each prime p and integer m. It is sufficient to assume that (ii) holds for all but one prime p, for an explanation of this see Borevich and Shafarevich (1966. p. 65).

Proof. For the case when one of the coefficients in (12) is zero, see Problem 2. Clearly if (12) is soluble in $\mathbb{Q}$, it is also soluble in $\mathbb{R}$ and $\mathbb{Q}_p$; thus by Theorem 1.3 it is sufficient to show that (i) and (ii) imply conditions (2) and (6) above. The arguments justifying (3) and (4) also apply here. Now f represents zero in $\mathbb{R}$ if and only if it is indefinite and this is equivalent to condition (2). To prove the remaining conditions (6) let $p^k|c$, then by (ii) we can derive $-abRp^k$ using the same method as in the proof of Theorem 1.3. Similarly the other conditions in (6) follow. □

As a general ternary quadratic form can be transformed into one of the type (12) using a standard diagonalization argument, this theorem proves the Hasse Principle in the ternary quadratic case, see Problem 2 for the binary case. The Principle can also be derived for forms in four and five variables, the standard proofs use methods similar to those above but rely on Dirichlet's theorem (1.4, Chapter 13). Further, Meyer has shown that an indefinite quadratic form in five or more variables always represents zero non-trivially. Hence these results provide a proof of Hasse's principle in the general quadratic case, for more details see Borevich and Shafarevich (1966) or Cassels (1978).

14.2 Fermat's last theorem

Fermat's conjecture, or 'last theorem' as it is often called, states: the equation

$$x^n + y^n = z^n \tag{1}$$

has no solutions in integers with $xyz \neq 0$ and $n > 2$. (For the case $n = 2$, see Problem 12, Chapter 1.) In 1637 Fermat proposed this problem and claimed to have a proof which has not survived. To this day no complete proof is known but, on the other hand, no counter-example has been found. Many mathematicians have studied the properties of equation (1), now known as Fermat's equation, and these investigations have had a profound effect on the development of both number theory and algebra over the past two centuries. For the history and current state of this conjecture see Edwards (1977) or Ribenboim (1979).

The greatest contributions to the study of Fermat's equation were made by Kummer in the middle of the last century. He proved Fermat's conjecture for a subset of the primes called the regular primes. [It is sufficient to show that (1) is insoluble when n is an odd prime, see Problem 4.] A prime p is *regular* if it does not divide any of the numerators of the Bernoulli numbers B_n for $n \leq (p-3)/2$ (see Chapter 12). This condition can be expressed more succinctly in terms of the class number of the pth cyclotomic field. Unfortunately infinitely many primes are not regular; examples are 37, 59, and 67, and so different methods have to be employed in these cases. Using various techniques, including many hours of computer time, it is now known that Fermat's conjecture is true at least for $n \le 125000$. Also the recent work of Faltings on the Mordell conjecture (see Chapter 15) shows that even if one of these equations is soluble, it can only have finitely many solutions.

As a further illustration of Diophantine methods, we shall prove Fermat's conjecture when $n = 3$ and 4 using the descent method first discussed in Chapter 6.

THEOREM 2.1 *The equation*

$$x^4 + y^4 = z^2 \tag{2}$$

has no solution in integers, provided $xyz \ne 0$.

Note that Fermat's conjecture for $n = 4$ follows immediately.

Proof. Assume that (2) is soluble, and $\{x_1, y_1, z_1\}$ is a solution with $x_1y_1 \ne 0$ and z_1 positive and as small as possible. We shall construct a new solution $\{x_2, y_2, z_2\}$ with $0 < z_2 < z_1$ and so show that our assumption is false. (Note, if $z_1 = 1$ then x_1 or y_1 is zero.)

We may clearly suppose $(x_1, y_1, z_1) = 1$ and so $(x_1, y_1) = (y_1, z_1) = (z_1, x_1) = 1$. If x_1 and y_1 are both odd, then $z_1^2 \equiv 2 \pmod 4$ which is impossible. So we may assume that x_1 is odd and y_1 is even; this implies that z_1 is also odd. If t is four or an odd prime, we cannot have both $t \mid z_1 - x_1^2$ and $t \mid z_1 + x_1^2$, because this would give $t \mid 2z_1$ and $t \mid 2x_1^2$ which contradicts our assumption $(x_1, z_1) = 1$. Hence

$$(z_1 - x_1^2, z_1 + x_1^2) = 2. \tag{3}$$

As $y_1^4 = (z_1 - x_1^2)(z_1 + x_1^2)$, it follows that one of $z_1 - x_1^2$ and $z_1 + x_1^2$ is divisible by 2 (and not by 4) and the other is divisible by 8. Hence we can write $y_1 = 2ab$, and either

$$z_1 - x_1^2 = 2a^4, \qquad z_1 + x_1^2 = 8b^4 \tag{4}$$

or

$$z_1 - x_1^2 = 8b^4, \qquad z_1 + x_1^2 = 2a^4, \tag{5}$$

where in each case $a > 0$, a is odd and $(a, b) = 1$. (4) is impossible because it would imply $x_1^2 = -a^4 + 4b^4$, giving $1 \equiv -1 \pmod 4$ as both x_1 and a are odd. Thus (5) holds and we have $z_1 = a^4 + 4b^4$ with $0 < a < z_1$, and

$$4b^4 = (a^2 - x_1)(a^2 + x_1).$$

Now $(a, b) = 1$, therefore $(a, x_1) = 1$ and we see, as in the proof of (3), that $(a^2 - x_1, a^2 + x_1) = 2$. Consequently we can write $b = x_2y_2$, and then

$a^2 - x_1 = 2x_2^4$ and $a^2 + x_1 = 2y_2^4$. If we let $a = z_2$ these equations give

$$z_2^2 = x_2^4 + y_2^4$$

with $0 < z_2 < z_1$. As noted above this completes the proof. □

The proof of Fermat's conjecture for $n = 3$ depends on properties of the integral domain $\mathbb{Z}[\omega]$ discussed in Section 1 and Problem 6 of Chapter 5. We review its basic properties. The complex number $\omega = (-1 + \sqrt{-3})/2$ satisfies the equation

$$\omega^2 + \omega + 1 = 0. \tag{6}$$

If $\alpha \in \mathbb{Z}[\omega]$ then α can be written in the form $a + b\omega$ where a and b belong to $\mathbb{Z}$. $\mathbb{Z}[\omega]$ is a unique factorization domain and has six units ± 1, $\pm\omega$, and $\pm\omega^2$. The number $\lambda = 1 - \omega$ plays a special role (similar to that of $2 \in \mathbb{Z}$ or $1 + i$ in the Gaussian integers); we have, as λ^2 is an associate of 3,

$$\alpha \in \mathbb{Z}[\omega] \text{ implies } \alpha \equiv -1, 0, \text{ or } 1, \pmod{\lambda}. \tag{7}$$

We shall prove

THEOREM 2.2 *Let ε be a unit in $\mathbb{Z}[\omega]$. The equation*

$$x^3 + y^3 = \varepsilon z^3 \tag{8}$$

has no solution $\{x_1, y_1, z_1\}$ where x_1, y_1, and z_1 belong to $\mathbb{Z}[\omega]$ and $x_1 y_1 z_1 \neq 0$.

This theorem is due essentially to Euler; note that Fermat's conjecture for $n = 3$ follows. We begin with a definition and two lemmas.

DEFINITION If $\alpha \in \mathbb{Z}[\omega]$ and $\alpha = \varepsilon\lambda^n\beta$ where ε is a unit and $\lambda \nmid \beta$ we define ord $\alpha = n$.

LEMMA 2.3 *Let ε be a unit in $\mathbb{Z}[\omega]$. If $x^3 + y^3 = \varepsilon z^3$, $\lambda \nmid xy$ and $\lambda | z$, where x, y, $z \in \mathbb{Z}[\omega]$, then* ord $z \geq 2$.

Proof. We show first that if $\alpha \in \mathbb{Z}[\omega]$,

$$\alpha \equiv 1 \pmod{\lambda} \text{ implies } \alpha^3 \equiv 1 \pmod{\lambda^4}. \tag{9}$$

For if $\alpha = 1 + \lambda\beta$

$$\begin{aligned} \alpha^3 - 1 &= (\alpha - 1)(\alpha - \omega)(\alpha - \omega^2) \\ &= \lambda\beta(1 - \omega + \lambda\beta)(1 - \omega^2 + \lambda\beta) \\ &= \lambda^3\beta(\beta + 1)(\beta + \omega + 1) \qquad \text{as } 1 - \omega = \lambda \\ &= \lambda^3\beta(\beta + 1)(\beta - \omega^2) \end{aligned}$$

by (6). But $\omega^2 \equiv 1 \pmod{\lambda}$ and so (9) follows from (7).

Reducing the equation in the lemma we have, by (9),

$$\pm 1 \pm 1 \equiv \varepsilon z^3 \pmod{\lambda^4}.$$

If $\varepsilon z^3 \equiv \pm 2 \pmod{\lambda^4}$, then $\lambda | 2$ which is impossible. Hence $\varepsilon z^3 \equiv 0 \pmod{\lambda^4}$ giving $3 \operatorname{ord} z \geq 4$, and so finally $\operatorname{ord} z \geq 2$. □

The next lemma provides the descent step in the main proof.

LEMMA 2.4 *Suppose $x^3 + y^3 = \varepsilon z^3$ where ε is a unit in $\mathbb{Z}[\omega]$, $(x, y) = 1$, $\lambda \nmid xy$ and* $\operatorname{ord} z \geq 2$, *then we can find x', y', z' and a unit ε' in $\mathbb{Z}[\omega]$ to satisfy*

$$x'^3 + y'^3 = \varepsilon' z'^3, \ \lambda \nmid x'y', \text{ and } \operatorname{ord} z' = \operatorname{ord} z - 1.$$

Proof. As

$$(x + y)(x + \omega y)(x + \omega^2 y) = \varepsilon z^3 \tag{10}$$

and $\operatorname{ord} \varepsilon z^3 \geq 6$, at least one of the factors on the left-hand side of (10) is divisible by λ^2. So, substituting ωy or $\omega^2 y$ for y if necessary, we have $\operatorname{ord}(x + y) \geq 2$. As $\operatorname{ord}(1 - \omega)y = \operatorname{ord} \lambda y = 1$ by assumption, we have

$$\operatorname{ord}(x + \omega y) = \operatorname{ord}(x + y - (1 - \omega)y) = \min(\operatorname{ord}(x + y), \operatorname{ord} \lambda y) = 1.$$

Using a similar argument we also have $\operatorname{ord}(x + \omega^2 y) = 1$. Combining these three equations with (10) we derive

$$\operatorname{ord}(x + y) = 3 \operatorname{ord} z - 2. \tag{11}$$

We show next that $(x + y, x + \omega y) = \lambda$. For if π is an irreducible element not equal to λ, $\pi | x + y$ and $\pi | x + \omega y$, then $\pi | (1 - \omega)y$, that is $\pi | y$, and similarly $\pi | x$ contrary to assumption. Similar arguments show that $(x + y, x + \omega^2 y) = (x + \omega y, x + \omega^2 y) = \lambda$. Thus as $\mathbb{Z}[\omega]$ is a unique factorization domain we have, by (11),

$$x + y = \varepsilon_1 \alpha^3 \lambda^{3 \operatorname{ord} z - 2}$$
$$x + \omega y = \varepsilon_2 \beta^3 \lambda$$
$$x + \omega^2 y = \varepsilon_3 \gamma^3 \lambda,$$

where λ does not divide α, β, or γ; $(\alpha, \beta) = (\beta, \gamma) = (\gamma, \alpha) = 1$; and ε_1, ε_2, and ε_3 are units. These give by (6)

$$\varepsilon_1 \alpha^3 \lambda^{3 \operatorname{ord} z - 2} + \omega \varepsilon_2 \beta^3 \lambda + \omega^2 \varepsilon_3 \gamma^3 \lambda = 0$$

or, dividing by $\lambda \omega \varepsilon_2$,

$$\beta^3 + \varepsilon_4 \gamma^3 = \varepsilon_5 \alpha^3 \lambda^{3(\operatorname{ord} z - 1)} \tag{12}$$

where $\varepsilon_4 = \omega \varepsilon_2^{-1} \varepsilon_3$ and $\varepsilon_5 = -\omega^{-1} \varepsilon_1 \varepsilon_2^{-1}$ are both units. Now by (7) and (9) this gives, as λ does not divide α, β, or γ,

$$\pm 1 \pm \varepsilon_4 \equiv 0 \pmod{\lambda^2}.$$

[For by (9), $\pm\alpha \equiv 1 \pmod{\lambda}$ implies $\alpha^3 \equiv \pm 1 \pmod{\lambda^2}$.] Further, λ^2 is an associate of 3 and so this congruence is only possible if $\varepsilon_4 = \pm 1$. Thus if we let $x' = \beta$, $y' = \pm\gamma$, where the sign depends on the sign of ε_4, and $z' = \alpha\lambda^{\operatorname{ord} z - 1}$ in (12) the lemma follows. □

Returning to the proof of Theorem 2.2 there are four cases to consider.

Case 1. (8) has a solution $\{x_1, y_1, z_1\}$ with $\lambda \nmid x_1 y_1 z_1$. This gives by (9)

$$\pm 1 \pm 1 \equiv \pm\varepsilon \pmod{\lambda^4},$$

which is impossible as 9 is an associate of λ^4.

Case 2. (8) has a solution $\{x_2, y_2, z_2\}$ with $\lambda \nmid x_2 y_2$ and $\lambda \mid z_2$. In this case we can apply Lemmas 2.3 and 2.4 to obtain a new solution $\{x_3, y_3, z_3\}$ with $\operatorname{ord} z_3 < \operatorname{ord} z_2$ and $\lambda \nmid x_3 y_3$. Repeating this argument k times, where $k = \operatorname{ord} z_2 - 1$, we obtain a contradiction because the resulting solution would satisfy the conditions of Case 1.

Case 3. (8) has a solution $\{x_4, y_4, z_4\}$ with $\lambda \mid x_4$ and $\lambda \nmid y_4 z_4$. In this case (9) gives $\varepsilon \equiv \pm 1 \pmod{\lambda^4}$ and so $\varepsilon = \pm 1$. Hence the equation can be rewritten as $(\pm z_4)^3 + (-y_4)^3 = x_4^3$ and Case 2 now applies.

Case 4. (8) has a solution $\{x_5, y_5, z_5\}$ with $\lambda \mid y_5$ and $\lambda \nmid x_5 z_5$. This is similar to Case 3.

As Cases 1–4 cover all possibilities, the theorem is proved. □

14.3 Skolem's method

Skolem developed a method for showing that a wide range of Diophantine equations have only finitely many integer solutions, and in some cases these solutions can be found explicitly. It uses p-adic analysis which was considered briefly in Chapter 3. A full discussion of Skolem's method is beyond the scope of this book, the reader should consult Borevich and Shafarevich (1966, Chapter 4). Here we shall describe the basic method and consider an example. The method relies on Dirichlet's unit theorem which states that the group of units in an algebraic number field is finitely generated. For example we have shown, using Pell's equation, that the group of units in the field $\mathbb{Q}(\sqrt{d})$ is an infinite cyclic group when $d > 0$ and not square.

Suppose we are given a two variable irreducible form F with integer coefficients and degree $n \geq 3$, and we wish to show that the equation

$$F(x, y) = 1 \tag{1}$$

has only finitely many integer solutions. (Note that the method can be applied in more general situations.) In many cases equation (1) can be rewritten as

$$N(x\alpha + y\beta) = 1 \tag{2}$$

where α and β are integral basis elements for some algebraic number field K of degree n, and N is the norm function for K. It follows that $x\alpha + y\beta$ is a unit in K, so by Dirichlet's unit theorem

$$x\alpha + y\beta = \varepsilon_1^{u_1} \ldots \varepsilon_k^{u_k} \tag{3}$$

where ε_i are 'fundamental' units in K (that is, generators of the unit group) and $u_i \in \mathbb{Z}$. Now if we let $u_1, \ldots, u_k$ range over the p-adic integers and show that (3) has only finitely many p-adic solutions, then clearly (3) can only have finitely many rational integer solutions. Borevich and Shafarevich (1966) gives the p-adic definition of exponentiation and shows that the usual properties of this operation apply p-adically, a proof of Dirichlet's unit theorem is also given.

Let us consider an example. We shall show that the only integer solutions of

$$x^4 - 2y^4 = 1 \tag{4}$$

are $x = \pm 1$, $y = 0$. Note first that if this equation is soluble then x is odd and y is even (as $x^4 - 1$ is congruent to 0 or 7 modulo 8). Thus we may replace (4) by

$$x^4 - 32y^4 = 1. \tag{5}$$

Let $\theta = \sqrt[4]{2}$ and $K = \mathbb{Q}(\theta)$. This field has the integral basis $\{1, \theta, \theta^2, \theta^3\}$ and fundamental units $1 + \theta$ and $1 + \theta^2$. Now, in K, $N(x + 2y\theta) = x^4 - 32y^4$ and so, by (5), $x + 2y\theta$ is a unit, and integers r and s exist satisfying

$$x + 2y\theta = (1 + \theta)^r(1 + \theta^2)^s, \tag{6}$$

changing the signs of both x and y if necessary. We show that $r = s = 0$ by deriving a contradiction from the assumption that either r or s is non-zero. We do this in the 2-adic field, and so it will also follow in $\mathbb{Z}$. As noted above, we can use standard analytic methods in the 2-adic field, and so we proceed as follows. Expanding the right-hand side of (6) as a binomial series we have, using the equation $\theta^4 = 2$,

$$\begin{aligned}
&(1+\theta)^r(1+\theta^2)^s \\
&= \left\{1 + 2\binom{r}{4} + \ldots + \theta\left[r + 2\binom{r}{5} + \ldots\right] + \theta^2\left[\binom{r}{2} + 2\binom{r}{6} + \ldots\right]\right. \\
&\quad \left. + \theta^3\left[\binom{r}{3} + 2\binom{r}{7} + \ldots\right]\right\}\left\{1 + 2\binom{s}{2} + \ldots \theta^2\left[s + 2\binom{s}{3} + \ldots\right]\right\} \\
&= 1 + 2\left[\binom{r}{4} + \binom{s}{2} + s\binom{r}{2}\right] + 2^2[\ldots] + \ldots \qquad (7) \\
&\quad + \theta\left\{r + 2\left[\binom{r}{5} + \binom{s}{2} + s\binom{r}{2}\right] + 2^2[\ldots] + \ldots\right\} \\
&\quad + \theta^2\left\{\binom{r}{2} + s + 2\left[\binom{r}{6} + s\binom{r}{4} + \binom{r}{2}\binom{s}{2} + \binom{s}{3}\right] + 2^2[\ldots] + \ldots\right\} \\
&\quad + \theta^3\left\{rs + \binom{r}{3} + 2\left[\binom{r}{7} + s\binom{r}{5} + \binom{r}{3}\binom{s}{2} + r\binom{s}{3}\right] + 2^2[\ldots] + \ldots\right\}.
\end{aligned}$$

As this expression equals $x+2y\theta$, we can equate coefficients of θ, θ^2, θ^3. Note first that, by considering the coefficient of θ, we have $r\equiv 0 \pmod 2$. Further the coefficients of θ^2 and θ^3 are both zero; hence, multiplying the coefficient of θ^2 by $-r$ and adding to the coefficient of θ^3, we have the 2-adic equation

$$\frac{(r-1)r(r+1)}{3}+2\left[\binom{r}{7}-r\binom{r}{6}+s\left(\binom{r}{5}-r\binom{r}{4}\right)+\left(\binom{r}{3}-r\binom{r}{2}\right)\binom{s}{2}\right]$$
$$+2^2[\ldots]+\ldots=0. \quad (8)$$

Note that in this expression the binomial coefficients $\binom{r}{2n}$ always occur as a factor in the product $r\binom{r}{2n}$; this is because the binomial coefficients with even, lower entries arise from the θ^2 terms in (7). Suppose $2^\sigma | r$ but $2^{\sigma+1} \nmid r$, and $\sigma>0$. We have

$$\binom{r}{2n+1}=\frac{r}{2n+1}\binom{r-1}{2n}\equiv 0 \pmod{2^\sigma}.$$

Thus each term in square brackets in (8) is divisible by $2^{\sigma+1}$ except the first. This contradicts (8) and so $r=0$ as r is even.

Returning to (6) we now have

$$s+2\binom{s}{3}+2^2\binom{s}{5}+\ldots=0,$$

and a similar argument shows that $s=0$ is the only possibility. For a full justification of this proof see Borevich and Shafarevich (1966) and Mordell (1969).

Many other equations can be treated in a similar way. For example, using this method it can be shown that the equation

$$x^3+dy^3=1,$$

where $d>0$, has at most two integer solutions. The proof uses the 3-adic field over $\mathbb{Q}(\sqrt[3]{d})$, see Mordell (1969. p. 203).

14.4 Mordell's equation

For our last example of Diophantine methods we consider the Mordell equation

$$y^2=x^3+k, \quad (1)$$

and look for both integer and rational number solutions. This equation is an important example of an elliptic curve, the subject of our final chapter. Here we shall discuss the rational case only briefly at the end of the section.

No general condition is known for the solubility of Mordell's equation (1) in integers. Some trivial solutions exist when k is a square or a cube, see Problem 11; we shall exclude these from now on. Using the method discussed in the Introduction to Chapter 8, Baker (1975) has shown that a bound, depending on k, exists on the size of integer solutions to (1). Thus for each k this equation has only finitely many solutions in integers, and in many cases these can be found with the aid of modern computer techniques using Baker's result and some results on cubic forms, see Ellison *et al.* (1972).

We shall describe two approaches. In the first, congruence considerations are used to show that (1) is insoluble when k has various special forms. In the second approach solutions are sought in the domain $\mathbb{Z}[(1+\sqrt{-k})/2]$. A third approach has been given in which equation (1) is replaced by a binary cubic form. For details of this, and further examples and methods, see Mordell (1969, Chapter 26). A list of solutions of Mordell's equation for $|k|\leq 50$ is given in Tables 6 and 7 on pp. 344–5.

The basic idea of our first approach is to write (1) as

$$y^2+jn^2=x^3+m^3=(x+m)(x^2-xm+m^2),$$

and show that if j, m, and n satisfy certain congruence conditions then the equation is insoluble. Note that the term $x^2-xm+m^2\geq 0$.

THEOREM 4.1 *The equation*

$$y^2=x^3+m^3-jn^2 \tag{2}$$

has no solution in integers in either of the following cases:

(i) $j=4$, $m\equiv 3 \pmod 4$, *and* $p\not\equiv 3 \pmod 4$ *whenever* $p|n$.

(ii) $j=1$, $m\equiv 2 \pmod 4$, n *is odd, and* $p\not\equiv 3 \pmod 4$ *whenever* $p|n$.

Proof. Writing (2) in the form

$$y^2+jn^2=(x+m)(x^2-xm+m^2), \tag{3}$$

we derive a contradiction from the assumption that a solution exists.

(i) Working modulo 4 we see that the left-hand side of (3) is congruent to 0 or 1 modulo 4, whilst the right-hand side is congruent to 0, 2, or 3 modulo 4, so the only possibility is $2|y$ and $x\equiv 1 \pmod 4$. It follows that

$$z=x^2-xm+m^2\equiv 3 \pmod 4$$

and so z has a prime factor p congruent to 3 (mod 4). But then $y^2\equiv -4n^2 \pmod p$, and so the Legendre symbol $(-4n^2/p)=0$ or 1. The first case implies $p|n$ which is ruled out by our hypothesis, and the second case is impossible as $(-1/p)=-1$.

(ii) This is similar to (i). We again have $x\equiv 1 \pmod 4$ and $x^2-xm+m^2\equiv 3 \pmod 4$ leading to a contradiction. □

This result shows that equation (1) has no non-zero integer solutions $\{x, y\}$ for $|k| \leq 50$, if $k = -41, -33, -17, -9, -5, -1, 7, 11, 23, 39$, and 47. Further examples of this type are given in the problem section.

In our second example we use cubic residues (see p. 83). Note that if $p \equiv 2 \pmod 3$ then $x^3 \equiv 2 \pmod p$ is always soluble, and if $p \equiv 1 \pmod 3$ then $x^3 \equiv 2 \pmod p$ is soluble if and only if $p = t^2 + 27u^2$ where $t, u \in \mathbb{Z}$, see Problem 3, Chapter 6, and Ireland and Rosen (1982).

THEOREM 4.2 *The equation*

$$y^2 = x^3 + 2a^3 - 3b^2 \tag{4}$$

has no solution in integers if $ab \neq 0$, $a \not\equiv 1 \pmod 3$, $3 \nmid b$, *a is odd if b is even, and* $p = t^2 + 27u^2$ *is soluble in integers t and u if* $p|a$ *and* $p \equiv 1 \pmod 3$.

Note that if $|a| < 7$ then the last condition is superfluous.

Proof. Let $X = x^3 + 2a^3$ and $Y = y^2 + 3b^2$. We derive a contradiction by showing that if (4) is soluble then X has no prime factors.

(i) Suppose $2|X$, then $2|x$ and $y \equiv b \pmod 2$. If b is even then, by hypothesis, a is odd and so $X \equiv 2 \pmod 4$ which is impossible since $Y \equiv 0 \pmod 4$ in this case. If b is odd then $X \equiv 4 \pmod 8$ giving $a^3 \equiv 2 \pmod 4$ which again leads to a contradiction.

(ii) Suppose $3|X$, then $3|y$ and $X \equiv 3 \pmod 9$. This is also impossible as $a^3 \equiv 0, 8 \pmod 9$ by hypothesis.

(iii) Now consider the case $p|X$ and $p > 3$. $p|X$ implies $(-3b^2/p) = 1$ or 0. If the former holds, then $(-3/p) = 1$ which is only possible when $p \equiv 1 \pmod 3$. If the latter holds, then $p|b$, and so $p^2|Y$. Let T denote the product of all prime factors p of this type. As $T|y$ and $T|b$, we have $4T^2 \leq Y$. Hence $Y = Y_1 T^2$ where $Y_1 > 1$ and Y_1 is a product of primes $\equiv 1 \pmod 3$.

Let $p|Y_1$. If $p|a$ then, by hypothesis, $p = t_1^2 + 27u_1^2$. If $p \nmid a$ then we can find a' to satisfy $aa' \equiv 1 \pmod p$ and so, as $X = Y$, $(-xa')^3 \equiv 2 \pmod p$. This again shows that $p = t_2^2 + 27u_2^2$ is soluble. Hence using these cases for all $p|Y_1$ we can find integers t and u satisfying $Y_1 = t^2 + 27u^2$.

Now we can factorise Y in the unique factorization domain $\mathbb{Z}[\omega]$ where $\omega = (-1 + \sqrt{-3})/2$. If $y = y'T$ and $b = b'T$, we have

$$y'^2 + 3b'^2 = t^2 + 27u^2$$

or

$$\pm(y' + b'\sqrt{-3})\omega^\alpha = t + 3u\sqrt{-3} \tag{5}$$

where $\alpha = 0, 1$, or 2. If $\alpha = 0$ then $3|b'$, and so $3|b$ contrary to hypothesis. If $\alpha = 1$ or 2, the left-hand side of (5) is

$$\pm\frac{1}{2}(-y' \mp 3b' + (\pm y' - b')\sqrt{-3}),$$

so in either case $y' + b'$ is even. Hence $2|Y$ and so $2|X$, but we have excluded this in (i). The result follows. □

We can use this theorem to show that (1) has no non-zero integer solutions if $k = -50, -21, -14, -5, 6, 13$, and 42.

In our final example we illustrate the second approach mentioned at the beginning of this section. Note that $\mathbb{Z}[\xi]$, where $\xi = (1 + \sqrt{-19})/2$, is a unique factorization domain with units ± 1. (This can be deduced from the fact that the class group of binary quadratic forms with discriminant -19 is trivial, see for example Hua, 1982, p. 442.)

THEOREM 4.3 *The only integer solutions of the equation* $y^2 = x^3 - 19$ *are* $x = 7$, $y = \pm 18$.

Proof. Note first that x is odd as $y^2 \not\equiv -19 \pmod 8$. In the domain $\mathbb{Z}[\xi]$ our equation can be written

$$(y + \sqrt{-19})(y - \sqrt{-19}) = x^3.$$

If $y + \sqrt{-19}$ and $y - \sqrt{-19}$ have a common factor σ in $\mathbb{Z}[\xi]$ then $\sigma | 2\sqrt{-19}$, and so 2 and 19 divide x, but this is impossible by the note above and Problem 12. Thus we may write

$$y + \sqrt{-19} = \tau^3$$

where $\tau \in \mathbb{Z}[\xi]$ (if the sign is negative replace τ by $-\tau$). Because $\tau = a + \frac{b}{2}(1 + \sqrt{-19})$, where $a, b \in \mathbb{Z}$ (see p. 78), we have

$$8(y + \sqrt{-19}) = (2a + b + b\sqrt{-19})^3. \tag{6}$$

Equating coefficients of $\sqrt{-19}$ this gives

$$8 = 3b(2a + b)^2 - 19b^3 = 4b(3a^2 + 3ab - 4b^2).$$

Thus $b = \pm 1$ or ± 2 and it is a simple matter to check that $b = 1$, with $a = 1$, is the only possibility. Substituting these values in (6) we obtain $y = -18$ and so $x = 7$. Repeating this construction with $y - \sqrt{-19}$ instead of $y + \sqrt{-19}$ gives $y = 18$, $x = 7$, and the result follows. □

This method can be used to solve Mordell's equation whenever $\mathbb{Z}[(1 + \sqrt{-k})/2]$ is a unique factorization domain and $k \not\equiv 0$, 1, or 4 (mod 8); for example when $k = 3, 7, 11, 27$, and 43, see Problem 16.

Rational solutions

As with the integer case, it is not known precisely for which k Mordell's equation

$$y^2 = x^3 + k \tag{7}$$

has rational solutions. A number of partial results have been developed; their proofs use properties of the field $\mathbb{Q}(\sqrt{k})$ and the related $\mathbb{Q}(\sqrt{-3k})$, and some theorems on cubic forms. The reader can find details in Mordell (1969), Cassels (1950) and Birch and Swinnerton–Dyer (1963). This last reference also contains a table of those k for which (7) has rational solutions when $|k| \leq 400$.

The rational and integer cases of Mordell's equation differ in one important respect, this concerns the number of solutions. Using the 'chord–tangent method' to be discussed in the next chapter it had long been assumed that, for most k, if (7) has a rational solution then it has infinitely many. This was finally proved by Fueter in 1930 and we have

THEOREM 4.4 *Suppose k has no sixth power factors and $k \neq 1$ or -432. If equation* (7) *has a rational solution $\{r, s\}$ with $rs \neq 0$, then it has infinitely many rational solutions.*

The exceptional cases are considered in Problems 13, 17, and 18.

Mordell has given a relatively short proof of Fueter's theorem as follows. With the intention of generating a new rational solution of (7) let

$$x = r + z, \qquad y = s + 3r^2z/2s.$$

If $\{x, y\}$ is a solution of (7) then

$$(s + 3r^2z/2s)^2 = (r + z)^3 + k$$

or, as $s^2 = r^3 + k$,

$$(3r^2z/2s)^2 = 3rz^2 + z^3.$$

This gives, as $z \neq 0$ if $x \neq r$,

$$z = 9r^4/4s^2 - 3r$$

and

$$x = \frac{9r^4 - 8rs^2}{4s^2}, \qquad y = \frac{27r^6 - 36r^3s^2 + 8s^4}{8s^3}.$$

Clearly x and y are rational numbers. If $x = 0$, then $9r^3 = 8s^2$ and so, for some t, $r = 2t^2$, $s = 3t^3$, and $k = t^6$, but we have excluded this case. By considering congruences modulo 2^n it also follows that $y \neq 0$. The proof is completed by showing that this process always leads to a new solution unless $k = 1$ or -432; for the details see Mordell (1966).

14.5 Problems 14

1. Which of the following equations have integer solutions? If the equation has no solution find a modulus for which the corresponding congruence is also insoluble.

(i) $3x^2+4y^2-5z^2=0$
(ii) $3x^2+5y^2-17z^2=0$
(iii) $3x^2-7y^2+43z^2=0$.

2. Using Theorem 3.3, Chapter 4, show that

$$ax^2+by^2=0$$

has a non-trivial solution in $\mathbb{Q}$ if and only if it has non-trivial solutions in $\mathbb{R}$ and $\mathbb{Q}_p$ for all primes p.

3. Let a and d be positive non-square integers such that $(a, 2d)=1$. Show that the equation

$$w^2=dy^2+az^2$$

has a non-trivial solution $\{w, y, z\}$, satisfying (i) $(w, d)=1$ and (ii) z is odd, if and only if aRd and dRa. Note, we have not assumed that a or d are square-free. [*Hint.* Adapt the proof of Legendre's theorem given in Section 1.]

4. Show that to prove Fermat's conjecture it is sufficient to prove it for all odd primes p.

5. Use Theorem 2.1 to show that the only rational solutions of the equation

$$y^2=x^4+1$$

are $x=0,\ y=\pm 1$.

☆6. Prove that the equation

$$x^4-y^4=z^2$$

has no solution with $(x, y)=1$ and $xyz\neq 0$. [Hint. Consider the cases y even and y odd separately, and use Problem 12, Chapter 1.] Hence show that the only solutions of the equation

$$x^4+y^4=2z^2,$$

where $(x, y)=1$, are given by $x^2=y^2=1$.

7. Use the method of proof of Theorem 2.2 to show that the equation

$$x^3+y^3=3z^3$$

has no solutions with $xyz\neq 0$. Begin by noting that $3=\varepsilon\lambda^2$ for some unit ε in $\mathbb{Z}[\omega]$. For the descent step consider the equation $x^3+y^3+\varepsilon\lambda^{3n+2}z^3=0$ in $\mathbb{Z}[\omega]$.

8. Is the equation $x^3+y^3=9z^3$ soluble?

9. Germain's theorem states: if p is an odd prime such that $q=2p+1$ is also prime (see p. 217) then

$$x^p+y^p+z^p=0 \qquad (*)$$

has no solution in integers with $xyz \neq 0$. Assuming that $(*)$ has a solution $\{x_1, y_1, z_1\}$ with $x_1y_1z_1 \neq 0$ and $(x_1, y_1, z_1) = 1$, prove the following propositions in turn.

(i) There are integers a and t such that $(a, t) = 1$,

$$y_1 + z_1 = a^p,$$
$$y_1^{p-1} + y_1^{p-2}z_1 + \ldots + z_1^{p-1} = t^p.$$

Note that, by symmetry, b and c exist satisfying

$$x_1 + y_1 = b^p \quad \text{and} \quad z_1 + x_1 = c^p.$$

(ii) Use the assumption to show that $q \mid x_1y_1z_1$, note $q > 5$.
(iii) By (ii) assume that $q \mid x_1$ and so $q \nmid y_1z_1$. Show that

$$b^{(q-1)/2} + c^{(q-1)/2} - a^{(q-1)/2} \equiv 0 \pmod q$$

and so deduce $q \mid a$.
(iv) Now using (i) again prove

$$t^p \equiv py_1^{p-1} \pmod q,$$

and so derive a contradiction which proves Germain's theorem.

10. Use Skolem's method to show that the equation

$$x^4 - 8y^4 = 1$$

has no integer solutions with $y \neq 0$.

11. Show that Mordell's equation always has integer solutions if k is a square or a cube. What is special about these solutions when considered as points on the curve in the real plane?

12. Suppose k is square-free. Show that if $y^2 = x^3 + k$ is soluble in $\mathbb{Z}$ then $(k, x) = 1$.

13. Prove that $y^2 = x^3 + kz^6$ has integer solutions if and only if $y^2 = x^3 + k$ is soluble in the rational field.

14. Show that the equation

$$y^2 - 2b^2 = x^3 - a^3$$

is not soluble in integers if $a \equiv 2$ or $4 \pmod 8$, b is odd, and $p \mid b$ implies $p \equiv 1$ or $7 \pmod 8$. [Hint. Show that $x \equiv -1$ or $3 \pmod 8$ and consider these cases separately.] Which Mordell equations, with $|k| \leq 50$, can be shown to be insoluble using this result?

15. Using the method of Theorem 4.2, show that the equation

$$y^2 + 3b^2 = x^3 + 4a^3$$

has no integer solutions if $ab \neq 0$, $b \equiv 1$ or $5 \pmod 6$, $a \equiv 0$ or $2 \pmod 6$, and a has no prime factor p satisfying $p \equiv 1 \pmod 3$. Give some examples.

16. Using the method of Theorem 4.3, find the integer solutions, where they exist, of the equations

$$y^2 = x^3 - 11, \qquad y^2 = x^3 - 3.$$

[Note that $\mathbb{Z}[\omega]$ has six units where $\omega = (-1 + \sqrt{-3})/2$.]

17. Given the result: the equation $x^3 + y^3 + 2z^3 = 0$ has no integer solutions with $|x|$, $|y|$, or $|z|$ larger than one (see Mordell, 1969, p. 124), show that

$$y^2 = x^3 + 1$$

has no rational solution with $y > 3$. Explain geometrically why the solutions $\{-1, 0\}$, $\{0, \pm 1\}$, and $\{2, \pm 3\}$ do not generate any further solutions [see (i) on p. 290].

18. The equation

$$y^2 = x^3 - 432 \qquad (*)$$

has the solutions $\{12, \pm 36\}$. Let $\{x, y\}$ be another rational solution, note $x \neq 0 \neq y$. Further let a, b, and c be given by $y/36 = a/c$ and $x/12 = b/c$ with $a \equiv c \equiv 0 \pmod 2$ and $b > 0$. If $r = (a + c)/2$, $s = (c - a)/2$ show that

$$r^3 + s^3 = b^3, \qquad rsb \neq 0 \qquad (**)$$

and hence deduce that $\{12, \pm 36\}$ are the only solutions to $(*)$.

Conversely show that if $(**)$ is soluble in rational integers then $(*)$ has a solution distinct from $\{12, \pm 36\}$. [Hint. Consider the transformation $x = 12b/(r + s)$ and $y = 36(r - s)/(r + s)$.]

15

ELLIPTIC CURVES

Most of the older theory of Diophantine equations is an *ad hoc* collection of individual methods without general applicability. Some constructions can be used in a number of cases, for example the infinite descent method (Chapter 6) or Skolem's method (Chapter 14), but others apply only to single equations; examples are the congruence arguments used to show that particular Mordell equations are insoluble (also Chapter 14). Beginning with the work of Poincaré and using the methods of algebraic geometry, a more organized and substantial theory has been developed during this century which applies to a wide range of equations defined over fields. Here we shall discuss one important case, that is two-variable polynomial equations defined over the rational field $\mathbb{Q}$ (or, equivalently, homogeneous three-variable polynomial equations defined over $\mathbb{Z}$ or $\mathbb{Q}$). In this case the equations represent plane curves.

This collection of Diophantine equations falls into three main classes depending on the genus of the underlying curve. The genus is a non-negative integer which is invariant under birational transformations; it will be defined in Section 1 together with the other basic ideas from algebraic geometry we require. The first class contains polynomial equations whose curves have genus zero. All curves in this class can be reduced (by birational transformations) to conics or lines and can be parameterized using rational functions. The linear and quadratic form theory developed earlier applies to equations in this class. Many of the equations have infinite collections of integer solutions (for example, Pell's equation) and so have a great variety of rational solutions.

The elliptic curves and the polynomial equations defining them form our second main class, in this case the genus is one. The equations in this class can be reduced to cubics in a special form (see Theorem 1.4) and can be parameterized using elliptic functions. They have only finitely many solutions over $\mathbb{Z}$ but often have infinitely many rational solutions generated from a small finite set. There is an extensive and elegant theory for these equations based on the group structure of the set of rational points on the underlying curve, we shall discuss this below.

The last main class involves equations whose curves have genus larger than one. Until recently very little was known about these equations. In 1922, Mordell conjectured that each equation in this class has only finitely many rational solutions. This remarkable conjecture was established by Faltings in 1983 using advanced techniques from algebraic geometry. Faltings's results and the ideas associated with them are bound to have important and far-reaching consequences in the years to come. Below we summarize the basic properties of our three main classes.

Properties of two-variable Diophantine equations defined over $\mathbb{Q}$

		Maximum number of solutions	
Genus	Parameterization	in $\mathbb{Z}$	in $\mathbb{Q}$
0	By rational functions	Infinite	Infinite
1	By elliptic functions	Finite	Infinite but finitely generated
≥ 2	—	Finite	Finite

15.1 Geometric preliminaries

We begin by introducing some concepts from algebraic geometry. Some sketch proofs will be given; a clear and detailed account of the geometry of plane curves can be found in Fulton (1969). Let K be a field and let $A^n(K)$ denote the set of n-tuples $(x_1, \ldots, x_n)$ where each $x_i \in K$. $A^n(K)$ is called the n-dimensional *affine space* over K, it is the underlying space of the n-dimensional vector space over K. On the set $A^3(K) - \{(0, 0, 0)\}$ we define a relation $\sim$ by $(x, y, z) \sim (x', y', z')$ if and only if $x = tx'$, $y = ty'$, and $z = tz'$ for some non-zero $t \in K$. This is an equivalence relation and the set of equivalence classes is called the two-dimensional *projective space* $P^2(K)$ over K; points in $P^2(K)$ will be denoted by $(x: y: z)$. For example $(1: -3: 7)$ and $(4: -12: 28)$ represent the same point in $P^2(\mathbb{Q})$. The subset of $P^2(K)$ consisting of the points $(x: y: 1)$ is called the affine part of the plane whilst the set of points of the form $(x: y: 0)$ is called the line at infinity in $P^2(K)$.

A *curve* C in $P^2(K)$ is the set of points $(x: y: z) \in P^2(K)$ where x, y, and z satisfy an equation $f(x, y, z) = 0$ for some homogeneous polynomial f with coefficients in K. The *degree* of the curve C is the degree of f. Note that although $f(x, y, z)$ is not well defined (since x, y, and z are not determined by the equivalence class $(x: y: z)$) it does make sense in the context of the equation $f(x, y, z) = 0$ because if $t \neq 0$ we have

$$f(tx, ty, tz) = t^{\deg(f)} f(x, y, z).$$

For example $x^2+y^2-z^2=0$ is the equation of the unit circle. A point $\mathsf{X}=(x_0: y_0: z_0)$ lying on the curve C is called *singular* if

$$\frac{\partial f}{\partial x}(\mathsf{X})=\frac{\partial f}{\partial y}(\mathsf{X})=\frac{\partial f}{\partial z}(\mathsf{X})=0. \tag{1}$$

This is called a double point if at least one of the second partial derivatives is non-zero at X. If X is not a singular point then the tangent to the curve C at X can be constructed and its equation is given by

$$x\frac{\partial f}{\partial x}(\mathsf{X})+y\frac{\partial f}{\partial y}(\mathsf{X})+z\frac{\partial f}{\partial z}(\mathsf{X})=0.$$

When two curves C and C' meet or cross at a point P we associate a positive integer, called the *intersection number* $I_{\mathsf{P}}(C, C')$ (or, sometimes I_{P}), to P. The definition of I_{P} is complex, for details see Fulton (1969, pp. 74–83). One of the difficulties concerns flex points (in the affine plane these are points where the second derivative is zero, for example the origin on the curve $y=x^3$ is a flex point). We list below some cases to illustrate the concept.

(i) If curves C and C' intersect at P, P is not a singular point on C or C' and the tangents to C and C' at P are distinct, then $I_{\mathsf{P}}=1$.

(ii) If P is neither a singular point nor a flex point on C and T is the tangent to C at P, then $I_{\mathsf{P}}(C, T)=2$.

(iii) If C is a cubic curve, P is a flex point on C, and T is the tangent to C at P, then $I_{\mathsf{P}}(C, T)=3$.

(iv) If P is a double point on C and L is a line passing through P which is not a tangent to C at P, then $I_{\mathsf{P}}(C, L)=2$.

Intersection numbers play a vital role in Bezout's theorem which generalizes the fact that two distinct lines always meet in a single point in the projective plane.

THEOREM 1.1 (Bezout) *Let C and C' be curves of degree m and n, respectively, which meet in at most a finite number of points* $\mathsf{P}_1, \ldots, \mathsf{P}_k$ then

$$I_{\mathsf{P}_1}+\ldots+I_{\mathsf{P}_k}=mn$$

provided the underlying field is algebraically closed.

Note that both affine points and points at infinity must be counted using their intersection numbers. For example two conics defined over the complex numbers always meet in 'four' points. If the underlying field is not algebraicly closed we have the weaker result

COROLLARY 1.2 *Two curves of degree m and n, respectively, with no common component meet in at most mn points.*

For example distinct cubics cannot meet in more than 9 points. A proof of Bezout's theorem can be found in Fulton (1969).

Genus of a curve

A homogeneous polynomial f of degree n has at most $\frac{1}{2}(n+1)(n+2)$ coefficients and so the curve C defined by f is, in general, determined by $\frac{1}{2}n(n+3) = \frac{1}{2}(n+1)(n+2) - 1$ conditions. For example, a cubic C_1 can be drawn through 9 points, but if a second cubic C_2 can also be drawn through these points then any cubic of the form $C_1 + kC_2$ will also pass through these points; thus in general 10 conditions are needed to define a cubic. If we assume that f has rational coefficients then the equations determining the singular points will also have rational coefficients and so (via symmetric function theory) the singular points will have rational coordinates. Also we shall assume that all singular points are double points; using birational transformations (see below), it can be shown that this is a reasonable assumption.

The genus of a curve is defined in terms of N, the number of singular points, and the degree. We have

THEOREM 1.3 *If the curve C is defined by an irreducible polynomial of degree $n > 1$, and has N singular points, then $N \leq \frac{1}{2}(n-1)(n-2)$.*

Proof. Suppose $N > \frac{1}{2}(n-1)(n-2)$. A curve C' of degree $n-2$ can be constructed to pass through $\frac{1}{2}(n-1)(n-2)+1$ of these singular points and any other $n-3$ points on C, for (see preceding paragraph)

$$\tfrac{1}{2}(n-2)(n+1) = \tfrac{1}{2}(n-1)(n-2) + 1 + (n-3).$$

Note, by hypothesis C is not the union of two or more curves of lower degree. The sum of the intersection numbers for the common points of C and C' is

$$(n-3) + (n-1)(n-2) + 2 > n(n-2).$$

The result follows as this contradicts Bezout's theorem. □

DEFINITION The *genus* g of a curve C of degree n with N singular points is given by

$$g = \tfrac{1}{2}(n-1)(n-2) - N.$$

By Theorem 1.3, g is a non-negative integer. We give some examples.

(i) If the degree of C is 2, then $N=0$ (no conic has a singular point) and so $g=0$; all conics have genus zero.

(ii) If the degree of C is 3, then $g=0$ or 1. For instance if C is given by $y^2z=x^3-3z^3$ then $N=0$ and $g=1$, but if C is given by $y^2z=x^3$, then $N=1$ and $g=0$ (the origin is a singular point).

(iii) If C_n is the Fermat curve $x^n+y^n-z^n=0$, then $N=0$ and so $g=\frac{1}{2}(n-1)(n-2)$. For example C_4 has genus 3.

Many properties of a curve are unaltered by rational transformations provided their inverses are also rational. For instance if a curve has a rational point (point with rational coordinates) then its transform also possesses a rational point.

DEFINITION A curve C is said to be *birationally equivalent* to a curve C' if C and C' are transformed into one another by the mappings

$$x'=a(x,y,z),\ y'=b(x,y,z),\ z'=c(x,y,z)$$

and

$$x=a_1(x',y',z'),\ y=b_1(x',y',z'),\ z=c_1(x',y',z')$$

where $a, b, \ldots, c_1$ are rational functions (that is, one polynomial divided by another), and the correspondence is one to one and onto except for a finite set of points. The functions a, b, and c may be undefined at a finite number of points of C, and likewise a', b', and c' may be undefined at a finite number of points of C'. See the examples given in Problem 4.

Birationally equivalent curves C and C' may have different degrees and different numbers of singular points, but it can be shown that they always have the same genus. This is a consequence of a theorem of Riemann (see Fulton, 1969). Hilbert and Hurwitz have shown that a curve C_0 of genus zero is birationally equivalent to a line or a conic, whilst Poincaré showed that a curve C_1 of genus one which contains at least one rational point is birationally equivalent to a cubic with a rational point; proofs of these results can be found in Mordell (1969, Chapter 17). As an example of the methods used we give below a proof, due to Nagell, which shows that the cubic in Poincaré's result above can be taken in a special form.

DEFINITION A plane cubic curve C, defined by a polynomial with rational coefficients, which possesses at least one rational point and has genus 1 is called an *elliptic curve*.

As noted in the Introduction the term elliptic is used because C can be parameterized using Weierstrass elliptic functions.

THEOREM 1.4 *Suppose C is an elliptic curve then C is birationally equivalent to a curve with equation*

$$y^2z = x^3 + axz^2 + bz^3$$

where a and b are integers.

Proof. Let P be a rational point on C and let O be the intersection of the tangent to C at P with the curve C. The point O exists by Bezout's theorem (1.1), and it has rational coordinates. Note that $\mathsf{P} = \mathsf{O}$ if and only if P is a flex point. Now make a linear rational translation of the plane so that the new origin coincides with the point O on C; in this new system the polynomial defining C will have no term in z^3. Putting $y = tx$ the equation for C takes the form

$$x(f(t)x^2 + 2g(t)xz + h(t)z^2) = 0 \tag{1}$$

where f, g, and h are polynomials with rational coefficients having degrees 3, 2, and 1, respectively. Thus, if $x \neq 0$,

$$(f(t)x + g(t)z)^2 = (g(t)^2 - f(t)h(t))z^2 = G(t)z^2, \tag{2}$$

where G is a quartic polynomial. Let $y = t_0x$ be the tangent to C at P given in the first part of the proof. (If t_0 is infinite interchange x and y in the argument so far.) We have $G(t_0) = 0$ for, if $\mathsf{P} \neq \mathsf{O}$, the left-hand side of (1) is x times a square when $t = t_0$, and, if $\mathsf{P} = \mathsf{O}$, (1) is $f(t_0)x^3 = 0$ when $t = t_0$.

Let $t = t_0 + 1/u$, and so $u = 1/(t - t_0)$ and $G(t) = r(u)/u^4$ for some cubic polynomial r. Make the birational transformation

$$x' = (u - d)z/c \qquad [\text{and so } u = (cx' + dz)/z]$$
$$y' = u^2(f(t)x + g(t)z)/e$$

where c, d, and e are rational numbers to be chosen. In this new system (2) becomes

$$\frac{(ey')^2}{u^4} = \frac{r(u)}{u^4}z^2$$

or

$$y'^2z = \frac{r(u)}{e^2}z^3 = \frac{1}{e^2}r\left(\frac{cx' + dz}{z}\right)z^3 = \frac{1}{e^2}r'(x', z)$$

where r' is a second cubic polynomial whose coefficients involve c, d and the coefficients of r. Now by a suitable choice of c, d, and e the right-hand side of the above equation can be written in the form $x^3 + axz^2 + bz^3$. Finally we may assume that a and b are integers; if not, let k be the LCM of the denominators of a and b, and use the transformation $x'' = k^2x'$, $y'' = k^3y'$. The result follows as all transformations used in this proof are birational. $\square$

15.2 Rational points on elliptic curves

Let C be an elliptic curve with equation in standard form as given by Theorem 1.4, and let $C(\mathbb{Q})$ denote the set of points on C with rational coordinates. Using the results of the previous section all curves of genus one which are defined over $\mathbb{Q}$ and have at least one rational point are birationally equivalent to an elliptic curve in standard form. Suppose P_1 and P_2 belong to $C(\mathbb{Q})$ and L is the line $\mathsf{P}_1\mathsf{P}_2$; its equation has rational coefficients. The line L meets C in one further point P_3 (see Theorem 1.1) and $\mathsf{P}_3 \in C(\mathbb{Q})$, for if a cubic polynomial with rational coefficients has roots α, β, and γ, and α, $\beta \in \mathbb{Q}$, then $\gamma \in \mathbb{Q}$ because $\alpha + \beta + \gamma$ is rational. Note that if $\mathsf{P}_1 = \mathsf{P}_2$, then the line L is the tangent to C at P_1. This process for generating members of $C(\mathbb{Q})$ is called the *chord–tangent method.* It is usually attributed to Bachet, but it was probably invented by Newton, see Cassels (1986). We shall prove in the next section that all members of $C(\mathbb{Q})$ can be generated from a finite set of points by this method.

Using the chord–tangent method we define an addition operation on $C(\mathbb{Q})$, and so provide $C(\mathbb{Q})$ with a group structure. Let O be a fixed member of $C(\mathbb{Q})$, it will act as the identity of the group. Suppose P_1, $\mathsf{P}_2 \in C(\mathbb{Q})$ and the line $\mathsf{P}_1\mathsf{P}_2$ meets C again at Q, see Figure 5; as noted above $\mathsf{Q} \in C(\mathbb{Q})$. Now draw the line QO, this line meets C in one further

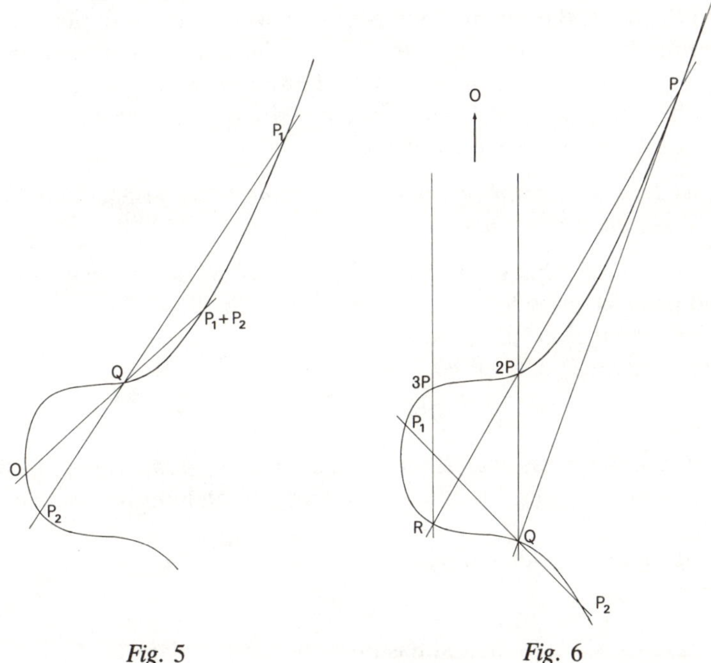

Fig. 5 *Fig.* 6

point (apart from Q and O) which is also a member of $C(\mathbb{Q})$. This new point is called the *sum* of P_1 and P_2 and is denoted by $P_1 + P_2$. Note that $P_2 + P_1 = P_1 + P_2$ and the set $C(\mathbb{Q})$ is closed under this operation.

As the point (0: 1: 0) belongs to $C(\mathbb{Q})$ for all elliptic curves in standard form, we shall take this point to be the identity O from now on. With this choice the inverse operation has a particularly simple form. We have $P + O = P$ for all $P \in C(\mathbb{Q})$, and if $P = (x: y: z)$ then $-P = (x: -y: z)$; the reader should check these facts by drawing a diagram. We illustrate some operations in Figure 6 where $Q = -2P = -(P + P) = -(P_1 + P_2)$ and $R = -3P$.

Associativity is more difficult to prove; we use the following result which is a consequence of Bezout's theorem. Suppose we have three cubic curves C_1, C_2, and C_3, at least one of which is non-degenerate (i.e. not a combination of lines or a line and a conic) and suppose C_1, C_2, and C_3 have eight mutually common points none of which are singular points on C_1, C_2, or C_3. By Bezout's theorem, let P be the ninth common point of C_1 and C_2, then it can be shown that P lies on C_3 and so the three cubic curves have nine common points. (For a proof see Fulton, 1969, p. 124; this result is the cubic analogue of Pappus's theorem.) The associativity of our addition operation follows immediately if we let C_1 be the elliptic curve C, C_2 be the triple of lines $P_1(P_2 + P_3)$, P_2P_3, $Q_{12}(P_1 + P_2)$, and C_3 be the triple of lines $P_3(P_1 + P_2)$, P_1P_2, $Q_{23}(P_2 + P_3)$; see Figure 7. The eight common points are O, P_1, $P_2 + P_3$, P_2, Q_{12}, $P_1 + P_2$, Q_{23} and P_3 (note, the singular points of C_2 and C_3 are not in this set). The ninth intersection point of C_1 and C_2 is R [the intersection of C with the line $P_1(P_2 + P_3)$] and the above result shows that R lies on C_3; that is R lies on the line $P_3(P_1 + P_2)$. Now $P_1 + P_2 + P_3$ is given by the point S on the line RO. Hence we have proved

THEOREM 2.1 *The set of points $C(\mathbb{Q})$ with the addition operation defined above forms an Abelian group.* □

We also denote this group by $C(\mathbb{Q})$. Let us consider some examples, detailed proofs can be found in Cassels (1950) or Mordell (1969). Further examples are given in the problem section.

(i) The curve C_1 defined by

$$y^2z = x^3 - 3z^3.$$

This curve has no rational affine points (that is, points $(x: y: 1)$ where x, $y \in \mathbb{Q}$) and so $C_1(\mathbb{Q})$ is the trivial group containing the single point (0: 1: 0).

(ii) The curve C_2 defined by

$$y^2z = x^3 + z^3.$$

In this case $C_2(\mathbb{Q})$ is finite and has order 6 (see Problem 17, Chapter 14).

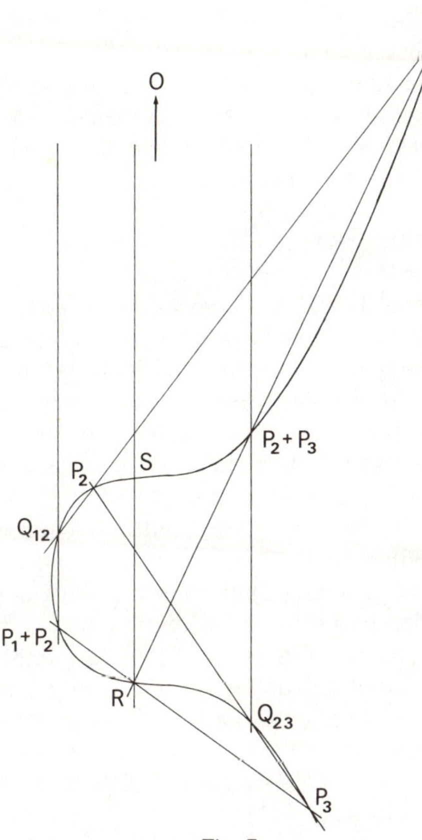

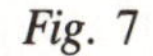
Fig. 7

(iii) The curve C_3 defined by

$$y^2z = x^3 - 7z^3.$$

Here $C_3(\mathbb{Q})$ is an infinite cyclic group generated by the point (2: 1: 1).

(iv) The curve C_4 defined by

$$y^2z = x^3 + 36z^3.$$

The group $C_4(\mathbb{Q})$ has both finite and infinite order elements. It is the direct product of the cyclic group (of order 3) generated by (0: 6: 1) and the infinite cyclic group generated by (−3: 3: 1). Tables 6 and 7 on pp. 344–5 list the generators for the groups whose curves have the equation $y^2z = x^3 + kz^3$ where $|k| \leq 50$ (see also Theorem 4.4, Chapter 14).

From these examples we see that $C(\mathbb{Q})$ can be finite or infinite but, as we shall show in the next section, $C(\mathbb{Q})$ is always finitely generated. In the cases considered up to the present time the *rank* (the number of infinite order generators) is usually small (0, 1, or 2), and no example of an elliptic curve has been given with rank larger than 14. We shall return to this topic in the final section.

A surprising application has been found recently for the groups $C(\mathbb{Q})$. In Chapter 1 we discussed computer factorization of large numbers. Using some properties of $C(\mathbb{Q})$, Lenstra has discovered a very efficient factorization method which is effective in many cases, see Lenstra (1986).

15.3 Mordell–Weil theorem

A proof of the Mordell–Weil theorem will be given in this section, it is quite difficult and the reader will need time to grasp the details. Also it is only complete in one special case; we shall indicate, in square brackets, the extensions required to derive the full result. These extensions do not alter the structure of the proof, they involve generalizations to arbitrary number fields of some of the definitions and lemmas given below in the rational case. Full proofs can be found in Lang (1978), Silverman (1986), and Husemöller (1987), and a recent account of the history of this topic is given in Cassels (1986).

The definition of the group $C(\mathbb{Q})$ given in the previous section does not rely on any particular properties of $\mathbb{Q}$ except that it forms a field of characteristic zero. Hence, if K is an algebraic number field, we can reconstruct the definition and define the group $C(K)$ of points on the elliptic curve C whose coordinates lie in K. The group operation is the same for both $C(\mathbb{Q})$ and $C(K)$. Note that, as $\mathbb{Q}$ is a subfield of K, we have $C(\mathbb{Q})$ is a subgroup of $C(K)$, and if $C(K)$ is finitely generated then so is $C(\mathbb{Q})$; see Scott (1964, p. 113).

The Mordell–Weil theorem states that, for an algebraic number field K, the group of rational points $C(K)$ on the elliptic curve C is finitely generated, that is every rational point (over K) can be obtained by the chord–tangent method from a finite number of points. Conjectured by Poincaré in 1901, the first proof was given by Mordell in 1922; the proof given here is due to Weil (see Mordell, 1969; Lang, 1978).

Outline of the proof

Let K be an algebraic number field, and let C be an elliptic curve given in standard form by

$$y^2 = x^3 + ax + b = (x - \alpha_1)(x - \alpha_2)(x - \alpha_3) \qquad (1)$$

where $a,\ b \in \mathbb{Z}$. For the first part of the proof we shall use the affine coordinate system. Note that α_1, α_2, and α_3 are distinct as C has no singular points (see Problem 5). We shall assume that α_1, α_2, and α_3 belong to K; if not, we can make a finite algebraic extension of K to include these numbers. Thus $C(K)$ has exactly three points of order two, they are $(\alpha_i, 0)$ for $i = 1, 2, 3$. It is this requirement that necessitates the use of general number fields in the full proof, even when a and b are rational and we are considering $C(\mathbb{Q})$.

In the main proof we shall define a 'height' function for points on C, and prove two inequalities for this function. The finiteness of the generating set will follow using these inequalities and

(i) the factor group $C(K)/2C(K)$ is finite;

(ii) the set of points of bounded height on $C(K)$ is finite.

Proposition (i) is sometimes called the weak Mordell–Weil theorem, we shall derive this result first.

Many of the lemmas in this section rely on the following three elementary coordinate geometry results concerning points on $C(K)$.

LEMMA 3.1 *Let P_1 and P_2 be distinct affine points on $C(K)$ with coordinates (x_1, y_1) and (x_2, y_2), respectively. Let $\mathsf{Q}_{12} = (x_{12}, y_{12})$ be the third intersection point of the line $\mathsf{P}_1\mathsf{P}_2$ with C. We have*

$$x_{12} = \frac{(x_1x_2 + a)(x_1 + x_2) + 2b - 2y_1y_2}{(x_2 - x_1)^2} \tag{2}$$

and, for $i = 1, 2, 3$,

$$x_{12} - \alpha_i = \frac{1}{(x_1 - \alpha_i)(x_2 - \alpha_i)}\left\{\frac{y_1(x_2 - \alpha_i) - y_2(x_1 - \alpha_i)}{x_2 - x_1}\right\}^2 \tag{3}$$

Proof. The line $\mathsf{P}_1\mathsf{P}_2$ has equation

$$y - y_1 = \frac{y_2 - y_1}{x_2 - x_1}(x - x_1).$$

Hence the numbers x_1, x_2, and x_{12} are the roots of

$$x^3 + ax + b - \left\{y_1 + \frac{y_2 - y_1}{x_2 - x_1}(x - x_1)\right\}^2 = 0. \tag{4}$$

By considering the coefficient of x^2 we have

$$x_1 + x_2 + x_{12} = \left\{\frac{y_2 - y_1}{x_2 - x_1}\right\}^2,$$

and so

$$\begin{aligned} x_{12} &= \frac{(y_2 - y_1)^2 - (x_1 + x_2)(x_2 - x_1)^2}{(x_2 - x_1)^2} \\ &= \frac{(x_1^3 + ax_1 + b) + (x_2^3 + ax_2 + b) - 2y_1y_2 - x_1^3 - x_2^3 + x_1x_2^2 + x_1^2x_2}{(x_2 - x_1)^2} \end{aligned}$$

and (2) follows. Note, a similar rational expression can be given for y_{12}.

To prove (3) write $x = \alpha_i + \bar{x}$, $x_1 = \alpha_i + \bar{x}_1$, $x_2 = \alpha_i + \bar{x}_2$ in (4) and we have, as α_i is a root of $x^3 + ax + b = 0$,

$$\bar{x}^3 + a_1\bar{x}^2 + b_1\bar{x} - \left\{y_1 + \frac{y_2 - y_1}{\bar{x}_2 - \bar{x}_1}(\bar{x} - \bar{x}_1)\right\}^2 = 0$$

for some suitably chosen integers a_1 and b_1. As the roots of this equation are $x_1 - \alpha_i$, $x_2 - \alpha_i$, and $x_{12} - \alpha_i$ we have

$$(x_1 - \alpha_i)(x_2 - \alpha_i)(x_{12} - \alpha_i) = \left\{\frac{y_1\bar{x}_2 - y_2\bar{x}_1}{x_2 - x_1}\right\}^2,$$

and this gives (3). □

LEMMA 3.2 *Let P_1 be an affine point on C with coordinates (x_1, y_1) where x_1 and y_1 belong to K and $y_1 \neq 0$, and let the tangent to C at P_1 meet C again at $\mathsf{Q}_{11} = (x_{11}, y_{11})$. We have*

$$x_{11} = \frac{x_1^4 - 2ax_1^2 - 8bx_1 + a^2}{4(x_1^3 + ax_1 + b)} \tag{5}$$

and, for $i = 1, 2, 3$,

$$x_{11} - \alpha_i = \left\{\frac{x_1^2 + a - 2\alpha_i x_1 - 2\alpha_i^2}{2y_1}\right\}^2. \tag{6}$$

Proof. This is similar to the proof above. In (4) replace the term $(y_2 - y_1)/(x_2 - x_1)$ by $(3x_1^2 + a)/2y_1$, the slope of the tangent at P_1, and proceed as in Lemma 3.1. □

LEMMA 3.3 *Suppose α_i, α_j, and α_k are distinct. If $y \neq 0$ and $x_{\alpha i}$ is the first coordinate of the point $(\alpha_i, 0) + (x, y)$ then, for $i = 1, 2$ or 3,*

$$x_{\alpha i} - \alpha_i = (x - \alpha_i)(\alpha_i - \alpha_j)(\alpha_i - \alpha_k)s^2$$

for some non-zero $s \in K$.

Proof. Note first that, as $\alpha_i + \alpha_j + \alpha_k = 0$,

$$\begin{aligned}(\alpha_i - \alpha_j)(\alpha_i - \alpha_k) &= 3\alpha_i^2 - 2\alpha_i(\alpha_i + \alpha_j + \alpha_k) + \alpha_i\alpha_j + \alpha_j\alpha_k + \alpha_k\alpha_i \\ &= 3\alpha_i^2 + a. \end{aligned} \tag{7}$$

Now $x \neq \alpha_i$ as $y \neq 0$, and so by (2) in Lemma 3.1 we have, if $s = (x - \alpha_i)^{-1}$,

$$\begin{aligned} x_{\alpha i} - \alpha_i &= \{\alpha_i x^2 + (a + \alpha_i^2)x + a\alpha_i + 2b - \alpha_i(x - \alpha_i)^2\}s^2 \\ &= \{x(3\alpha_i^2 + a) + \alpha_i(\alpha_i\alpha_j + \alpha_j\alpha_k + \alpha_k\alpha_i) - 2\alpha_i\alpha_j\alpha_k - \alpha_i^3\}s^2 \\ &= (x - \alpha_i)(\alpha_i - \alpha_j)(\alpha_i - \alpha_k)s^2 \qquad \text{by (7).} \end{aligned}$$ □

Our next objective is to show that $C(K)/2C(K)$ is finite. A point Q belongs to $2C(K)$ if there is $\mathsf{P} \in C(K)$ which satisfies $-\mathsf{Q} = \mathsf{P} + \mathsf{P} = 2\mathsf{P}$;

that is Q is constructed using the tangent at P. Further the points P_1 and P_2, where $P_1, P_2 \in C(K)$, belong to the same element of the group $C(K)/2C(K)$ [that is, they belong to the same coset of $2C(K)$ in $C(K)$] if and only if $P_1 + P_2 = 2P$ for some $P \in C(K)$. Geometrically this is equivalent to: if Q is constructed using the chord P_1P_2 then it can also be constructed using the tangent at P (see Figure 6 on p. 289). Thus the weak Mordell–Weil theorem states that there exists a finite set of points [of $C(K)$] $\mathbf{P} = \{P_1, \ldots, P_n\}$ such that every point $Q \in C(K)$ can be expressed in the form $Q = P_i + 2R$ where $P_i \in \mathbf{P}$ and $R \in C(K)$, that is every element of $C(K)$ is the sum of a member of $\mathbf{P}$ and a 'square'.

We let K^* denote the multiplicative group of the non-zero elements of K. Note that if t_1 and t_2 belong to the same coset of K^{*2} in K^* then $t_1 = t_2 s^2$ for some $s \in K^*$. We prove our result by defining three homomorphisms θ_i and applying the first homomorphism theorem for groups to the intersection of their kernels.

Definition For $i = 1, 2, 3$, let θ_i be the mapping $C(K) \to K^*/K^{*2}$ given by (where an element of K^*/K^{*2} is denoted by sK^{*2})

$$\theta_i(O) = K^{*2}$$
$$\theta_i((x, y)) = (x - \alpha_i)K^{*2} \qquad \text{if } x \neq \alpha_i$$
$$\theta_i((\alpha_i, 0)) = (\alpha_i - \alpha_j)(\alpha_i - \alpha_k)K^{*2}$$

where O is the identity of $C(K)$ and i, j, and k are distinct.

Lemma 3.4 *For $i = 1, 2, 3$, θ_i is a homomorphism.*

Proof. Assume throughout that α_i, α_j, and α_k are distinct. As $(\alpha_j, 0) + (\alpha_k, 0) = (\alpha_i, 0)$ we have

$$\begin{aligned}\theta_i((\alpha_j, 0) + (\alpha_k, 0)) &= \theta_i((\alpha_i, 0)) = (\alpha_i - \alpha_j)(\alpha_i - \alpha_k)K^{*2} \\ &= \theta_i((\alpha_j, 0))\theta_i((\alpha_k, 0)),\end{aligned}$$

and

$$\begin{aligned}\theta_i((\alpha_i, 0) + (\alpha_j, 0)) = (\alpha_k - \alpha_i)K^{*2} &= -(\alpha_i - \alpha_k)(\alpha_i - \alpha_j)^2 K^{*2} \\ &= \theta_i((\alpha_i, 0))\theta_i((\alpha_j, 0)).\end{aligned}$$

This gives the result for the points of order 2. The result also follows in the remaining cases using Lemma 3.1 for the points (x_1, y_1) and (x_2, y_2) when $y_1 y_2 \neq 0$, and Lemma 3.3 for the points (x, y) and $(\alpha_i, 0)$ when $y \neq 0$. □

Lemma 3.5 $\bigcap_{i=1}^{3} \ker \theta_i \subset 2C(K)$.

Proof. To prove this result we must show that if θ_i maps a point $P_1 = (x_1, y_1) \in C(K)$ onto a square in K^* when $i = 1, 2$, and 3, then we can find another point $P_2 = (x_2, y_2) \in C(K)$ such that $P_1 = 2P_2$. Let L be the tangent to the curve C at P_2 and let $-P_{22} = (x_{22}, y_{22})$ be the remaining point where L intersects C. We shall prove the result by showing that $P_1 = -P_{22} = 2P_2$.

First consider the case $y_1 \neq 0$. By assumption we have

$$x_1 - \alpha_i = s_i^2$$

for some $s_i \in K$, $i = 1$, 2, and 3. The set of equations

$$u + \alpha_i v + \alpha_i^2 w = s_i$$

has a unique solution $\{u, v, w\}$ in K as the coefficient matrix has a Vandermonde determinant and the α are distinct. Hence

$$(u + \alpha_i v + \alpha_i^2 w)^2 = x_1 - \alpha_i,$$

and, as $\alpha_i^3 + a\alpha_i + b = 0$, we have, multiplying out,

$$-x_1 + \alpha_i + u^2 + v^2\alpha_i^2 - w^2(a\alpha_i^2 + b\alpha_i) + 2uv\alpha_i + 2uw\alpha_i^2 - 2vw(a\alpha_i + b) = 0.$$

Equating coefficients we have

$$\begin{aligned} -x_1 + u^2 - 2bvw &= 0 \\ 1 - bw^2 + 2uv - 2avw &= 0 \\ v^2 - aw^2 + 2uw &= 0. \end{aligned} \tag{8}$$

The last two equations give, eliminating u,

$$w - bw^3 - avw^2 - v^3 = 0$$

or, as $w \neq 0$ [if $w = 0$ then $v = 0$ contradicting (8)],

$$\left(\frac{v}{w}\right)^3 + a\frac{v}{w} + b = \left(\frac{1}{w}\right)^2.$$

Let $x_2 = v/w$, $y_2 = 1/w$ and $P_2 = (x_2, y_2)$. It follows that P_2 lies on $C(K)$ and, using the last equation of (8) again,

$$x_2^2 + 2uy_2 = a.$$

Thus $u = (a - x_2^2)/2y_2$ and the first equation of (8) gives

$$x_1 = \left(\frac{a - x_2^2}{2y_2}\right)^2 - \frac{2bx_2}{y_2^2} = \frac{x_2^4 - 2ax_2^2 - 8bx_2 + a^2}{4y_2^2}.$$

The first part of the result follows because, by Lemma 3.2(5), $x_{22} = x_1$, and so $P_1 = 2P_2$.

For the second part suppose α_i, α_j, and α_k are distinct and $P_1 = (\alpha_i, 0)$. If $P_1 \in \bigcap \ker \theta_i$ then

$$\alpha_i - \alpha_j = s^2 \quad \text{and} \quad \alpha_i - \alpha_k = t^2$$

for $s, t \in K$. Note, by (7), $(\alpha_i - \alpha_j)(\alpha_i - \alpha_k) = 3\alpha_i^2 + a = s^2t^2$. The tangent to C at a point (x_1, y_1) has the equation

$$2y_1 y = (3x_1^2 + a)x + (2b + ax_1 - x_1^3). \tag{9}$$

If this intersects C at P_1 we have

$$x_1^3 - 3\alpha_i x_1^2 - ax_1 - (a\alpha_i + 2b) = 0$$

which gives, as α_i is one root of this equation,

$$x_1 = \alpha_i \pm \sqrt{(3\alpha_i^2 + a)} = \alpha_i \pm st \in K.$$

By (9), $y_1 \in K$ and so $\mathsf{P}_1 = 2(x_1, y_1) \in 2C(K)$. This completes the proof. □

We can now prove the weak Mordell–Weil theorem.

THEOREM 3.6 *If $C(K)$ is the group of points defined over an algebraic number field K on an elliptic curve C then $C(K)/2C(K)$ is finite.*

Proof. As $\bigcap \ker \theta_i$ is a subgroup of $C(K)$, we have by Lemmas 3.4 and 3.5

$$C(K)/2C(K) \subset C(K)/\bigcap \ker \theta_i. \tag{10}$$

Suppose first $K = \mathbb{Q}$, and so $\alpha_1, \alpha_2, \alpha_3 \in \mathbb{Z}$ as $a, b \in \mathbb{Z}$. Let $(x, y) = (r/t^2, s/t^3) \in C(\mathbb{Q})$ where $r, s, t \in \mathbb{Z}$ and $(r, t^2) = (s, t^3) = 1$. By (1) we have

$$(r - \alpha_1 t^2)(r - \alpha_2 t^2)(r - \alpha_3 t^2) = s^2,$$

If $u|s$, $u|r - \alpha_i t^2$ and $u|r - \alpha_j t^2$, then $u|t^2(\alpha_i - \alpha_j)$ and $u|r(\alpha_i - \alpha_j)$, and so $u|\alpha_i - \alpha_j$ and there are only finitely many choices for u. Hence for each i and each $(x, y) \in C(\mathbb{Q})$

$$x - \alpha_i = z^2 u$$

where $u, z \in \mathbb{Q}$ and u belongs to a fixed finite set. It follows that there are only finitely many cosets of $\bigcap \ker \theta_i$ in $C(\mathbb{Q})$, and so by (10) the proof is complete in this case.

[The proof above relies on unique factorization (of $\mathbb{Q}$), but in the general case K may not have this property and so we proceed as follows. The ring of algebraic integers in K is a Dedekind domain in which ideals have unique factorization into prime ideals. Also the group of units of this ring is finitely generated (Dirichlet's unit theorem). Using these two results the proof of the general case can be completed easily using an argument very similar to the one above. Details of the ideal theory and of Dirichlet's result can be found in, for example, Hua (1982) or Stewart and Tall (1979).] □

THEOREM 3.7 *Suppose $K = \mathbb{Q}(\xi)$ is a simple algebraic extension of $\mathbb{Q}$. If $C(K)/2C(K)$ is finite then $C(\mathbb{Q})/2C(\mathbb{Q})$ is also finite.*

Proof. Let $\mathcal{G}$ be the (finite) set of conjugate operations for K over $\mathbb{Q}$ (see p. 75; $\mathcal{G}$ is the Galois group of K over $\mathbb{Q}$); and if $\mathsf{Q} = (x, y) \in C(K)$ and

$i \in \mathcal{G}$, let $Q_{(i)} = (x_{(i)}, y_{(i)})$ where $x_{(i)}, y_{(i)}$ are the ith field conjugates of x and y, respectively, and where $Q = Q_{(1)}$. Let C_2 denote the set of points of order 2 on C (i.e. points P satisfying $2P = O$, by Lemma 3.2 C_2 has at most four points). Finally let J be the kernal of the natural homomorphism $C(\mathbb{Q})/2C(\mathbb{Q}) \to C(K)/2C(K)$, that is

$$J = (C(\mathbb{Q}) \cap 2C(K))/2C(\mathbb{Q}). \tag{11}$$

As $C(K)/2C(K)$ is finite the result follows from: J is finite.

If $P \in J$, there is a $Q^P \in C(K)$ such that $P = 2Q^P$ (Q^P may not be unique, but there are only finitely many possibilities). Now associate with each $P \in J$ a mapping f_P: $\mathcal{G} \to C_2$ given by

$$f_P(i) = Q^P_{(i)} - Q^P.$$

It is a simple, if rather tedious, matter to show that

$$2(Q^P_{(i)} - Q^P) = (2Q^P)_{(i)} - 2Q^P = P_{(i)} - P = O$$

as $P \in C(\mathbb{Q})$. Hence each f_P is well defined. Suppose further $P, P' \in C(\mathbb{Q}) \cap 2C(K)$ and $f_P = f_{P'}$. It follows that, as the range of each f_P is contained in C_2,

$$(Q^{P'} - Q^P)_{(i)} = Q^{P'} - Q^P$$

for all $i \in \mathcal{G}$, and so $Q^{P'} - Q^P \in C(\mathbb{Q})$. Hence

$$P' - P = 2Q^{P'} - 2Q^P \in 2C(\mathbb{Q}),$$

that is P and P′ belong to the same coset of J. Therefore the association of P with f_P is one to one, and so the result follows by (11) as there are only finitely many maps between $\mathcal{G}$ and C_2. □

This result completes the proof of the weak Mordell–Weil theorem over $\mathbb{Q}$; see Problem 17 where a special case is treated.

For the second part of the derivation of the Mordell–Weil theorem we shall revert to the projective coordinate system and consider points defined over $\mathbb{Q}$ only (see Theorem 3.7). We begin with a

DEFINITION Let $P \in C(\mathbb{Q})$ have projective coordinates $(x\!:\!y\!:\!z)$ where x, $z \in \mathbb{Z}$ and $(x, z) = 1$, and let the function h from $C(\mathbb{Q})$ to the non-negative reals be given by

$$h(O) = 0 \qquad \text{where O is the point at infinity on C}$$
$$h(P) = \ln \max(|x|, |z|) \qquad \text{for } P \neq O.$$

h(P) is called the *height* of P. Note that it is not necessary for y to be an integer but, as $y^2z = x^3 + axz^2 + bz^3$, we have y^2z is an integer and

$$\ln y^2z \leq 3h(P) + c_1 \tag{12}$$

where c_1 is a constant determined by a and b.

[In the general number field K case we replace the absolute value in this definition by a product, over all primes, of p-adic valuations defined over K; see Section 3, Chapter 3 and Lang (1978, pp. 77 ff) The proofs of Lemmas 3.8, 3.9 and 3.10 in the general case follow closely those given below for $\mathbb{Q}$ using the p-adic properties of the new height function.]

We shall now prove the three basic properties of the function h.

LEMMA 3.8 *Given a constant k, there are only finitely many points on $C(\mathbb{Q})$ whose height is at most k.*

Proof. Let $\mathsf{P}=(x:y:z)\in C(\mathbb{Q})$, $x, z\in\mathbb{Z}$, $(x,z)=1$, and $h(\mathsf{P})\le k$. Clearly there are only finitely many choices for x and z. Further $(z, y^2z)=(z, x^3+axz^2+bz^3)=1$, and so if $y=y_1/z_1$ where y_1, $z_1\in\mathbb{Z}$ then $z=z_1^2$. The lemma follows by (12). □

LEMMA 3.9 *For a fixed point P_1 on the curve $C(\mathbb{Q})$ there is a constant c_2 (dependent on P_1) such that, if $\mathsf{P}\in C(\mathbb{Q})$,*

$$h(\mathsf{P}+\mathsf{P}_1)\le 2h(\mathsf{P})+c_2.$$

Proof. Let $\mathsf{P}_1=(x_1: y_1: z_1)$, $\mathsf{P}=(x_2: y_2: z_2)$, and $\mathsf{P}+\mathsf{P}_1=(x_{12}: -y_{12}: z_{12})$ where in each case the first and last coordinates are coprime integers. Writing Lemma 3.1(2) in projective coordinates we have

$$\frac{x_{12}}{z_{12}}=\frac{(x_1x_2+az_1z_2)(x_1z_2+x_2z_1)+2bz_1^2z_2^2-2y_1y_2z_1z_2}{(x_1z_2-x_2z_1)^2}.$$

Let the right-hand side of this equation be abbreviated to e_1/e_2. By (12) we have, as x_1, y_1, and z_1 are fixed,

$$\ln|e_1|\le 2h(\mathsf{P})+c' \quad\text{and}\quad \ln|e_2|\le 2h(\mathsf{P})+c''$$

where c' and c'' are constants dependent upon x_1, z_1, a, and b. This gives, as $(x_{12}, z_{12})=1$,

$$\begin{aligned}h(\mathsf{P}+\mathsf{P}_1)=\ln\max(|x_{12}|, |z_{12}|)&\le\ln\max(|e_1|, |e_2|)\\ &\le 2h(\mathsf{P})+\max(c', c'')\end{aligned}$$

and the result follows if we let $c_2=\max(c', c'')$. □

LEMMA 3.10 *A non-negative constant c_3 exists with the property*

$$3h(\mathsf{P})-c_3\le h(2\mathsf{P})$$

for all points P lying on $C(\mathbb{Q})$.

Proof. The result is obvious when $\mathsf{P}=\mathsf{O}$, so we may suppose P is an affine point. Writing the equation for x_{11} in Lemma 3.2 in projective coordinates we have if $\mathsf{P}=(x_1: y_1: z_1)$ and $2\mathsf{P}=(x_{11}: -y_{11}: z_{11})$,

$$\frac{x_{11}}{z_{11}}=\frac{f_1(x_1, z_1)}{f_2(x_1, z_1)}$$

where

$$f_1(x, z) = x^4 - 2ax^2z^2 - 8bxz^3 + a^2z^4,$$
$$f_2(x, z) = 4(x^3 + axz^2 + bz^3).$$

We can find an integer Δ and homogeneous polynomials g_i ($i = 1, \ldots, 4$), in x and z, with integer coefficients having the following properties: g_1 and g_3 have degree 3, g_2 and g_4 have degree 4, and

$$\begin{aligned} f_1(x, z)g_1(x, z) + f_2(x, z)g_2(x, z) &= \Delta x^7, \\ f_1(x, z)g_3(x, z) + f_1(x, z)g_4(x, z) &= \Delta z^7. \end{aligned} \tag{13}$$

The coefficients of the polynomials g_i can be found by equating the coefficients of $x^7, x^6z, \ldots, z^7$ in the equation above. In each case Δ is the determinant of the coefficient matrix and equals $4(4a^3 + 27b^2)^2$; as C has no singular points, $\Delta \neq 0$ (see Problems 6 and 10). If we let c^* be the maximum of the absolute values of the coefficients of the polynomials g_i ($i = 1, \ldots, 4$), the equations (13) give

$$\begin{aligned} \Delta \max(|x_1|^7, |z_1|^7) &\leq 2 \max(|f_1(x_1, z_1)|, |f_2(x_1, z_1)|) \max_{1 \leq i \leq 4} |g_i(x_1, z_1)| \\ &\leq 8c^* \max(|f_1(x_1, z_1)|, |f_2(x_1, z_1)|) \max(|x_1|^4, |z_1|^4). \end{aligned}$$

Taking logarithms we have

$$\ln \Delta + 7h(\mathsf{P}) \leq \ln 8c^* + h(2\mathsf{P}) + 4h(\mathsf{P}),$$

and the result follows if we let $c_3 = \max(0, \ln(8c^*/\Delta))$. □

THEOREM 3.11 *The Mordell–Weil theorem for* $\mathbb{Q}$.
The group of rational points $C(\mathbb{Q})$ *on an elliptic curve* C *is finitely generated.*

Note that our proof of this result is only complete in the case when α_1, α_2, and $\alpha_3 \in \mathbb{Z}$, that is when $K = \mathbb{Q}$. We have indicated the extensions required when α_1, α_2, and $\alpha_3 \notin \mathbb{Z}$, these involve two basic results from algebraic number theory. But the full result can only be derived using some p-adic analysis.

Proof. By Theorems 3.6 and 3.7 we can choose a fixed finite set of points $\mathsf{Q}_1, \ldots, \mathsf{Q}_k$ on $C(\mathbb{Q})$ such that each point $\mathsf{P} \in C(\mathbb{Q})$ can be expressed in the form

$$\mathsf{P} = \mathsf{Q}_i + 2\mathsf{R}$$

where $\mathsf{R} \in C(\mathbb{Q})$ and $i \in \{1, \ldots, k\}$. Hence for an arbitrary point $\mathsf{P}_0 \in C(\mathbb{Q})$ we can choose a sequence of points $\mathsf{P}_0, \mathsf{P}_1, \ldots$ on $C(\mathbb{Q})$ such that

$$\mathsf{P}_n = \mathsf{Q}_{i_n} + 2\mathsf{P}_{n+1} \quad \text{or} \quad 2\mathsf{P}_{n+1} = \mathsf{P}_n - \mathsf{Q}_{i_n}. \tag{14}$$

Applying Lemmas 3.9 and 3.10 we obtain, as $h(\mathsf{P}) = h(-\mathsf{P})$,

$$-c_3 + 3h(\mathsf{P}_{n+1}) \leq h(2\mathsf{P}_{n+1}) - h(\mathsf{P}_n + \mathsf{Q}_{i_n}) \leq 2h(\mathsf{P}_n) + c'_n,$$

where c'_n is dependent on Q_{i_n}. If we let $c_2^* = \max\limits_{1 \leq j \leq k} c'_j$, then we can replace c'_n by the absolute constant c_2^* in the above inequality. Thus, if $c_4 = c_2^* + c_3$, we have

$$h(\mathsf{P}_{n+1}) \leq \frac{2}{3} h(\mathsf{P}_n) + \frac{c_4}{3}$$

$$\leq \left(\frac{2}{3}\right)^n h(\mathsf{P}_0) + c_4\left(\frac{1}{3} + \ldots + \frac{1}{3^n}\right).$$

Hence for large n the height of P_n is bounded by some fixed constant c_5. By Lemma 3.8, there are only finitely many points of height less than c_5, and so if we add the set of points with height bounded by c_5 to the set $\{\mathsf{Q}_1, \ldots, \mathsf{Q}_k\}$ we obtain a finite set G of generators for $C(\mathbb{Q})$. Every point on $C(\mathbb{Q})$ can be constructed from the points in G using equation (14), that is by the chord–tangent method. □

It is possible for some of the generators of $C(\mathbb{Q})$ to be of finite order, see Table 6 on p. 344 and Problem 17, Chapter 14. If $\mathsf{P} \in C(\mathbb{Q})$ has finite order, that is if $n\mathsf{P} = \mathsf{O}$ for some integer n, then P is called an *exceptional* point. For example if C is the curve $y^2z = x^3 + 4z^3$, then the points $(0: \pm 2: 1)$ are exceptional as they have order 3 (they are flex points). Nagell has proved

THEOREM 3.12 *If* (x_1, y_1) *is an exceptional affine point on the curve*

$$y^2 = x^3 + ax + b, \tag{15}$$

where $a, b \in \mathbb{Z}$, *then* x_1 and y_1 *are integers and either* $y_1^2 | 4a^3 + 27b^2$ *or* $y_1 = 0$.

More recently Mazur (1977) has given a complete list of possible torsion subgroups (subgroups of points of finite order) that can occur on elliptic curves over $\mathbb{Q}$. They are cyclic groups of order $1, 2, \ldots, 10$ or 12, or direct products of a group of order 2 and a cyclic group of order $2m$ where $m = 1, \ldots, 4$. If $a = 0$ in (15), the possible orders are 1, 2, 3, or 6; see Olsen (1974).

15.4 Rank and the zeta function of a curve

One of the major problems remaining in the theory of rational points on elliptic curves defined over $\mathbb{Q}$ concerns the number g_C of generators of infinite order in the group $C(\mathbb{Q})$. The number g_C is called the *rank* of $C(\mathbb{Q})$ and is the smallest number (finite by the Mordell–Weil theorem) of rational points on the elliptic curve C from which all other rational points of infinite order can be obtained using the chord–tangent method. For many curves the rank is zero, one, or two, but some examples have been given in which g_C is at least 14. It is not known if the rank can be arbitrarily large or, conversely, if an absolute constant m exists such that $g_C \leq m$ for all elliptic curves C. This problem has been investigated extensively and some remarkable conjectures relating the rank to the behaviour of the 'zeta function of the curve' have been proposed. We shall consider these briefly now. To simplify the discussion we only treat curves of the form

$$y^2z = x^3 + bz^3, \tag{1}$$

where b is a non-zero integer.

So far in this chapter we have considered points on elliptic curves defined over fields of characteristic zero, but it is possible to study points defined over fields of any characteristic $\neq 2, 3$; in particular, the set of points $C(\mathbb{Z}/p\mathbb{Z})$ defined over the finite field $\mathbb{Z}/p\mathbb{Z}$. If $\mathsf{P} = (x: y: z) \in C(\mathbb{Z}/p\mathbb{Z})$ then

$$y^2z \equiv x^3 + bz^3 \pmod{p}.$$

As before $C(\mathbb{Z}/p\mathbb{Z})$ is an Abelian group (see Problem 13) with identity $(0: 1: 0)$ and is necessarily finite. Let the order of $C(\mathbb{Z}/p\mathbb{Z})$ be N_p, we write

$$N_p = p + 1 - a_p \tag{2}$$

where it can be shown that $|a_p| \leq 2\sqrt{p}$ (see Problem 12). For curves of the type (1) a_p can be expressed in terms of sixth-power residues (see Ireland and Rosen, 1982, p. 305).

Following some work of Siegel extending the class number formulas for binary quadratic forms (of Chapter 10) to general quadratic forms, Birch and Swinnerton–Dyer considered the product

$$\prod_p \frac{N_p}{p},$$

defined over the primes, and its associated zeta function ζ_C. As some primes require special treatment, it turns out that the most suitable definition for ζ_C is, for complex s,

$$\zeta_C(s) = \zeta(s)\zeta(s-1)L^*(C, s)^{-1} \tag{3}$$

where ζ is the Riemann zeta function (see p. 224) and where

$$L^*(C, s) = \prod_{p \nmid 6b} (1 - a_p p^{-s} + p^{1-2s})^{-1}$$

Note that, if $p \nmid 6b$, $1 - a_p p^{-s} + p^{1-2s}$ at $s = 1$ equals N_p/p. The infinite product in (3) converges if the real part of s is larger than $3/2$; this follows from the result

$$1 - a_p p^{-s} + p^{1-2s} = (1 - \gamma p^{-s})(1 - \bar{\gamma} p^{-s})$$

where γ satisfies $|\gamma| = |\bar{\gamma}| = \sqrt{p}$ [and this is a consequence of the result quoted below (2)]. Hasse conjectured that the domain of definition of ζ_C can be extended (by analytic continuation) to the whole plane for all elliptic curves C.

Hasse's conjecture has been established for curves of the type (1). This is achieved by showing that

$$L^*(C, s) = L(s, \chi) \tag{4}$$

where L is similar to the L function defined in Chapter 13 and χ is a character (called a Hecke character) defined over the ideals in $\mathbb{Z}[\omega]$ where ω is a complex cube root of unity. Equation (4) implies that the behaviour of $L^*(C, s)$ can be studied near $s = 1$. This study and a considerable amount of numerical evidence led Birch and Swinnerton–Dyer to postulate the following

CONJECTURE For an elliptic curve C, the rank of the group $C(\mathbb{Q})$ equals the order of the zero of $L^*(C, s)$ at $s = 1$.

Some progress has been made on this conjecture for it has been shown (by Coates and Wiles) that g_C is zero if $L^*(C, 1)$ is non-zero. The extensive tables given by Birch and Swinnerton–Dyer suggest that their conjecture is true for a large class of elliptic curves [including most of those of type (1) above], but it is possible that some exceptions exist. For further details the reader should consult Ireland and Rosen (1982, Chapter 18) for a good introduction to this topic, Birch and Swinnerton–Dyer (1963, 1965) for a more detailed discussion and extensive tables, and Husemöller (1987) for an up-to-date and comprehensive treatment of the whole subject.

It is remarkable here, as so often in number theory, how ideas and theorems from different areas come together to give new insights and results, and to suggest new lines of enquiry. And it is not only completed theories and proofs that are important, conjectures and new 'angles' on established concepts all play a vital role. Progress in number theory, as well as science in general, would come to a halt if this was not so.

15.5 Problems 15

1. Find the singular points and draw the curves of the equations
   (i) $y^2z^3 = x^2(x - z)(x - 2z)^2$,
   (ii) $y^2z^2 = x(x - z)^2(x - 2z)$,
   (iii) $y^2z = x(x - z)(x - 2z)$,
   (iv) $x^2y^2 = (x^2 + y^2)z^2$.

2. Check that Bezout's theorem holds in the following cases.
   (i) The parabolas $y^2 = 4xz$, $y^2 = 8xz$.
   (ii) The cubics $y^2z = x^3$, $y^2z = 8x^3$.
   (iii) The line $y = z$ and the cubic $y^2z = x^3 + z^3$.

3. Find the genus of the curves given in Problem 1.

4. Show that the following pairs of curves are birationally equivalent, and check that the genus is preserved in both cases.
   (i) $x^2 + y^2 - xz = 0$ and $x^3 + x^2z - y^2z = 0$.
   (ii) $x^2 + y^2 = z^2$ and $x^2y^2 = (x^2 + y^2)z^2$.
[Hint. Work in the affine plane and try the mappings

$$x' = (1 - 2x)/x,\ y' = y(1 - 2x)/x^2 \quad \text{and} \quad x' = 1/x,\ y' = 1/y.]$$

5. What is the genus of the curve $y^2z = x^2(x - az)$?

6. Let C be the curve $y^2z = x^3 + axz^2 + bz^3$, a, $b \in \mathbb{Z}$; show that C is elliptic if and only if $4a^3 + 27b^2 \neq 0$.

7. Explain geometrically why the point $(0:1:0)$ on the curve defined in the previous problem does not generate any new points on $C(\mathbb{Q})$.

8. Find as many points (x, y) as possible on the curve $y^2 = x^3 + 36$ where both x and y are integers.

9. Let C be an elliptic curve, let W_C denote the order of $C(\mathbb{Q})/2C(\mathbb{Q})$, and let R_C equal the number of generators (finite and infinite) of $C(\mathbb{Q})$. Give examples of curves C where (i) $W_C - 1 = R_C$, and (ii) $W_C - 1 < R_C$.

10. Complete the proof of Lemma 3.9 by finding Δ and the coefficients of the polynomials g_i $(i = 1, \ldots, 4)$.

11. Let $C_t(\mathbb{Q})$ be the torsion subgroup of the group $C(\mathbb{Q})$ where C is an elliptic curve in standard form. Give examples to show that $C_t(\mathbb{Q})$ can be cyclic of order 1, 2, 3, 4, or 6, or a direct product of two cyclic groups of order 2. [Hint. Consider the cases $a = 0$ and $b = 0$.]

12. Suppose p is an odd prime satisfying $p \equiv 2 \pmod 3$. Using the theory of Jacobi sums given in Chapter 6, show that

$$N(y^2 \equiv x^3 + b(p)) = p + 1$$

where one of the points counted is the point at infinity on the curve.

13. Show that the set of points $C(\mathbb{Z}/p\mathbb{Z})$ on the elliptic curve C defined over the finite field $\mathbb{Z}/p\mathbb{Z}$ forms an Abelian group.

14. Let C be the curve $y^2 = x^3 + b$ and let p be a prime congruent to 2 (mod 3) such that $p \nmid 6b$. Assuming the result: $C_t(\mathbb{Q})$ is isomorphic to a subgroup of $C_t(\mathbb{Z}/p\mathbb{Z})$, where C_t is defined in Problem 11, show that the order of $C_t(\mathbb{Q})$ can only be 1, 2, 3, 4, or 6. ($C_t(\mathbb{Q})$ cannot have order 4 but this does not follow from the method given here.) [Hint. Use Problem 12, Dirichlet's theorem (1.4, Chapter 13), and a density argument to show that if $r \mid p + 1$ then $\phi(r) \leq 2$.]

15. (i) Let C_1 be the curve $y^2 = x^3 + k$ and let C_2 be the curve $y^2 = x^3 - 27k$.

Show that if $(x_1, y_1) \in C_1(\mathbb{Q})$ then $(x_2, y_2) \in C_2(\mathbb{Q})$ where

$$x_2 = (x_1^3 + 4k)/x_1^2, \qquad y_2 = y_1(x_1^3 - 8k)/x_1^3,$$

and if $(x_3, y_3) \in C_2(\mathbb{Q})$ then $(x_4, y_4) \in C_1(\mathbb{Q})$ where

$$x_4 = (x_3^3 - 108k)/9x_3^2, \qquad y_4 = y_3(x_3^3 + 216k)/27x_3^3.$$

(ii) Show that if $x_3 = x_2$ and $y_3 = y_2$ in (i), and if $\mathsf{P}_1 = (x_1, y_1)$ and $\mathsf{P}_4 = (x_4, y_4)$, then $\mathsf{P}_4 = 3\mathsf{P}_1$. Deduce that

$$C_1(\mathbb{Q}) \supset C_2(\mathbb{Q}) \supset 3C_1(\mathbb{Q}),$$

where $3C_1(\mathbb{Q})$ is the subgroup of $C_1(\mathbb{Q})$ containing the points $\mathsf{P}_1 + \mathsf{P}_1 + \mathsf{P}_1$ for $\mathsf{P}_1 \in C_1(\mathbb{Q})$.

(iii) Suppose $\mathsf{P}_1 = (x_1, y_1)$, $\mathsf{P}_2 = (x_2, y_2)$ belong to $C_1(\mathbb{Q})$ and suppose $\mathsf{P}_1 + \mathsf{P}_2 = (x_{12}, y_{12})$. Show that

$$(y_1 + \sqrt{k})(y_2 + \sqrt{k}) = \alpha^3(y_{12} + \sqrt{k})$$

where

$$\alpha = (x_1 y_2 - x_2 y_1 + (x_1 + x_2)\sqrt{k})/x_{12}.$$

Further show that if $3\mathsf{P}_1 = \mathsf{P}_3 = (x_3, y_3)$ then

$$y_1 + \sqrt{k} = \beta^3$$

for some $\beta \in \mathbb{Q}(\sqrt{k})$.

(iv) Finally show that the point (x_{12}, y_{12}) defined in (iii) lies on $C_1(\mathbb{Q})$ if and only if

$$(y_1 + \sqrt{k})(y_2 + \sqrt{k}) = \gamma^3$$

for some $\gamma \in \mathbb{Q}(\sqrt{k})$.

These results can be used to re-prove the Mordell–Weil theorem for C_1, the proof uses the quadratic field $\mathbb{Q}(\sqrt{k})$ rather than the cubic field of the original proof, see Mordell (1969).

16. A rational number is called *congruent* if it is the area of a right-angled triangle with rational sides. For example, 6 is congruent being the area of a 3,4,5 triangle. Suppose n is congruent and is associated with a right-angled triangle having sides of length x_1, y_1, and z_1. Using the numbers n, x_1, y_1, and z_1 construct a rational point on the elliptic curve C given by

$$y^2 = x^3 - n^2 x.$$

[Hint. Show first $((x_1 \pm y_1)/2)^2 = (z_1/2)^2 \pm n$.]

Conversely suppose $(x, y) \in C(\mathbb{Q})$ where x is a square with even denominator, show that n is congruent [Hint. Reverse the argument above and use Problem 12, Chapter 1.]

What is special about the points on $C(\mathbb{Q})$ which are associated with right-angled triangles? (This problem is treated in great depth by Koblitz, 1984.)

☆17. In this problem we give a proof of the weak Mordell–Weil theorem over $\mathbb{Q}$ in the case when the cubic is a product of linear and quadratic factors.

Let $a \neq 0 \neq c^2 - 4a$ and suppose a has r distinct prime factors and $c^2 - 4a$ has s distinct prime factors. Let the elliptic curves C_1 and C_2 be given by

$$C_1 : y^2 = x^3 + cx^2 + ax$$
$$C_2 : y^2 = x^3 - 2cx^2 + (c^2 - 4a)x$$

where O is the point at infinity on both C_1 and C_2. For $i = 1, 2$, let $f_1 : C_1(\mathbb{Q}) \to C_2(\mathbb{Q})$ and $f_2 : C_2(\mathbb{Q}) \to C_1(\mathbb{Q})$ be given by $f_i(\mathsf{O}) = \mathsf{O}$, $f_i((0, 0)) = \mathsf{O}$ and, for $x \neq 0$,

$$f_1((x, y)) = (y^2/x^2, y(x^2 - a)/x^2)$$
$$f_2((x, y)) = (y^2/4x^2, y(x^2 - (c^2 - 4a))/8y^2),$$

and let $\theta_i : C_i(\mathbb{Q}) \to \mathbb{Q}^*/\mathbb{Q}^{*2}$ be given by

$$\theta_i(\mathsf{O}) = \mathbb{Q}^{*2}$$
$$\theta_i((x, y)) = x\mathbb{Q}^{*2} \qquad \text{if } x \neq 0$$
$$\theta_1((0, 0)) = a\mathbb{Q}^{*2} \quad \text{and} \quad \theta_2((0, 0)) = (c^2 - 4a)\mathbb{Q}^{*2}.$$

Using elementary coordinate geometry show that

(i) $f_2(f_1((x, y))) = 2(x, y)$,

(ii) $\theta_2(f_1((x, y))) = \mathbb{Q}^{*2}$,

(iii) if $\theta_2((x, y)) = \mathbb{Q}^{*2}$ then x_1 and y_1 can be found to satisfy $f_1((x_1, y_1)) = (x, y)$. [Hint. Using $x = t^2$, let $2x_1 = t^2 - c + y/t$, $2x_2 = t^2 - c - y/t$ and $y_i = x_i t$, and show that $x_1 x_2 = a$.]

By considering $\mathbb{Q}^*/\mathbb{Q}^{*2}$ as the additive part of the vector space over the two element field generated by -1 and the primes, show that

(iv) θ_1 and θ_2 are group homomorphisms and the image of θ_1 (θ_2) has cardinality at most 2^{r+1} (2^{s+1}, respectively).

Finally using the above and the homomorphism theorem deduce that

(v) the cardinality of $C_1(\mathbb{Q})/2C_1(\mathbb{Q})$ is at most 2^{r+s+2}. (For further details see Husemöller (1987).)

ANSWERS AND HINTS TO PROBLEMS

We give here answers to numerical exercises and hints or sketch solutions to problems. More details are given if the problem is hard or tricky. In many cases it is possible that shorter, easier, or more elegant solutions can be found.

Problems 1

1. Use the division algorithm.

2. Consider $S_n 2^{k-1} \cdot 3 \cdot 5 \cdot 7 \cdot 9 \cdot \ldots$, where $2^k \le n < 2^{k+1}$.

3. (i) $\{14 + 5t, 7 + 3t\}$; (ii) no solution;
(iii) $\{55 - 127t, -42 + 97t\}$; (iv) $\{8 + 11t, 5 + 7t\}$.

4. Use Theorem 1.3 and induction to show that $a_1x_1 + \ldots + a_nx_n = (a_1, \ldots, a_n)$ is soluble. As $(a_1, \ldots, a_n) | a_1x_1 + \ldots + a_nx_n$, for any $x_1, \ldots, x_n$, these show $c \le, \ge (a_1, \ldots, a_n)$.

5. If $n = a + (a+1) + \ldots + (a + k - 1)$ then $n = k(2a + k - 1)/2$ if k is even, and $n = k(a + (k-1)/2)$ if k is odd. In either case n has an odd factor. Conversely let $n = 2^x(2y+1)$ with $y > 0$. If $y < 2^x$ then $n = (2^x - y) + (2^x - y + 1) + \ldots + (2^x + y)$, and if $2^x \le y$ then $n = (y - 2^x + 1) + \ldots + (y - 2^x + 2^{x+1})$.

6. Use Theorem 1.5 and note that if $ak > -a$ then $k \ge 0$.

7. For $k = 1$, $bt + c = av$ has solution $\{t_1, v_1\}$ with $0 \le t_1 < a$, so let $u_1 = t_1^2 + v_1$. If $at_k^2 + bt_k + c = a^k u_k$ with $0 \le t_k < a^k$, try $t_{k+1} = t_k + a^k x$. So solve $a^{2k+1}x^2 + (2a^{k+1}t_k + ba^k)x + a^k u_k = a^{k+1}u_{k+1}$, i.e. solve $(2at_k + b)x - av = u_k$, and then $u_{k+1} = a^k x^2 + v$.

8. (i) If $(m+n, m-n) = d$ then $d | 2m$ and $d | 2n$. (ii) If $d | m$ and $d | m + n$ then $d | n$. (iii) Use contradiction.

9. (i) $(2n+1)^2 - 1 = 4n(n+1)$; (ii) $n^2 - 1 = (n-1)(n+1)$, both terms are even, and one divisible by 3; (iii) $n^5 - n = n(n-1)(n+1)(n^2+1)$, so use 5 divides $(5k+2)^2 + 1$ and $(5k+3)^2 + 1$; (iv) $n^7 - n = n(n-1)(n+1)(1 + n + n^2)(1 - n + n^2)$, so use $7 | (7k+2)^2 + (7k+2) + 1$, etc.; (v) $504 = 7 \cdot 8 \cdot 9$ so check factors seperately with 7 as in (iv). For 8 we have $2 | n \rightarrow 8 | n^3$, and $2 \nmid n \rightarrow 2 | n^3 - 1$ and $4 | n^3 + 1$ or vice versa. For 9 use $3 | (3k+1)^2 + (3k+1) + 1$, etc.

10. Suppose $x \in X$ possesses t of the properties $P_1, \ldots, P_r$, then x is counted once in n, t times in $n_1 + \ldots + n_r$, $\binom{t}{2}$ times in $n_{12} + n_{13} + \ldots + n_{r-1,r}$, and so on.

Hence the total contribution of x to m is

$$1-\binom{t}{1}+\binom{t}{2}-\ldots=(1-1)^t=0.$$

11. Use Problem 10.

12. Divide out by $k=(a, b, c)$, so assume $k=1$ and then $(a, c)=1$. One of a or b is even, suppose b. Now $b^2=(a+c)(c-a)$ and 2 is the only common factor of this product, i.e. $(a+c)/2=m^2$, $(c-a)/2=n^2$.

13. If $2|n$ and $n>2$ then $(n-1)!$ is even, but n^k-1 is odd. If $2\nmid n$ and $n>5$ then $n-1$ is even and $n-1=2((n-1)/2)\,|(n-2)!$, so $(n-1)^2|\,(n-1)!$. Now $n^k-1=(n-1)^k+k(n-1)^{k-1}+\ldots+k(n-1)$, so $n-1|k$ iff $(n-1)^2|n^k-1$. Hence $k\geq n-1$, but $n^{n-1}>(n-1)!$ if $n>5$.

14. (i) If $p\leq n$ then p, $2p, \ldots$ divide $n!$, with $[n/p]$ terms; similarly $[n/p^2]$ terms p^2, $2p^2, \ldots$ divide $n!$, add up.
(ii) $\text{ord}_p\, n!+\text{ord}_p(m-n)!=[n/p]+\ldots+[(m-n)/p]+\ldots\leq[m/p]+\ldots=$ $\text{ord}_p\, m!$.

15. Treat cases $x<, \geq[x]+\frac{1}{2}$ and $y<, \geq[y]+\frac{1}{2}$ separately and use Problem 14.

16. Consider $m!\,k+2$, $m!\,k+3, \ldots, m!\,k+m$.

17. Suppose contrary, as in Theorem 1.8, and use $(4m+1)(4n+1)=4k+1$, etc.

18. (i) and (ii) As $\zeta(s)=\prod_{p|n}(1-1/p^s)^{-1}$, if the number of primes is finite with product n, and $s\in\mathbb{Z}$, then this is finite and rational. (iii) Use Lemma 2.2. (iv) Use $\sqrt[n]{n!}\rightarrow\infty$ as $n\rightarrow\infty$.

19. (i) Use induction, noting that $x^{n+2}=x^{n+1}+x^n$ and $x-y=\sqrt{5}$. (ii) Use defining equations and induction. (iii) Use induction on n. (iv) Use (iii) and induction on r. (v) Use Euclidean algorithm: if $m=nq+r$, $u_m=u_{nq-1}u_r+u_{nq}u_{r+1}$ and so $(u_m, u_n)=(u_n, u_r)$, now repeat.

20. If d is odd, let e^2 be the largest squared factor of d. If d is even, let $d=rs^2$ (s largest), now $2|s$. If $4|r-2$ or $4|r-3$ let $d_0=4r$. Note that if d is fundamental, and $4|d^*$ or $4|d^*-1$, it will have an even number of prime factors p satisfying $4|p-3$.

21. (i) 0/1, 1/4, 1/3, 1/2, 2/3, 3/4, 1/1, etc. (ii) $a/b<(a+c)/(b+d)<c/d$, so $b+d\leq n$ contradicts hypothesis. (iii) Use $a/b<a/(b-1)<(a+1)/b$. (iv) Let $x=x_0+ta$, $y=y_0+tb$ be general solution, choose t: $n-b<y\leq n$, so $x/y\in\mathsf{F}_n$ and $x/y\geq c/d$. If not equal, then $x/y-c/d\geq 1/dy$, $c/d-a/b\geq 1/bd$ so $1/by=(bx-ay)/by=x/y-a/b\geq 1/dy+1/bd=(b+y)/bdy>n/bdy\geq 1/by$, impossible. (v) $bc-ad=1$, $de-cf=1$ so solve for c and d. (vi) 4/9, 5/11.

22. We have $n(d^2-b^2)=a^2d^2-b^2c^2$, so $(n, ad-bc)$ and $(n, ad+bc)$ are factors of n. Now look at $10b^2-n$ and check for squareness. $34889=143^2+10\cdot 38^2=157^2+10\cdot 32^2=139\cdot 251$.

Problems 2

1. (i) Use $\sum n^3 = (\sum n)^2$. (ii) If $n = \prod p^{\alpha}$ then $\prod_{d|n} d = \prod p^{1+2+\ldots+\alpha} = \prod p^{\alpha(1+\alpha)/2}$, use formula for τ. (iii) If $n = p^{\alpha}$ then $\tau(n^2) = 2\alpha + 1$, and if $\mathrm{LCM}(a, b) = p^{\alpha}$ then $a = p^{\alpha}$ and $b = p^{\beta}$ for $0 \le \beta \le \alpha$ or vice versa, i.e. $2\alpha + 1$ pairs, use multiplicativity. (iv) Let $n = (2 \cdot 3 \cdot \ldots \cdot p_r)^m$ (r fixed) so $\tau(n) > m^r$, but $m = \ln n / \ln(2 \cdot 3 \cdot \ldots \cdot p_r)$ hence $\tau(n) > C \ln^r n$, C a constant independent of n. (v) Let $f(n) = \tau(n)/n^s$, f is multiplicative. Now $f(p^m) = (m+1)/p^{ms} \le c \ln p^m / p^{ms} \to 0$ as p or $m \to \infty$, so $f(n) \to 0$ as $n \to \infty$.

2. (i) Let e be maximum such that $e^2 | n$ then $\sum_{d^2|n} \mu(d) = \sum_{d|e} \mu(d)$, use Lemma 1.4 as $|\mu(n)|$ is zero unless n is square-free. (ii) Use Lemmas 2.1 and 1.4, and so $\left|\sum_{i \le n} n\mu(i)/i\right| \le n$. (iii) The first sum counts the number of combinations of distinct prime power factors of n, and the second counts the number of square-free factors of n.

3. (i) $\displaystyle\sum_m \frac{1}{m^s} \sum_n \frac{1}{n^s} = \sum_{m,n} \frac{1}{m^s n^s} = \sum_i \frac{1}{i^s} \sum_{d|i} 1 = \sum_n \frac{\tau(n)}{n^s}$.

(ii) $\displaystyle\sum_m \frac{1}{m^s} \sum_n \frac{n}{n^s} = \sum_i \frac{1}{i^s} \sum_{d|i} d$.

(iii) As above using Theorem 1.8(iii) and $(**)$ in the proof of Lemma 2.4.

4. We have $\sigma(n) = (1 + 2 + 2^2 + \ldots)(1 + p + p^2 + \ldots)\ldots$, the first term is always odd, and the second, ... is odd if it has an odd number of summands.

5. Let $n = 2^k(2m+1)$ and $e | (2m+1)$, then $-\sum_{d|n} (-1)^{n/d} d = \sum' [-e - 2e - \ldots - 2^{k-1}e + 2^k e]$, where $\sum'$ is the sum over all odd divisors e of n; note that the sum in the square bracket equals e. Secondly let $\sigma(n) = (2^{k+1} - 1)X$ then $\sigma(n) - 2\sigma(n/2) = ((2^{k+1} - 1) - 2(2^k - 1))X = X$.

6. (i) $\sigma(220) = 504 = \sigma(284)$; (ii) 120 is 3-perfect.

7. Proof similar to Theorem 2.5, we have by DSI

$$\sum_{i \le n} \sigma(i) = \sum_{i \le n} \sum_{d \le n/i} d = \frac{1}{2} \sum_{i \le n} \left(\left[\frac{n}{i}\right]^2 + \left[\frac{n}{i}\right] \right) = \frac{n^2}{2} \sum_{i \le n} \frac{1}{i^2} + O(n \ln n).$$

8. Use a similar method to that in Lemma 1.7 and Theorem 1.8 as there are n^k sets involved.

9. (i) Consider each prime power separately. (ii) If $n > 2$, $\phi(n)$ is even and $(n, t) = 1$ implies $(n, n - t) = 1$, so add up corresponding pairs. (iii) Use Lemma 1.7 and DSI. (iv) Use $\sum_{d|n} \mu(d) f(d) = \prod_{p|n} (1 - f(p))$ if f is multiplicative, with $f = \mu/\phi$, and Theorem 1.8. (v) Use (iv) and DSI to obtain $\displaystyle\sum_{n \le x} \frac{1}{\phi(n)} = \sum_{d \le x} \frac{\mu^2(d)}{d\phi(d)} \left(1 + \ldots + \frac{1}{[x/d]}\right) \sim \sum' \frac{1}{n\phi(n)} (\ln x - \ln d)$, etc. (vi) If $n = \prod p^{\alpha}$, $\sigma(n)\phi(n) = n^2 \prod (1 - p^{-(\alpha+1)})$. But $\prod_{p|n} (1 - p^{-2}) > \prod_{2 \le k \le n} (1 - k^{-2}) > 1/2$.

10. Let $g(i, j) = 1$ if $j|i$ and 0 otherwise, so by Lemma 1.7 $(i, j) = \sum_{d \le n} g(i, d)g(j, d)\phi(d)$. Now $\det((i, j)) = \det(g(i, d))\det(g(j, d)\phi(d)) = \prod_{d \le n} \phi(d)$ as matrices are triangular.

11. (ii) If $\delta_n > 0$, there is an $s : s^2 \le n$ and $(s^2, n) = 1$, so D_n is false. (iii) Let $M(i, m) = \sum_{d|(i,m)} \mu(d)$, by Lemma 1.4(ii), $\phi(m) = \sum_{i \le m} M(i, m)$, now use DSI. (iv) Use Theorem 1.8(iii). (v) If $n = \prod_{i \le r} p_i^{\alpha_i}$ we need

$$2\prod_{i \le r} p_i^{(1/2)(\alpha_i - 1)}\left(\frac{p_i - 1}{2\sqrt{p_i}}\right) \ge 1 + \frac{1}{2^{r-1}}.$$

Work numerically, if $2 \nmid n$ then $4 \nmid n$, if $3 \nmid n$ then $9 \nmid n$, and if $5 \nmid n$ then $25 \nmid n$, so if $n > 25$ consider only integers divisible by 30, i.e. $r \ge 3$, $p_1 = 2$, $p_2 = 3$, and $p_3 = 5$. If $r > 3$ inequality gives $7 \le p_r \le 39$, so $n \ge 30 \cdot 7 > 13^2$, etc., hence $p_4 = 7$, $p_5 = 11$, and $p_6 = 13$, but now $n > 41^2$ contradicting $41 \nmid n$. If $n = 2^\alpha \cdot 3^\beta \cdot 5^\gamma$ and $\max(\alpha, \beta, \gamma) > 1$ then $n \ge 60 > 7^2$.

12. Use Lemma 1.9(ii) and DSI, both terms count the number of squares less than x, secondly use this and note $|\lambda(n)| = 1$.

13. (i) Use $\sum_{d|p^\alpha} \Lambda(\mathrm{d}) = \alpha \ln p$ and $\ln n = \alpha_1 \ln p_1 + \alpha_2 \ln p_2 + \ldots$ if $n = p_1^{\alpha_1} p_2^{\alpha_2} \ldots$. (ii) Use the Möbius inversion formula and Lemma 1.4(ii).

14. For the first part note that both equalities hold iff $b_n = \sum_{d|n} a_d$. So for (i) use Lemma 1.4(ii), for (ii) use Problem 3(i), and for (iii) use Lemma 1.7 as $x/(1-x)^2 = \sum (n+1)x^n$.

15. (i) See (2) on p. 17; (ii) multiply out; (iii) use $\ln n = \ln d + \ln n/d$.

16. Use Theorem 2.5.

Problems 3

1. Use divisibility properties. We have $a \equiv b\ (m)$ gives $a = b + mt$, and so $(b, m)|a$, etc., also $a - b|a^k - b^k$.

2. (i) No solution; (ii) $x \equiv 9\ (21)$; (iii) x arbitrary, $y \equiv 6 - 12x\ (15)$.

3. (i) Use $10^k \equiv 1 \pmod 9$; (ii) use $10 \equiv -1 \pmod{11}$.

4. (i) If $k > 0$ we have $2^{4+k} \equiv 2^k \pmod{10}$, and so consider $2^{k-1}(2^k - 1)$ for $k = 2$, 3, and 5, using Lemma 1.3 of Chapter 2. (ii) Use Problem 3(i) and $2^6 \equiv 1 \pmod 9$ to reduce $2^{2k}(2^{2k+1} - 1)$ to the cases $k = 2$, 4, and 6.

5. $7^{139} \equiv (7^4)^{34} \cdot 7^3 \equiv 7^3 \equiv 3\ (10)$ and $13^{2001} \equiv 3^{2001} \equiv 3\ (10)$.

6. Use Euler's theorem and note that $a - 1|a^{k\phi(m)} - 1$. Now put $a = 10$.

7. If $c^2 \equiv 1$ then $(n - c)^2 \equiv 1$, so check $c \ne n - c$. Given a_i, let a_j satisfy $a_i a_j \equiv 1\ (n)$; if $i \ne j$ then $a_i a_j$ has no affect on product, and if $i = j$ use first part.

8. Euler's theorem gives $(m^{\phi(n)}-1)(n^{\phi(m)}-1)\equiv 0 \pmod{mn}$.

9. (i) Note first $p\nmid(a,c)$ as $p|d$ and $(a,b,c)=1$. So assume $p\nmid a$. If $p>2$, $(p,4a)=1$ and $4af(x,y)=(2ax+by)^2-dy^2\not\equiv 0\ (p)$ is equivalent to $2ax+by\not\equiv 0$ (p), now count solutions. If $p=2$ then $2|b$, so consider $ax+cy\equiv 1\ (2)$. (ii) Use the Chinese remainder theorem and the multiplicativity of ϕ.

10. By Fermat's theorem we have $x\equiv xa^{p-1}\equiv ba^{p-2} \pmod{p}$. So $x\equiv 3^{27}\cdot 17\equiv(-2)^9\cdot(-12)\equiv 2^{10}\cdot 6\equiv 9\cdot 6\equiv 25 \pmod{29}$.

11. (i) Let h be the least positive integer such that $2^h\equiv 1\ (p)$. $p|F_n$ gives $p|2^{2^{n+1}}-1$ and $h=2^{n+1}$ (using Lemma 2.2, Chapter 1), but by Fermat's theorem $h|p-1$, and so $p=2^{n+1}k+1$. (ii) Use Fermat's theorem, $56759=211\cdot 269$ as $211|2^{7!}-1$.

12. (i) By Fermat's theorem $a=b+kp$, and so $a^p=(b+kp)^p$, etc. (ii) Use Euler's theorem, and Theorem 1.1 and Lemma 1.4 of Chapter 2.

13. First we have $a^{2^n}\equiv 1 \pmod{2^{n+2}}$ by Problem 9(i) of Chapter 1 and (iii) of the second proof of Theorem 1.12, so $(a,p)=1$ gives $a^{\chi(p^\alpha)}\equiv 1 \pmod{p^\alpha}$ for all p and α. For the next three parts we note that if m is prime then $\phi(m)=m-1$, $\chi(m)|\phi(m)$ and (iii) implies $2^{m-1}=2^{\chi(m)k}\equiv 1 \pmod{m}$ as m is odd. Now $341=11\cdot 31$, so $\chi(341)=30\nmid 340$, but $2^{340}\equiv 1 \pmod{341}$; similarly for 645, $\chi(645)=84\nmid 644$. Further $561=3\cdot 11\cdot 17$, so $\phi(561)=320\nmid 560$, but $\chi(561)=80|560$; similarly $\phi(1105)=768\nmid 1104$, but $\chi(1105)=48|1104$. Finally if $2^{k-1}\equiv 1 \pmod{k}$ then $2^k-2=nk$, but $2^k\equiv 1 \pmod{2^k-1}$, so $2^{nk}=2^{2^k-2}\equiv 1 \pmod{2^k-1}$.

14. 301.

15. $-2846265 \pmod{9699690}$, $47 \pmod{60}$.

16. Let $x=a+mt$ so solve $mt\equiv b-a \pmod{n}$, possible by condition. t is unique modulo $n/(m,n)$ and so x is unique modulo LCM(m,n).

17. $x\equiv c_i(m_i)$ jointly soluble iff $(m_i,m_j)|c_j-c_i$ for all i, j $(i\neq j)$. By 16 the condition is necessary as each pair of congruences must have a common solution. For the converse suppose $x\equiv c_1 \pmod{m_1}$ and $x\equiv c_2 \pmod{m_2}$ have the common solution $x\equiv d$ (mod LCM(m_1,m_2)), then we need $3\leq i\leq k$ implies $M=(m_i,\text{LCM}(m_1,m_2))|c_i-d$. Consider each prime factor separately and suppose α_i is the exponent of p in m_i, then the exponent of p in M is $\min(\alpha_i,\max(\alpha_1,\alpha_2))=\max(\min(\alpha_i,\alpha_1),\ \min(\alpha_i,\alpha_2))=\max(a_1,a_2)$. By assumption $p^{a_j}|c_j-c_i$ and $p^{\alpha_j}|c_j-d$, hence $p^{a_j}|c_i-d$ for $j=1,2$. Now take maxima. (See Theorem 1.10, Chapter 1.)

18. (i) 15, (ii) 11, (iii) 8, 23, 33, 48, 58, 69, 73, 83, 98, 108, and 123.

19. Use Theorem 2.4. (i) If m is not prime then $(m-1)!\equiv 0 \pmod{m}$. (ii) Use induction as $(p-k)!(k-1)!=(p-k)(p-k-1)!(k-1)!\equiv-(p-k-1)!k! \pmod{p}$.

20. Clearly if $p\nmid f'(x_i)$ then $p\nmid f'(x_i+p^et)$, so each solution $\pmod{p}$ gives rise to a unique solution $\pmod{p^e}$. Now use Theorem 2.3.

21. Use the division algorithm for $\mathbb{Z}/p\mathbb{Z}[x]$. Now by Fermat's result $x^p - x$ has p zeros in $\mathbb{Z}/p\mathbb{Z}[x]$ and so

$$x^p - x = x(x - \bar{1}) \dots (x - \overline{p-1})$$

in $\mathbb{Z}/p\mathbb{Z}[x]$. If

$$f(x) = (x - \bar{r}_1)^{e_1} \dots (x - \bar{r}_k)^{e_k} g(x)$$

then $\gcd(f(x),\ x^p - x) = (x - \bar{r}_1) \dots (x - \bar{r}_k)$ with degree k. (Note this result counts the number of distinct roots only.)

22. Let $f(x) = \sum a_i x^i$ and $g(x) = \sum b_j x^j$ (both primitive). For each p, we have i and j so that $p \mid a_t$ for $t < i$ and $p \nmid a_i$; $p \mid b_u$ for $u < j$ and $p \nmid b_j$. If the coefficients of the product are $c_0, c_1, \dots$ then $p \mid c_k$ for $k < i + j$ and $p \nmid c_{i+j}$.

23. (i) Use $\prod (x - i) \equiv x^{p-1} - 1\ (p)$. (ii) Note $a_{p-1} = (p-1)!$, and so $p^2 \mid a_{p-2}$.

24. Let $H = 1 - f^{p-1}$ and so $H(x_1, \dots, x_n) \equiv 1$ if $x_i \equiv a_{j,i}(p)$, and $\equiv 0$ otherwise. As in Theorem 2.7, we have, if H' is the reduction of H, $H' \sim H^*$. The term of degree $n(p-1)$ in H^* is $s(-1)^n(x_1x_2 \dots x_n)^{p-1}$, it does not occur in H', hence $s \equiv 0\ (p)$.

25. $x^2 - xy + y^2$; $\quad xyz + xy(1+z) + yz(1+x) + zx(1+y) + x(1+y)(1+z) + y(1+z)(1+x) + z(1+x)(1+y)$.

26. First remove any common factor of x, y, and z. Now if $3 \mid y$ then $4x^3 \equiv z^3$ (9), and so $x \equiv y \equiv z \equiv 0$ (3) as $x^3 \equiv 0, \pm 1$ (9), hence $3 \nmid xy$. Further, by Fermat, $z \equiv z^3 \equiv ax^3 + dy^3 \equiv ax + dy$ (3), and so

$$((a - a^3)/3)x^3 + (b - a^2d)x^2y + (c - ad^2)xy^2 + ((d - d^3)/3)y^3 \equiv 0\ (3)$$

and, using conditions, $0 \equiv x + (b-1)y + (c-1)x + y \equiv cx \equiv \pm x$ (3).

27. (i) By cases. (ii) Remove common factors, and modulo 7 we have $x^3 + 2y^3 + 4z^3 + xyz \equiv 0$ (7). If $z \equiv 0$ (7) then $x \equiv y \equiv 0$ (7) by (i). If $z \not\equiv 0$ (7) use Corollary 1.5 to obtain the congruence $X^3 + 2Y^3 + 4 + XY \equiv 0$ (7). Now consider cases X^3, $Y^3 \equiv 0, 1, -1$ (7), as $XY \not\equiv 0$ (7).

28. Modulo p the equation gives $x_1 \equiv y_1 \equiv z_1 \equiv 0\ (p)$, now divide by p twice to show all arguments $\equiv 0 \pmod p$.

29. Use unique factorization.

30. Use condition $x = 0$ iff $|x|_p = 0$, etc.

31. We have $\left| \sum_{i=k}^{m} a_i \right|_p \le \max_{i \ge k} (|a_i|_p) \to 0$ as $k, m \to \infty$.

32. $\{1, 7, 25, 79, 241, 727, \dots\}$ and $\{1, 1, 73, 73, 721, 721, \dots\}$.

Problems 4

1. (i) Note $x \ne y$ as a is a non-residue. (ii) Similar to (i) but consider z and $p - z$, where $z^2 \equiv a\ (p)$, separately; $z(p-z) \equiv -a\ (p)$. (iii) Use Theorem 2.3, Chapter 3.

2. 1, 1, 1.

3. (i) $(a/p)=(b/p)$ iff $(ab/p)=1$ iff $y^2 \equiv ab$ or $b(b^*y)^2 \equiv a\ (p)$ is soluble where $b^*b \equiv 1\ (p)$. (ii) For each prime $q|d$, $(q/p)=(4m+1/q)=(1/q)$. (iii) Factorize and note that $(2/p)$, $(3/p)$, or $(6/p)=1$ for all $p>3$. Also we have

$$x^{2^\alpha}-2^{2^{\alpha-1}}=(x^2-2)(x^2+2)((x-1)^2+1)((x-1)^2-1)(x^{2^3}-2^{2^2})\ldots$$

and $(2/p)$, $(-2/p)$, or $(-1/p)=1$ for all $p>2$. Use $x^2 \equiv a(p)$ has $1+(a/p)$ solutions.

4. Use Theorem 1.3 and the results from Section 2, Chapter 3 and Chinese remainder theorem. Note, if $x^2 \equiv -1\ (p)$ is soluble, so is $x^2 \equiv -1\ (2p)$.

5. If $p|4m^2+3$ then $(-3/p)=1$ and $p \equiv 1$ or 7 (12). Also, not all factors $\equiv 1$ (12). Now let m be product of all primes $\equiv 7$ (12).

6. We have $4(n^4-n^2+1)=4((n^2-1)^2+n^2)=(2n^2-1)^2+3$ so $(-3/p)=1=(-1/p)$. Hence $p \equiv 1$ (mod 3) and (mod 4).

7. (i) Use Theorem 2.2. (ii) If $p|6119$ then $p \equiv \pm 1$ (10) and we have $6119=29\cdot 211$. (iii) Use $4751=86^2-5\cdot 23^2$.

8. (i) Note $p-k \equiv (-1)k\ (p)$, and if k is odd then $p-k$ is even. (ii) We have $R_p \equiv 1^2 3^2 \ldots (p-2)^2\ (p)$ as different quadratic congruences have different solutions.

9. Note $x^2 \equiv y^2+a\ (p)$ has $1+(y^2+a/p)$ solutions x, so total number of solutions is $\sum_{y=0}^{p-1}(1+(y^2+a/p))$.

10. (i) Use Lemma 1.1. (ii) For each i, choose t_i: $it_i \equiv 1\ (p)$, we have

$$(i(i+1)/p)=(t_i^2/p)(i(i+1)/p)=(it_i(it_i+t_i)/p)=(t_i+1/p),$$

now use (i) noting that t_i ranges over the reduced residue system (p). (iii) Use result $x^2 \equiv n\ (p)$ has $1+(n/p)$ solutions. Now

$$4N=p-2+\sum_{i=1}^{p-2}(i/p)+\sum_{i=2}^{p-1}(i/p)+\sum_{i=1}^{p-2}(i(i+1)/p)=p-4-(-1/p).$$

11. The total number of points is $(p-1)(q-1)/4$. Write

$$S(q,p)=\sum_{s=1}^{(p-1)/2}\left[\frac{sq}{p}\right],$$

$S(q,p)$ is the number of points below $y=qx/p$ as there are no points on this line.

(i) We have $\sum r_i+\sum s_j=1+2+\ldots+(p-1)/2=(p^2-1)/8$.

(ii) u_k is a positive absolute least residue of kq, etc.

(iii) Summing (*) we find $(p^2-1)q/8=pS(q,p)+\sum u_k+\sum v_k$, and (i) and (ii) give $(p^2-1)/8=tp+\sum u_k-\sum v_k$.

(iv) As q is odd and $p^2-1 \equiv 0$ (8), (iii) gives $S(q,p) \equiv t$ (2). Now use first result.

12. (i) Use properties of the sine function.

(ii) Put $x = e^{2\pi iz}$ and $y = e^{-2\pi iz}$ and we have

$$f(nz) = \prod_{k=0}^{n-1} f\left(z + \frac{k}{n}\right) = f(z) \prod_{k=1}^{(n-1)/2} f\left(z + \frac{k}{n}\right) \prod_{k=(n+1)/2}^{n-1} f\left(z + \frac{k}{n} - 1\right), \text{ etc.}$$

(iii) Let $ja \equiv \pm m_j$ where $1 \leq m_j \leq (p-1)/2$ so $f\left(\frac{ja}{p}\right) = \pm f\left(\frac{m_j}{p}\right)$, use Gauss's lemma.

(iv) Using (ii) with $n = q$ and $z = j/p$, and (iii) with $a = q$ we have

$$(q/p) = \prod_{m=1}^{(q-1)/2} \prod_{j=1}^{(p-1)/2} f\left(\frac{j}{p} + \frac{m}{q}\right) f\left(\frac{j}{p} - \frac{m}{p}\right).$$

By symmetry we obtain a similar expression for (p/q), finally count minus signs.

13. There are 2^{t-1} odd numbers x: $0 < x < 2^t$ and, for each, $x^2 \equiv m$ (2^t) for some $m \equiv 1$ (8). Each m occurs exactly four times. For if $x^2 \equiv y^2(2^t)$, x odd, then $2^{t-2} | (x+y)/2$ or $(x-y)/2$, i.e. $x \equiv \pm y$ (2^{t-1}), that is four values of x. Now use pigeon-hole principle. For $p = 2$ the number of solutions is (i) 0 if $t = 2$ and $n \equiv 3$ (4), or $t > 2$ and $n \not\equiv 1$ (8); (ii) 1 if $t = 1$; (iii) 2 if $t = 2$ and $n \equiv 1$ (4); and (iv) 4 if $t > 2$ and $n \equiv 1$ (8). For $p > 2$ there are $1 + (n/p)$ solutions, see Section 2, Chapter 3.

14. Note aRm and aRn give $(x + km)^2 \equiv a(m)$ and $(y + jn)^2 \equiv a(n)$ both soluble (any k, j). As $(m, n) = 1$, choose k, j: $km - jn = y - x$.

15. For the first part use Problems 13 and 14 and the Chinese remainder theorem. From this and 14 we have, if $d \equiv 1$ (4)

$$v(d, 4k) = 2\prod_{p|k} (1 + (d/p)) = 2{\sum_{u|k}}' (d/u)$$

with a similar argument when $d \equiv 0$ (4).

16. $$\begin{aligned}(2/p) &= (8/p) = (8-p/p) = (-1/p)(p-8/p) \\ &= (p/p-8) \qquad \text{by reciprocity (two cases)} \\ &= (8/p-8) = (2/p-8) \text{ as } p \equiv 8 \pmod{p-8}.\end{aligned}$$

17. Let P denote the list of exceptional primes. The proof is as before except that each time k is chosen it must now also satisfy $(k, p) = 1$ for each p in P.

18. Interchange the order of summation in the middle sum, and then the new general term is, as $(2n - 1/p)^2 = 1$,

$$\frac{(2n-1/p)}{2n-1} \sum_{t=1}^{p-1} ((2n-1)t/p) \sin[2\pi t(2n-1)/p].$$

If $p \nmid 2n - 1$ then $(2n-1)t$ ranges over a reduced residue system (mod p) and so this term equals, by (ii),
$$\frac{(2n-1/p)}{2n-1} \sqrt{p},$$

and is zero if $p | 2n - 1$. Now apply (iii) to the inner Fourier series and (i) gives result.

19. Use $x^4+4=((x+1)^2+1)((x-1)^2+1)$.

20. (iii) If $q|a$ we have $b^2 \equiv p\ (q)$ is soluble, so $1=(p/q)=(q/p)$. (iv) $x^2 \equiv 2p\ (a+b)$ is soluble, so $(2/a+b)(p/a+b)=1$. (v) We have $b^2 \equiv a^2f^2 \equiv (p-b^2)f^2(p)$, now use $(b, p)=1$. By (iii) there exists x: $x^4 \equiv a^2\ (p)$, so $ab \equiv x^4f$ and

$$(2ab)^{(p-1)/4} \equiv (2f)^{(p-1)/4}\ (p).$$

Further as $(a+b)^2 \equiv 2ab\ (p)$ we have, by (iv) and Theorem 1.2,

$$(-1)^{((a+b)^2-1)/8} \equiv (a+b)^{(p-1)/2} \equiv (2ab)^{(p-1)/4}\ (p).$$

Hence by (iii) and Theorem 1.2, and as $f^2 \equiv -1\ (p)$ we have

$$2^{(p-1)/4} \equiv f^{3(p-1)/4}(-1)^{((a+b)^2-1)/8} \equiv f^{(3(a^2+b^2-1)+(a+b)^2-1)/4} \equiv f^{ab/2}f^{a^2+b^2-1} \equiv f^{ab/2}\ (p),$$

as a is odd and b is even. (vi) If $8|b$ (i.e. $p=a^2+64c^2$) then $f^{ab/2} \equiv 1\ (p)$, so use (v) and (i). If $x^4 \equiv 2\ (p)$ soluble, $1 \equiv f^{ab/2}(p)$ by (i) and (v), so $4|ab/2$ and $8|b$ as $2 \nmid a$.

21. (i) We can assume that $|d|=p^tg$, $p>2$ and t and g odd. Choose s, a quadratic non-residue (p), and $n>0$ so that $n \equiv s(p)$ and $n \equiv 1(g)$. Then $(d/n)=(n/p)^t(n/g)=-1$. (ii) Choose $n>0$ such that $n \equiv 5\ (8)$ and $n \equiv 1\ (|b|)$, b the odd part of d. Remaining parts similar.

22. Use Theorem 3.4. If $d=2^tb$, show that $(2/|d|-1)^t=1$ using $|d|-1 \equiv 7\ (8)$ if $t>2$. For the second part use $(d/n)=(d/m|d|-m)$ and the first part.

Problems 5

1. Follow the method given on p. 3, (i) 1, $-x^3-x$, $x-1$; (ii) $-\frac{1}{2}(x+1)$, $\frac{1}{2}(x^4+x^3-x^2-x+1)$, 1.

2. (i) If $f=g^2h$ then g divides f and Df. Suppose f has no squared factor, $f=gh$, $(g, h)=1$ and $g|Df$. Now $Df=(Dg)h+g(Dh)$, and so $g|Dg$, impossible if $\deg g>0$ for then $\deg Dg \geq \deg g$. (ii) f and Df coprime (for otherwise f has a squared factor) so by Lemma 1.1, t and u exist: $ft+(Df)u=1$ over K. This also holds over $\mathbb{C}$, so use Lemma 1.1.

3. (i) Equations are (a) $(x^2-3)^2-14(x^2+3)+49=0$, (b) $x^{17}+1=0$, and (c) $x^4-4x^2+8=0$. (ii) As $\beta+\bar{\beta}$ and $\beta\bar{\beta}$ are rational, consider $(x^2+\beta x+\gamma)(x^2+\bar{\beta}x+\gamma)(x^2+\beta x+\bar{\gamma})(x^2+\bar{\beta}x+\bar{\gamma})$.

4. As $\beta_i=\sum a_j\gamma_{j,i}$, Cramer's rule gives $a_k=\det A_k/\det(\gamma_{i,j})$ where A_k is the matrix $(\gamma_{i,j})$ with its kth column replaced by $(\beta_1, \ldots, \beta_n)'$. By Lemma 1.8, $\det(\gamma_{i,j}) \neq 0$ and A_k can be written as a linear expression in $\beta_1, \ldots, \beta_n$ with coefficients defined in terms of $\gamma_{i,j}$. Take absolute values.

5. (i) As in Lemma 1.8, $\Delta=\prod_{i<j}(\alpha_i-\alpha_j)^2$, α_i conjugate of α. Also $f(x)=\prod_i(x-\alpha_i)$, so $Df(x)=\sum_j\prod_{i\neq j}(x-\alpha_i)$ and $Df(\alpha_j)=\prod_{i\neq j}(\alpha_j-\alpha_i)$ giving $N(Df(\alpha))=\prod_j Df(\alpha_j)=\prod_{i,j\neq i}(\alpha_i-\alpha_j)$. Now note $\alpha_i-\alpha_j$ and $\alpha_j-\alpha_i$ occur in last expression. (ii) $\Delta=(-1)^{(p-1)/2}N(Df(\alpha))$ as $p-2$ is odd. As $f(x)=(x^p-1)/(x-1)$, we have $Df(\alpha)=p\alpha^{p-1}/(1-\alpha)$ and $N(Df(\alpha))=N(p)N(\alpha)^{p-1}/N(1-\alpha)=p^{p-2}$.

6. (ii) $(a+\omega b)(a+\omega^2 b)=a^2-ab+b^2$ by (i). (iii) If $\alpha=\beta\gamma+\delta$ then $N\delta\le 3N\beta/4$. Also $N\alpha\beta=N\alpha N\beta$. (iv) $\lambda^2=(1-\omega)^2=-3\omega$. Also $a+\omega b=a+b-\lambda b\equiv a+b\equiv -1, 0, 1 \pmod{\lambda}$. (v) Use $Np=p^2$.

7. As $p\equiv 1$ (4), there is an x: $p|x^2+1$ or $p|(x+i)(x-i)$. Now $p\nmid x\pm i$ as $x/p\pm i/p$ is not an integer. Hence p factorizes $p=\alpha\bar{\alpha}$ in $Z[i]$, and if $\alpha=a+ib$ then $p=a^2+b^2$.

8. (i) 11, 11, 22. (ii) 3, 5; 3, 5; 3, 5, 10, 12, 17, 24, 26, 33, 38, 40, 45, 47. [Note. Using the notation of Theorem 2.5, $g+kp$ is not a primitive root $\pmod{p^2}$ if $kg^{p-2}(p-1)\equiv(1-g^{p-1})/p$ (p), so 19 is not a primitive root (mod 49).]

9. As $(a/p)=-1$ Euler's criterion gives $a^{2^{m-1}}\equiv -1$ (p), hence the order $\pmod{p}$ of a is 2^m.

10. If $p-1|k$ use Fermat's Theorem. If not, choose primitive root g (p) and the sum is $\equiv g^k+g^{2k}+\ldots+g^{(p-1)k}\equiv(g^{pk}-1)/(g^k-1)-1\equiv 0$ (p).

11. If g is a primitive root so is g^t where $(t,\phi(m))=1$. If $m>6$, $\phi(\phi(m))$ is even, and if g^t is a primitive root so is $g^{\phi(m)-t}$, hence the exponent of the product $g^{t_1}g^{t_2}\ldots$ is divisible by $\phi(m)$.

12. If k exists we have $g^k(g-1)\equiv 1$ and $g^k(g^2-1)\equiv 2$ (p) which gives $g+1\equiv 2$ (p).

13. If $\text{ord}_p 2=t<4q$, as $t|4q$ (Lemma 2.1) we have $t=4$ or $t|2q$. $2^4\not\equiv 1$ (p) and $2^{2q}=2^{(p-1)/2}\equiv(2/p)\equiv -1$ (p) as $p\equiv 5$ (8). Second part similar.

14. If $a=b^2$ Euler's theorem gives $a^{\phi(m)/2}\equiv 1$ (m), also $(-1)^t=\pm 1$.

15. First we have, by induction, $10^k=p(10^{k-1}a_1+\ldots+a_k)+b_k$, now divide by $10^k p$ and take limits, also reduce mod p. We have $b_k=b_{k+(p-1)}$ using congruence and inequality, hence period is $p-1$ as 10 is a primitive root.

16. (i) Use $a\equiv g^{\text{ind}_g a}(m)$. (ii) $g^{\text{ind}_g ab}\equiv ab\equiv g^{\text{ind}_g a+\text{ind}_g b}(m)$. $x\equiv 5^3\equiv 17$ (18).

17. By Lemma 2.3 and induction we have $5^{2^{t-3}}\equiv 1+2^{t-1}$ (2^t), this shows $\text{ord}_{2^t}5=2^{t-2}$. Now the numbers $\pm 5^\alpha$, for $0<\alpha\le 2^{t-2}$, are all incongruent mod 2^t, and there are 2^{t-1} of them; note that $5^\alpha\equiv 1$ (mod 4).

18. 1, 10, 16, 18, and 37; 10, 16, 18, and 37. (Use Theorem 2.8.)

19. (i) By Theorem 2.8 $x^4\equiv a$ (p) soluble iff $a^{(p-1)/2}\equiv 1$ (p) iff $(a/p)\ge 0$. Now $(3/11)=-(11/3)=1$, solutions are 4 and 7. (ii) If $p\equiv 3$ (4) use (i), if $p\equiv 1$ (4), $x^4\equiv -1$ (p) soluble iff $(-1)^{(p-1)/4}\equiv 1$ (p) iff $8\mid p-1$ as $p>2$.

20.

y	1	2	3	4	5	6
χ_1	1	ω	$-\bar{\omega}$	$\bar{\omega}$	$-\omega$	-1
χ_2	1	$\bar{\omega}$	ω	ω	$\bar{\omega}$	1
χ_3	1	1	-1	1	-1	-1
χ_4	1	ω	$\bar{\omega}$	$\bar{\omega}$	ω	1
χ_5	1	$\bar{\omega}$	$-\omega$	ω	$-\bar{\omega}$	-1
χ_6	1	1	1	1	1	1

modulo 7

where $\omega=(-1+\sqrt{-3})/2$

y	1	3	5	7
χ_1	1	−1	1	−1
χ_2	1	−1	−1	1
χ_3	1	1	−1	−1
χ_4	1	1	1	1

modulo 8

21. If $m=p^u$ $(p>2)$ then χ_1^t is a character mod p^u, provided $(t, p^u)=1$. These are distinct if $0 \le t < p^u$, that is for $\phi(p^u)$ values of t. Hence by Theorem 3.4 all characters mod p^u have this form. We can argue similarly when $p=2$ using χ_2, χ_3, and χ_4. Now consider products over the prime power factors of m.

22. Use Cases 1 and 2 on p. 86. A typical character χ mod $2^{\alpha_0}p^{\alpha_1}q^{\alpha_2}$ has the form $\chi_3^b\chi_4^c\chi_{1,p}^d\chi_{1,q}^e$, where $\chi_{1,p}$ and $\chi_{1,q}$ are given by Case 1 and have moduli p^{α_1} and q^{α_2}, respectively; and b, c, d, and e are to be chosen. As in the proof of Theorem 3.6 we can replace χ_3, χ_4, $\chi_{1,p}$, and $\chi_{1,q}$ by Kronecker symbols, and so $\chi(a)=(-1/a)^{b'}(2/a)^{c'}(p^*/a)^{d'}(q^*/a)^{e'}$ where b', c', d', and e' are defined in terms of b, c, d, and e, respectively.

23. Yes. If $m=p$, let g be a primitive root (mod p), $\chi_t(a)=e(t \operatorname{ind} a/p-1)$, and define the character group (mod p) by $\psi(z)=\chi_{\operatorname{ind} z}$, ψ is an isomorphism. Remaining cases similar.

24. Proceed as in Problem 22.

25. If $a\equiv 0$ then $x\equiv 0$ (p) and $N=1$. If $a\not\equiv 0$, b exists such that $b^n\equiv a$ (p), and $\chi^n=\chi_0$, then $\chi(a)=(\chi(b))^n=1$ and $\sum_\chi^n \chi(a)=N$ in this case. Finally if $a\not\equiv 0$ and $x^n\equiv a$ (p) is not soluble we have a character ρ with the property $\rho(a)\neq 1$ (Lemma 3.3). Let $\psi=\rho^{(p-1)/n}$, so $\psi(a)\neq 1$ and $\psi^n=\chi_0$. Now $\sum_\chi^n \chi(a)=\sum_\chi^n \psi(a)\chi(a)=\psi(a)\sum_\chi^n \chi(a)=0$.

26. We have

$$\left|\sum_{n=r}^{s}\chi(n)f_n\right|=\left|\sum_{n=r}^{s}(Q(n)-Q(n-1))f_n\right|$$
$$=\left|-Q(r-1)f_r+\sum_{n=r}^{s-1}Q(n)(f_n-f_{n+1})+q(s)f_s\right|$$
$$\le\phi(k)\left[f_r+\sum_{n=r}^{s-1}(f_n-f_{n+1})+f_s\right]\le 2\phi(k)f_r.$$

Problems 6

1. Put $\alpha=-z/n$, $\beta=\sqrt{n}$ and $x=zy+nu$, then conditions give $x\equiv zy$ (n), $0<x^2+y^2<2n$ and $x^2+y^2\equiv 0$ (n). Now use $n=(zy+nu)^2+y^2$ to prove $(x, y)=1$. If $x<0$, let $x=-y$ and $y=x$. For uniqueness use product formula and $x\equiv zy$ (n).

2. (i) For $n>1$ we have $x\neq 0\neq y$ as $(x, y)=1$, so each solution has four associates $(\pm x, \pm y)$. Now use Problem 1 and result: $x\equiv zy$ (n) has a unique

solution if $(y, n) = 1$. (ii) Use (i), for if $(x, y) = d$ then $n/d^2 = x_1^2 + x_2^2$, $x_1 = x/d$, $y_1 = y/d$, and $(x_1, y_1) = 1$. (iii) Use Chinese remainder theorem to show that v is multiplicative, and so deduce $(n_1, n_2) = 1$ implies $u(n_1 n_2)/4 = (u(n_1)/4)(u(n_2)/4)$, also $\sum_{d|n} \chi_2(d)$ is multiplicative. Now use Problem 14 of Chapter 4. (iv) Use (iii) checking cases $p = 2$, $p \equiv 1$ (4), and $p \equiv 3$ (4) for $p|n$. (v) Use (iv).

3. If $x \equiv 2$ (3) replace x by $-x$. We have $x + 3y$ is even and $p = (\frac{1}{2}(x + 3y) + 3y\omega)(\frac{1}{2}(x + 3y) + 3y\omega^2)$ where each factor is an associate of a prime. Repeat this construction beginning with $4p = t^2 + 27u^2$ and use unique factorization.

4. (i) This is equivalent to finding all possible ways of decomposing $4n = 2r + 2s$ where $2r = u_1^2 + u_2^2$ and $2s = u_3^2 + u_4^2$ [as $u_1^2 + u_2^2 \equiv u_3^2 + u_4^2 \equiv 2$ (4)]. Now

$$a(n) = \sum_{r+s=2n} \frac{u(2r)}{4}\frac{u(2s)}{4} = \sum_{tx+uy=2n} \chi_2(xy)$$

where the last sum is taken over all odd t, u, x, and y satisfying the condition. If $x = y$ terms in sum give $\sigma(n)$, so we need: terms in sum with $x > y$ add up to zero. To do this define bijection ϕ: $(t, u, x, y) \rightarrow (t', u', x', y')$ and show $\chi_2(xy) + \chi_2(x'y') = 0$. ϕ is given by: let $q = [y/(x - y)]$, $t' = -qx + (q + 1)y$, $u' = (q + 1)x - (q + 2)y$, $x' = (q + 2)t + (q + 1)u$, and $y' = (q + 1)t + qu$.

(ii) If $2n = \sum x_i^2$ then two x_i are even and two odd, so the number of solutions with x_1, x_2 even and x_3, x_4 odd is $Q(2n)/6$. Now put $y_1 = (x_1 + x_2)/2$, $y_2 = (x_1 - x_2)/2$, $y_3 = (x_3 + x_4)/2$, $y_4 = (x_3 - x_4)/2$ and we have $n = \sum y_i^2$, $y_1 + y_2 \equiv 0$, $y_3 + y_4 \equiv 1$ (2) so $Q(2n)/6 \leq Q(n)/2$, and the converse is similar. For $Q(2n) = Q(4n)$ use substitution above. Finally if $4n = \sum x_i^2$ then all x_i even or all odd so use (i) and definition.

(iii) Use (ii). For (i)–(iii), see Landau (1958, pp. 146–51).

(iv) 1536 solutions generated by $\{10, 2, 1, 0\}$, $\{9, 4, 2, 2\}$, $\{8, 6, 2, 1\}$, $\{8, 5, 4, 0\}$, $\{8, 4, 4, 3\}$, and $\{7, 6, 4, 2\}$ for $n = 105$.

5. If $a > 2$ then each x_i is even and we have $8n = \sum x_i^2$ is equivalent to $2n = \sum (x_i/2)^2$. So if a is even ($a = 2c$) then $2^{2c} = 4(2^{c-1})^2 = (2^c)^2 + 0^2 + 0^2 + 0^2$ and $2^{2c+1} = (2^c)^2 + (2c)^2 + 0^2 + 0^2$, check with problem above. Remaining parts follow.

6. If n is odd let $x_5 = 0$, and if n is even let $x_5 = 1$, so assume n is odd and all solutions of $n = x_1^2 + x_2^2 + x_3^2 + x_4^2$ have at least $x_1 = x_2$. Now the number of solutions (for fixed x_1) of $n - 2x_1^2 = x_3^2 + x_4^2$ is $O(n^s)$ for any $s > 0$, by Problem 2 and Problem 1 of Chapter 2, so the number of solutions of $n = 2x_1^2 + x_3^2 + x_4^2$ is $O(n^{s+\frac{1}{2}})$. But by Problem 4, $Q(n) = 8\sigma(n) \geq (n + 1)/k$ for some fixed constant k and this gives a contradiction.

7. Let $n - 169 = x_1^2 + x_2^2 + x_3^2 + x_4^2$ with $x_1 \geq x_2 \geq x_3 \geq x_4 \geq 0$. If $x_4 > 0$ result follows as $169 = 13^2$, if $x_4 = 0$ and $x_3 > 0$ use $169 = 12^2 + 5^2$, if $x_3 = 0$ and $x_2 > 0$ use $169 = 12^2 + 4^2 + 3^2$, and if $x_2 = 0$ and $x_1 > 0$ use $169 = 10^2 + 8^2 + 2^2 + 1^2$. Exceptional set is 1, 2, 3, 4, 6, 7, 9, 10, 12, 15, 18, and 33.

8. Use the identities $2x + 1 = (x + 1)^2 - x^2$ and $2x = x^2 - (x - 1)^2 + 1^2$. Now $k = 3$ for $n = 6$ as 6 is not a sum of two squares, and there is no solution of $6 = x^2 - y^2 = (x + y)(x - y)$.

9. Using the identity and Theorem 1.4 we see that $6m^2$ can be written as a sum of 12 fourth powers, so by 1.4 again $6n$ is a sum of 48 fourth powers, now use $1=1^4$. Further $6n+2=6n+1^4+1^4$, $6n+3=6(n-13)+3^4$, and $6n+4=6(n-2)+2^4$.

10. $J(\psi,\chi)=\sum_{u=0}^{p-1}\chi(1-u)(\psi(u)+1)=\sum_{t=0}^{p-1}\chi(1-t^2)$ as $t^2\equiv u\ (p)$ has zero or two solutions. Let $v=2^*k(t+1)$.

11. (i) Twenty solutions generated by $\{0,1\}$, $\{2,4\}$, and $\{3,7\}$. (ii) Twenty-four solutions: $\{0,1\}$, $\{0,7\}$, $\{0,11\}$, $\{t,10\}$, $\{t,13\}$, $\{t,15\}$ for $t=2$, 3, and 14 and transposes.

12. If $p|a$ then $y=x$ or $p-x$. If $p\nmid a$ number of solutions N is $\sum_{t+u=a}N(x^2\equiv t\ (p))N(y^2\equiv u\ (p))$. Now choose c, d so that $t\equiv ac$ and $u\equiv ad\ (p)$, then $N=\sum_{c+d=1}(1+(ac/p))(1+(-ad/p))=p-(-1/p)^2$ as in proof of Theorem 2.7.

13. By Theorem 2.8, Chapter 5, $N(x^3\equiv a\ (p))=1$ so $N(x^3+y^3\equiv 1\ (p))=\sum_{a+b=1}1=p$. $\{0,1\}$, $\{2,2\}$, $\{3,4\}$; $\{0,1\}$, $\{2,3\}$, $\{4,11\}$, $\{5,6\}$, $\{7,9\}$, $\{8,16\}$, $\{10,13\}$, $\{12,14\}$, $\{15,15\}$; and transposes.

14. $N(x^n+y^n\equiv 1\ (p))=\sum_{a+b=1}N(x^n\equiv a(p))N(x^n\equiv b\ (p))=\sum_{i=0}^{n-1}\sum_{j=0}^{n-1}J(\chi^i,\chi^j)$. If $i+j=n$, $J(\chi^i,\chi^j)=-\chi(-1)$, and if $i\neq 0\neq j$ and $i+j\neq n$ then $|J(\chi^i,\chi^j)|=\sqrt{p}$.

15. (i) Use $\chi^3=\bar{\chi}$ and Theorem 2.4. (ii) Note $2|\psi(a)-1$ and $1+i$ divides 0, -2, $-1-i$ and $i-1$, so $\sum_{a+b=1}(\psi(a)-1)(\chi(b)-1)\equiv 0\ (2+2i)$, expand and get $p\equiv z\ (2+2i)$ and note that $2+2i|4$. (iii) We have $a+bi\equiv 1\ (2)$. Now as $1+i|4$, we have $a+bi\equiv a\equiv 1\ (2+2i)$, take conjugates and so $(2+2i)(2-2i)=8|(a-1)^2$, etc. (iv) As on p. 106 we have $N(x^2+y^4\equiv 1(p))=p+J(\psi,\chi)+J(\psi,\chi^2)+J(\psi,\chi^3)$, but $\chi^2=\psi$ so $J(\psi,\chi^2)=-1$. (v) The transformation is 1 to 1 but not onto as the inverse is not defined for $\{0,\pm y\}$ when $y^2\equiv -1\ (p)$, so there are two fewer points. (vi) $17=1^2+4^2$. First congruence: $N=14$ and solutions generated by $\{0,1\}$, $\{1,0\}$, $\{0,4\}$, $\{6,2\}$, and $\{6,8\}$, where $\{0,4\}$ and $\{0,13\}$ have no corresponding solutions. Second congruence: $N=12$ and solutions generated by $\{0,1\}$ and $\{2,8\}$.

16. (i) Use $(\zeta_{2^n})^2=\zeta_{2^{n-1}}$. (ii) As $p-1=2^n$ characters have order 2^m, $m\leq n$. Now note each $J(\chi,\chi^j)$ is constructible by (i). (iii) We have $\sum_{\chi}g(\chi)=\sum_{t=0}^{p-1}\left(\sum_{\chi}\chi(t)\right)\zeta_p^t=1+(p-1)\zeta_p$.

17. Use $\zeta^{(j+n/2)^2}=-\zeta^{j^2}$.

18. (i) Assume d_1 odd, so $(d_1/|d_2|)(d_2/|d_1|)=(|d_2|/|d_1|)(d_2/|d_1|)$ by Theorem 3.4, Chapter 4. Now treat $d_2>0$ and $d_2<0$ separately. (ii) We have $\sum(d_1/j_1)(d_2/j_2)e_{|d_1d_2|}(n(j_1|d_2|+j_2|d_1|))=(d_1/n)(d_2/n)(\sqrt{d_1})(\sqrt{d_2})$. Now multiply by $(d_1/|d_2|)(d_2/|d_1|)$ and note that $(d_1/j_1|d_2|)=(d_1/j_1|d_2|+j_2|d_1|)$, etc. (iii) We have, if $e(\cdot)=e_{|d|}(\cdot)$,

$$(d/n)\sum(d/j)e(j)=(d/n)\sum(d/nj)e(nj)=\sum(d/j)e(nj).$$

(iv) Do first three cases by direct calculation. If $d=(-1)^{(p-1)/2}p$, $\sum (d/j)e(j) = \sum (j/p)e(j)$ so use Theorem 3.1.

Problems 7

1. e $=[2, 1, 2, 1, 1, 4, 1, 1, 6, 1, \ldots]$ and $\pi=[3, 7, 15, 1, 292, 1, 1, 1, 21, 31, \ldots]$. The third convergent to π is 355/113, now use Lemma 1.2(i).

2. (i) Expand by last column and use induction. Delete first row and column to obtain determinant for b_n, for $n>0$. (ii) Show by induction $\alpha_k = q_k + 1/\alpha_{k+1}$.

3. By Theorem 1.1 $[q_0, \ldots, q_{n-1}, q_n+\beta] = \dfrac{a_n+\beta a_{n+1}}{b_n+\beta b_{n+1}}$, this lies between a_n/b_n and a_{n+1}/b_{n+1}. So inequality is reversed if n is even.

4. (i) If a_n/b_n is the nth convergent of $\alpha=(1+\sqrt{5})/2$ then $a_0=u_2$, $a_1=u_3$, $b_0=1$, and $b_1=u_2$, use induction as $\alpha=[1, 1, 1, \ldots]$. (ii) Assume $p>5$. Expanding $\alpha^n-\alpha'^n$ we have ($\alpha'=$ conjugate of α)

$$2^{n-1}u_n = n + \binom{n}{3}5 + \binom{n}{5}5^2 + \ldots + 5^{(n-1)/2}.$$

So using Fermat's theorem and Euler's criterion we have $u_p \equiv (5/p) = \pm 1 \pmod p$. By (i) and Lemma 1.2, $u_p^2 - u_{p-1}u_{p+1} = 1$, so $p \mid u_{p-1}u_{p+1}$. Using equation above with $n=p+1$ we get $2^p u_{p+1} \equiv 1 + (5/p) \pmod p$, so $p \mid u_{p+1}$ if $(5/p)=-1$, etc. Now note $u_3=2$, $u_4=3$, and $u_5=5$ so use (iii) of Problem 19, Chapter 1.

5. (i) (a) If $\alpha=[0, q_1, \alpha_2]$ and $q_1 \geq m$ then $0<\alpha\leq 1/m$, and probability is $1/m$. (b) Probability is $1/m(m+1)$. (ii) If $\alpha=[0, q_1, q_2, \alpha_3]$ and $q_2\geq m$ then $1/q_1 < \alpha \leq 1/(q_1+1/m)$, sum over q_1. (iii) Probability of $q_2 \geq 2$ is $\sum_x (1/x(2x+1)) = 2\sum_x (1/2x - 1/(2x+1)) = 2(1-\ln 2)$, so probability of $q_2=1$ is $2\ln 2 - 1 \doteq 0.3$.

6. $167/43=[3, 1, 7, 1, 1, 2]$, so the third convergent is 35/9 giving solution $x=-9$ and $y=35$, see (i) in problem below.

7. (i) Use Lemma 1.2. (ii) and (iii) Use induction and ($*$) in Theorem 1.1.

8. $20929/86400=[0, 4, 7, 1, 3, 1, 16, \ldots]$; convergents 1/4, 7/29, 8/33, 31/128, ... Eight leap years in 33 better.

9. (i) Use $1/\alpha'_{n+1} = \alpha'_n - q_n$ and induction. (ii) Use (i). (iii) As α is quadratic, $\alpha_j=\alpha_k$ for some j, k, use $\alpha_{j-1} = [-1/\alpha'_j] + 1/\alpha_j$. (iv) Begin with $\alpha=(\alpha a_{n-1} + a_{n-2})/(\alpha b_{n-1}+b_{n-2})$. (v) Show that $f(0)<0$ and $f(-1)>0$.

10. We have $q_{n+2}(a_n+a_{n+1}t) = (q_{n+2}-t)a_n + ta_{n+2}$, etc., so inequality follows. First choose T: $b_n < T \leq b_{n+2}$, then choose t: $b_n + b_{n+1}t < T \leq b_n + b_{n+1}(t+1)$. Now $779/207=[3, 1, 3, 4, 2, 5]$ so convergents are 3, 4, 15/4, 64/17, and 143/38. Best lower estimate is 143/38 and best upper is 207/55, but former is closer.

11. 52518/16717.

12. Given α $(0<\alpha<1)$ choose consecutive a/b, $c/d \in \mathsf{F}_n$ so that α lies

between a/b and $(a+c)/(b+d)$, hence

$$\left|\alpha - \frac{a}{b}\right| < \left|\frac{a+c}{b+d} - \frac{a}{b}\right| = \frac{1}{b(b+d)} < \frac{1}{bn} \leq \frac{1}{b^2}.$$

13. (i) Suppose result false, as α lies between a_n/b_n and a_{n+1}/b_{n+1} we have

$$\frac{1}{b_n b_{n+1}} = \left|\frac{a_{n+1}}{b_{n+1}} - \frac{a_n}{b_n}\right| = \left|\alpha - \frac{a_n}{b_n}\right| + \left|\alpha - \frac{a_{n+1}}{b_{n+1}}\right| \geq \frac{1}{2b_n^2} + \frac{1}{2b_{n+1}^2},$$

but this gives $(b_{n+1} - b_n)^2 \leq 0$.

(ii) By Theorem 2.1 and Lemma 1.2 we have $b < b_{n+1}$ gives $|a - \alpha b| \geq |a_{n+1} - \alpha b_{n+1}|$, and the first part follows. Now use similar argument to (i).

14. Use $\dfrac{x}{y} - d^{1/3} = (x^3 - dy^3)/y^3\left[\left(\dfrac{x}{y} + \dfrac{d^{1/3}}{2}\right)^2 + \dfrac{3}{4}d^{2/3}\right]$ and Problem 13.

15. We have $f(c, x) = f(c+1, x) + \dfrac{x}{c(c+1)} f(c+2, x)$ and so

$$\frac{f(c+1, x)}{f(c, x)} = \left(1 + \frac{x}{c(c+1)} \frac{f(c+2, x)}{f(c+1, x)}\right)^{-1},$$

put $x = y^2$. For given substitution $\alpha_{n+2} \geq 1$ for all n. Finally use $e^z + e^{-z} = 2f(1/2, z^2/4)$ and $e^z - e^{-z} = 2zf(3/2, z^2/4)$. (See Davis, 1945.)

16. For $n > 1$, $s_n = (2 + 4n)s_{n-1} + s_{n-2}$, $q_{3n-m} = q_{3n-m-1} + q_{3n-m-2}$ if $m = 3$, 2, 0, or -1, and $q_{3n-1} = 2nq_{3n-2} + q_{3n-3}$. Adding we get $q_{3n+1} = (2+4n)q_{3n-2} + q_{3n-5}$, etc., and so

$$\frac{q_{3n+1}}{r_{3n+1}} = \frac{s_n/t_n + 1}{s_n/t_n - 1} \to \gamma \quad \text{as} \quad n \to \infty.$$

17. {1520, 273}, no solution, {27, 5}.

18. As $x^2 - dy^2 = 1$ is always soluble, we have $(2x)^2 - d(2y)^2 = 4$. Now proceed as in the proof of Theorem 3.2.

19. If $p \equiv 3$ (4), use $(-1/p) = -1$. If $p \equiv 1$ (4) then x' is odd and y' is even, and so $(x'+1, x'-1) = 2$. Then $x' \mp 1 = 2z^2$ and $x' \pm 1 = 2pt^2$. If lower signs apply start again.

20. (i) Each α_i is purely periodic, so $\alpha = \alpha_{nk}$ and $\alpha \neq \alpha_i$ if $i \nmid k$, and $(c_{nk} + \sqrt{d})/e_{nk} = (c_0 + \sqrt{d})/e_0 = [\sqrt{d}] + \sqrt{d}$. This gives $e_{nk} = 1$. Now if $e_i = 1$, $\alpha_i = c_i + \sqrt{d}$, which implies $c_i = [\sqrt{d}]$ by Problem 9. (ii) This is similar to (i). (iii) Use (**) on p. 119 for by above $e_t \geq 2$.

21. For first part write $\sqrt{d} = -[\sqrt{d}] + \alpha$ and use Problem 20. For second part use $\sqrt{d} = [q_0, \ldots, 1/(-q_0 + \sqrt{d})]$ to obtain quadratic for $\sqrt{d}$, deduce that $a_{k-2} + b_{k-2}q_0 - b_{k-1} = 0$, and use Problem 7.

22. (i) $\sqrt{d} = [a, \overset{*}{2}a]$, so $d = a^2 + 1$. (ii) $\sqrt{d} = [a, \overset{*}{b}, b, \overset{*}{2}a]$, this gives $(b^2+1)(a+\sqrt{d}) = a(b^2+1) + \sqrt{\{a^2(b^2+1)^2 + (b^2+1)(2ab+1)\}}$. So $2ab+1 = (b^2+1)c$ and $d = a^2 + c$, b is even and $c > 1$. (iii) $\sqrt{d} = [a, \overset{*}{b}, \overset{*}{2}a]$, so $d = a^2 + c$ where $bc = 2a$. $\sqrt{41} = [6, \overset{*}{2}, 2, \overset{*}{12}]$, $\sqrt{370} = [19, \overset{*}{4}, 4, \overset{*}{38}]$, etc.

23. Solve $(2n+1)^2 - 8z^2 = 1$.

24. (i) Yes. For each collection (with t elements) $\frac{0}{t}, \ldots, \frac{t-1}{t}$, $[(b-a)t]$ or $[(b-a)t]+1$ elements lie in the interval $[a, b)$, so sum over t and take limits. (ii) No. We have $e = 1 + 1 + \frac{1}{2!} + \ldots + \frac{1}{n!} + \frac{a_n}{(n+1)!}$ where $a_n < e$, hence $((n!\,e)) = \frac{a_n}{n+1} \to 0$ as $n \to \infty$. (iii) Yes. Use Theorem 4.2 as θ is irrational.

25. First show that the problem is equivalent to considering a ray of light passing across the plane which has been ruled off into squares, and then plot the mirror images of P. The path is periodic if it passes through one of these images otherwise use Theorem 4.1.

Problems 8

1. If a_n/b_n is the nth convergent for α we have

$$\left|\alpha - \frac{a_n}{b_n}\right| < \frac{1}{b_n b_{n+1}} < \frac{1}{b_n^2 a^{(n+1)!}} < \frac{1}{(a^{n!})^n} < \frac{1}{b_n^{n/2}}.$$

For the last inequality show that $b_n < a^{2(n!)}$.

2. Let $c_1, c_2, \ldots$ be an arbitrary sequence of 1's and 2's and consider the sum $\sum_{i=1}^{\infty} c_i 3^{i!}$.

3. Assume the contrary and represent each real α_n in the interval $[0, 1)$ by its infinite decimal $0.a_{n1}a_{n2}\ldots$. Show that the decimal $0.b_1b_2\ldots$ where $b_i = a_{ii} + 1$ if $a_{ii} < 9$ and $b_i = 0$ if $a_{ii} = 9$ is a 'new' real.

4. We have

$$0 < |g(\alpha_1)| = \left|x_0^v \prod_{i=1}^{n} f(\beta_i)\right| \Big/ \left\{a_0^n h^{v-1} \prod_{j=2}^{v} (|g(\alpha_j)|/h)\right\}$$

By the symmetric function theorem $\prod_i f(\beta_i)$ is an integer so

$$|g(\alpha_1)| \geq 1 \Big/ \left\{a_0^n h^{v-1} \prod_{j=2}^{v} |g(\alpha_j)/h|\right\} = 1/Th^{v-1}.$$

By definition of h, $|g(\alpha_j)/h|$ is bounded so $1/T$ has a positive lower bound $A(\alpha, n)$ depending only on α and n. So if the inequality has infinitely many solutions with n fixed then α is not algebraic of degree $< t+1$, but t is arbitrary.

5. For example consider the equations $x^2 - 2y^2 = 1$, $x^3 - y^3 = 0$ and $(x^2 - 2y^2)^2 = 1$.

6. See (ii) on p. viii.

7. Use induction, the symmetric function theorem, and Problem 6.

8. If $\gamma = a/b$ and $\sinh \gamma = c$ (algebraic) then $e^{2a/b} - 2ce^{a/b} - 1 = 0$, hence $e^{a/b}$ is algebraic, and so e^a is also. Now consider the polynomial in e^a as a polynomial in e.

9. (i) Use $e^{i\pi} + 1 = 0$. Multiply out and collect together terms with zero exponent, there are 2^c terms altogether. (ii) We have from the definitions

$$L = -(2^c - n)\sum_{j=0}^{m} f^{(j)}(0) - \sum_{j=0}^{m}\sum_{k=1}^{n} f^{(j)}(\alpha_k)$$

where $m = (n+1)p - 1$. By the symmetric function theorem each $S_j = \sum_{k=1}^{n} f^{(j)}(\alpha_k)$ is a rational integer. Also $f^{(j)}(\alpha_k) = 0$ when $j < p$, so $p!|S_j$ and $p!|f^{(j)}(0)$ unless $j = p - 1$, and then $f^{(p-1)}(0) = (p-1)!(-t)^{np}(\alpha_1 \ldots \alpha_n)^p$. Now suppose $p > 2^c - n$.

10. (i) Use $\int_0^\infty x^k e^{-x}\,dx = k\int_0^\infty x^{k-1}e^{-x}\,dx = k!$.

(ii)

$$a_k e^k I_k^\infty = \begin{cases} a_0 \displaystyle\int_0^\infty y^r f_0(y) e^{-y}\,dy & \text{if } k = 0, \\ a_k \displaystyle\int_0^\infty y^{r+1} f_r(y) e^{-y}\,dy & \text{if } 0 < k \le n, \end{cases}$$

for some polynomials f_i with integer coefficients. Therefore all terms of P_1 are integers and all but the first are multiples of $(r+1)!$. So $P_1 \equiv a_0(-1)^{n(r+1)}(n!)^{r+1}r!$ $(\text{mod}(r+1)!)$. Choose r so that $(r+1, a_0 n!) = 1$.

(iii) If $X = \max_{0 \le x \le n} |x(x-1)\ldots(x-n)|$ and $Y = \max_{0 \le x \le n} |(x-1)\ldots(x-n)e^{-x}|$, then $|a_k I_0^k| \le k|a_k|\, X^r Y$. Now use $X^r = o(r!)$.

(iv) We have $P_1 + P_2 = 0$.

11. If β is rational then the numbers θ_i are not all distinct, and so t and n are bounded.

12. $\alpha \ln \beta + \gamma \ln \delta = 0$ gives $\beta^{(-\alpha/\gamma)} = \delta$.

Problems 9

1. By relabelling if necessary, suppose $A_1, \ldots, A_r$ are the linearly independent rows of A, and $A_s = c_{1,s}A_1 + \ldots + c_{r,s}A_r$ for $s = r+1, \ldots, n$. Let B_i be the vector formed by taking the first r coordinates of A_i and let B be the matrix with ith row B_i; again, relabelling if necessary, we have $\det B \neq 0$. Now the ith coordinate of $\mathbf{y}$ is $x_i + c_{i,r+1}x_{r+1} + \ldots + c_{i,n}x_n$.

2.

$$A = \begin{pmatrix} a_1 & c_1 & e_1c_2 & f_1c_3 & \cdots & h_1c_{n-1} \\ a_2 & d_1 & e_2c_2 & f_2c_3 & \cdots & h_2c_{n-1} \\ a_3 & 0 & d_2 & f_3c_3 & \cdots & h_3c_{n-1} \\ \cdot & \cdot & \cdot & \cdot & \cdots & \cdot \\ a_{n-1} & 0 & 0 & 0 & \cdots & h_{n-1}c_{n-1} \\ a_n & 0 & 0 & 0 & \cdots & d_{n-1} \end{pmatrix}.$$

The numbers $e_1, e_2, f_1, \ldots, h_{n-1}$ are given by linear equations with coprime coefficients built up using the integers c_i and d_i, for example we have $f_1d_1d_2 - f_2c_1d_2 + f_3c_2 = 1$.

3. Second and fifth forms are equivalent using matrix $\begin{pmatrix}1 & 0\\ 1 & 1\end{pmatrix}$. Also these do not represent 1 whereas the third form does. The remaining forms have determinant -6, use Pell's equation to show that they are inequivalent.

4. One, all forms are equivalent to $x^2 + 2y^2$.

5. F and G have determinant 68, F represents 3 but G does not represent any positive number less than 7, show this by completing the square: $7G(x, y) = (7x + 3y)^2 + 68y^2$. We have $z^2 - G(x, y) = (x - 14s + 5t)^2 - (207s^2 - 140st + 24t^2)$. [Solution due to J. W. S. Cassels.]

6. If a permutation exists use a transposition matrix. Now assume the contrary and $a \le b \le c$, $a' \le b' \le c'$, if $a < a'$ then one form represents a while the other does not, so assume $a = a'$ *and* $b > b'$. Now show directly for any J

$$\begin{pmatrix} a & 0 & 0\\ 0 & b' & 0\\ 0 & 0 & c'\end{pmatrix} \ne J\begin{pmatrix} a & 0 & 0\\ 0 & b & 0\\ 0 & 0 & c\end{pmatrix}J'$$

7. Rewrite the form as a sum of squares beginning with $\frac{1}{2}(2x_1 - x_3)^2$.

8. As $p \equiv 1$ (4), b and c exist satisfying $b^2 = -1 + cp$, consider the form $px^2 + 2bxy + cy^2$ and use Lemma 2.1.

9. (i) Note that if $m \equiv 7$ (8), then $m - 1$ is a sum of three squares. (ii) Solve $x_1^2 + x_2^2 + x_3^2 = 8m + 3$, show that x_1, x_2, and x_3 are odd, and consider $\sum y_i(y_i + 1)$ where $x_i = 2y_i + 1$.

10. If $8 \nmid n$ then $n - 1$ is a sum of three squares, and if $8|n$ then $2|x_i$ for all i.

11. If A is the matrix of F then unimodular $T = (t_{ij})$ and $U = (u_{ij})$ exist satisfying $TT' = A$, $UAU' = I$, $U = T^{-1}$ and $t_{31}^2 + t_{32}^2 + t_{33}^2 = m$ (Theorem 1.1), so if $(t_{31}, t_{32}, t_{33}) = k$ then $k|(u_{11}, u_{12}, u_{13})$, but $F(u_{11}, u_{12}, u_{13}) = 1$.

12. Write $n = x^2 + y^2$; $n \equiv 1$ (4) so x is even and y is odd, also $n \equiv 2$ (3) so $3 \nmid x$ and $3 \nmid y$. Hence (changing signs if necessary) $x \equiv 1$ (3) and $y \equiv 2$ (3). Note also if $2y - x = 0$, $5y^2 = n$, but n is square-free etc.

13. There is only one equivalence class of ternary forms with determinant 2 so method is valid. Proceed exactly as in the proof of Theorem 2.2. There are four cases: (i) $m \equiv 3$ (4), so $(8m, m - 2) = 1$, $p = 8mt + m - 2$ and $c = 8t + 1$; (ii) $m \equiv 1$ (4), so $(8m, 5m - 2) = 1$, $p = 8mt + 5m - 2$ and $c = 8t + 5$; (iii) $m \equiv 4$ (8), so $(m - 2)/2$ is odd, $(4m, (m - 2)/2) = 1$, $p = 4mt + (m - 2)/2 \equiv 1$ (4) and $2p = cm - 2$ where $c = 8t + 1$; and (iv) $m \equiv 2$, 6, or 10 (16), let $e = 3$ if $m \equiv 2$ (8), $e = 1$ if $m \equiv 6$ (16) so $(em - 2)/4$ is odd. Now proceed as in the second case.

14. Forms are $x^2 + y^2 + 3z^2$ and $x^2 + 2y^2 + 2yz + 2z^2$, the first represents 5 whilst the second does not, note the second cannot represent any integer $\equiv 5$ (8). Use the method of Theorem 2.2 to find a form equivalent to one of the above, as

$m \equiv 5$ (8) it must be the first one. $3 \nmid m$, so $(24m, 4m - 3) = 1$ and a prime exists: $p = 4c'm - 3$ where $c' \equiv 1$ (6).

15. If a solution exists then $t = (x_1 - x_2)^2 + (x_1 - x_3)^2 + (x_1 - x_4)^2 + (x_2 - x_3)^2 + (x_2 - x_4)^2 + (x_3 - x_4)^2 \geq 0$, if $t = 0$ and $4|m$, let $x_1 = x_2 = x_3 = x_4 = m/4$, and if $t > 0$ and $t = u^2 + v^2 + w^2$, let $4x_1 = m + u + v + w$, $4x_2 = m + u - v - w$, $4x_3 = m - u + v - w$, and $4x_4 = m - u - v + w$.

16. Consider the cases d even and odd separately, note $d \equiv b$ (2).

17. (i) No solution as $m^2 \not\equiv -35$ (8). (ii) Solutions are $\{1, -1\}$ and $\{-1, 1\}$. (iii) Primary solution $\{3, -1\}$ with automorph $\begin{pmatrix} 2 & 1 \\ 5 & 3 \end{pmatrix}$. (iv) Two primary solutions $\{3, 1\}$, for $m = 3$, and $\{4, -1\}$, for $m = 7$, automorph $\begin{pmatrix} 11 & 5 \\ 35 & 16 \end{pmatrix}$.

18. Use Theorem 3.7 as class number is one in all cases. For (i) use $(-3/2) = -1$ and $(-3/p) = \pm 1$ iff $p \equiv \pm 1$ (6), other cases similar.

19. (i) As $-ac > 0$, we have $(v - bw)(v + bw) = 4(1 - acw^2) > 0$. So choose sign of v so that $v - bw$ and $v + bw$ are both positive. Now choose sign of w so that $aw > 0$. (ii) If r and s negative use $\begin{pmatrix} -1 & 0 \\ 0 & -1 \end{pmatrix}$. If $rs < 0$ we have $4r's' = 4rs + 8(-ac)rsw^2 + 2r^2aw(v - bw) - 2cws^2(v + bw) = 4rs + X$; hence use $((r^2aw(r - bw))^{1/2} - (-s^2cw(v + bw))^{1/2})^2 \geq 0$ to show X is positive, now by (i) choose v and w large (in absolute value) so that $r's' > 0$.

20. Use $\begin{pmatrix} 0 & 1 \\ -1 & 0 \end{pmatrix}$, $\begin{pmatrix} 1 & 0 \\ t & 1 \end{pmatrix}$, and $\begin{pmatrix} 1 & u \\ 0 & 1 \end{pmatrix}$.

21. First and third equivalent by $\begin{pmatrix} 3 & -1 \\ 4 & -1 \end{pmatrix}$, second and fifth equivalent, and sixth equivalent to $\langle 1, 0, 21 \rangle$ by Problem 20 ($t = 3$), see tables.

22. If equivalence holds then r and s exist: $r^2 - ps^2 = -1$, and so $p \equiv 1$ (4). If $p \equiv 1\,(4)$ we can find r and s to satisfy this equation, so use $\begin{pmatrix} r & s \\ -sp & -r \end{pmatrix}$.

23. $d = -67$: so $a \leq 4$ and $|b| = 1$ or 3. If $|b| = 3$ then $a = 1$, impossible so $|b| = 1$ and $a = 1$, $c = 17$. As $|b| = a$, $b = 1$ and $h(-67) = 1$. $d = -68$: so $a \leq 4$ and $|b| = 0$, 2, or 4. If $|b| = 4$ then $a < 4$, so $|b| = 0$ or 2 giving $h(-68) = 4$. $d = 29$: so $|a| \leq 2$ and $|b| = 1$, this gives $|a| = 1$ and $|c| = 7$. Now $(-1 + \sqrt{29})/2 = [2, \overset{*}{5}]$ giving solution $\{2, 1\}$, and $2^2 + 2.1 - 7(-1)^2 = -1$. Matrix $\begin{pmatrix} 2 & 1 \\ -5 & -2 \end{pmatrix}$ transforms $\langle 1, 1, -7 \rangle$ into $\langle -1, -1, 7 \rangle$. As $\langle 1, -1, 7 \rangle \sim \langle 1, 1, 7 \rangle$ we have $h(29) = 1$. $d = 69$: so $|a| \leq 4$ and $|b| = 1$ or 3. $|b| = 3$ gives $|a| = 3$ and $|c| = 5$, and $|b| = 1$ gives $|a| = 1$ and $|c| = 17$. Now $(-1 + \sqrt{69})/2 = [3, \overset{*}{1}, 1, 1, \overset{*}{7}]$ so possible solution $\{4, 1\}$ gives $4^2 + 4 - 17 = 3$ and $\begin{pmatrix} 4 & 1 \\ 3 & 1 \end{pmatrix}$ transforms $\langle 1, 1, -17 \rangle$ into $\langle 3, -3, -5 \rangle$. But similarly $\langle 1, 1, -17 \rangle \not\sim \langle -1, -1, 17 \rangle$, so $h(69) = 2$.

24. Solving $ru - st = 1$ and $(2ar + bs)t + (br + 2cs)u = b'$ we have $t = (b'r - br - 2cs)/2a'$ and $u = (2ar + bs + b's)/2a'$. It follows by direct calculation that $at^2 + btu + cu^2 = c'$. Now we have $4aa' = (2ar + bs)^2 - ds^2$ and, as $d = b'^2 - 4a'c'$, $4a' \mid (2ar + bs + b's)(2ar + bs - b's)$, so use conditions as b and b' have the same parity.

Problems 10

1. Choose t: $t \det T \equiv 1\ (p)$, then $t(\det T)T^{-1}$ is integral with determinant coprime to p.

2. Part (ii), all congruences mod 8. We show $k \equiv ax^2 + 2bxy + cy^2$ and $b^2 - ac \equiv 2$ implies $k \equiv \pm a$, and so $(2/k) = (2/a)$. Case $1: 4 \mid b$, so a and x odd, c and y even which gives $a \equiv 1$ or 5 and $c \equiv 6$, or $a \equiv 3$ or 7 and $c \equiv 2$. So if x odd and y even we have $k \equiv a$, and if x and y odd we have $k \equiv -a$. Other cases similar.

3. This is similar to Problem 2.

4. $x^2 + xy - 5y^2 = k$ soluble only if $k \equiv 1$, 25, or 37 (42), and $-x^2 + xy + 5y^2 = k$ soluble only if $k \equiv 5$, 17, or 41 (42). Solutions: $k = 1$, $\{1, 0\}$; $k = 5$, $\{0, 1\}$, $\{-3, 2\}$; $k = 17$, $\{3, 2\}$, $\{-4, 3\}$; $k = 25$, $\{5, 0\}$, $\{5, 1\}$, $\{7, 3\}$; $k = 37$, $\{6, 1\}$, $\{9, 4\}$; $k = 41$, $\{-1, 3\}$, $\{-7, 5\}$.

5. Equation soluble only if $k \equiv 13$, 37, 43, or 67 (120). Solutions $k = 13$, $\{\pm 1, \pm 1\}$; $k = 37$, $\{\pm 3, \pm 1\}$; $k = 43$, $\{\pm 1, \pm 2\}$; $k = 67$, $\{\pm 3, \pm 2\}$; $k = 133$, no solution; $k = 157$, $\{\pm 7, \pm 1\}$; $k = 163$, $\{\pm 1, \pm 4\}$; $k = 187$, $\{\pm 3, \pm 4\}$ and $\{\pm 7, \pm 2\}$.

6. $d = -20$, $x^2 + 5y^2 = k$ only if $k \equiv 1$ or 9 (20), $2x^2 + 2xy + 3y^2 = k$ only if $k \equiv 3$ or 7 (20). $d = -84$, $x^2 + 21y^2 = k$ only if $k \equiv 1$, 25, or 37 (84), $5x^2 + 4xy + 5y^2 = k$ only if $k \equiv 5$, 17, or 41 (84), $2x^2 + 2xy + 11y^2 = k$ only if $k \equiv 11$, 23, or 71 (84), $3x^2 + 7y^2 = k$ only if $k \equiv 19$, 31, or 55 (84). $d = 96$, $x^2 - 24y^2 = k$ only if $k \equiv 1$ (24), $-x^2 + 24y^2 = k$ only if $k \equiv 23$ (24), $3x^2 - 8y^2 = k$ only if $k \equiv 19$ (24), $-3x^2 + 8y^2 = k$ only if $k \equiv 5$ (24).

7. Note that $x^2 - dy^2$ belongs to the principal genus, i.e. the character system is all plus ones.

8. Choose a_1, a_2, a_3: $(a_1, a_2) = 1$ and $(a_3, a_1a_2) = 1$. Let $\langle a_j, b, - \rangle \in C_j$ $(j = 1, 2, 3)$ and consider $\langle a_1a_2a_3, b, - \rangle$.

9. If $f(r, s) = z$ with $(r, s) = 1$, then t and u exist satisfying $ru - st = 1$. Now use Lemma 1.5.

10. If $f \in C_0$, then $f \sim \langle 1, -, - \rangle$ and so f represents 1. Use Problem 9 for the converse.

11. $\langle\langle 3, -2, 4 \rangle\rangle$, $\langle\langle 4, 1, 9 \rangle\rangle$, $\langle\langle 5, 56, 150 \rangle\rangle \sim \langle\langle 3, 2, -11 \rangle\rangle$.

12. This is similar to Lemma 2.5.

13. (i) f is improperly equivalent to its inverse in $\mathcal{G}$. (ii) Use Problem 20, Chapter 9. (iii) $\begin{pmatrix} 1 & 0 \\ k & -1 \end{pmatrix}$ transforms $\langle a, ka, c \rangle$ into itself. (iv) Evaluate

coefficients. (v) If $T=\begin{pmatrix} e & f \\ g & h \end{pmatrix}$ then the top right-hand entry of TST^{-1} is $(es-fr)^2-f^2$, so let $f(1+r)=es$, that is $f=s/j$, $e=(1+r)/j$ where $(s, 1+r)=j$. (vi) Take $r=0$ and $s=1$ in the above.

14. Use: discriminant of $\langle a, 0, c\rangle$ is $-4ac$, and of $\langle a, a, c\rangle$ is $a(a-4c)$. Exclude the negative definite case.

15. If T is proper and S is improper then TS is improper and $TSTS=I$, so $TST^{-1}=T^2S$. Now use Problem 13.

16. Use Theorem 2.9 and Problem 13.

17. Use Problem 12 and consider the non-identity values of f_0.

18. As f represents c^n, $f\sim\langle c^n, b, -\rangle$. Check that $(b, c)=1$ and let $g=\langle c, b, -\rangle$.

19. $h(12)=2$, $h(-23)=3$, and $h(-575)=6h(-23)=18$.

20. Let $d=-3\cdot 5^{2\alpha}$ so $h(d)=h(-3)5^{\alpha-1}(5-(-3/5))=6\cdot 5^{\alpha-1}$.

21. We have $1\le h(-p)=[z/(2-d/2)]\sum_{t=1}^{p/2}(-p/t)$ but $(-p/t)=(t/p)$.

22. $S(n)-2S(n-1)+S(n-2)=(d/n)$ so $K(d)=\sum S(n)\left[\frac{1}{n}-\frac{2}{n+1}+\frac{1}{n+2}\right]$. Now $|T_1|\le\sum_{n=1}^{k-1}1/(n+2)\le\ln(k-1)+\gamma-3/2+1/k+1/(k+1)$ and $|T_2|\le\sum_{n=k}^{\infty}2\sqrt{d}/(n+1)(n+2)=2\sqrt{d}/(k+1)$. Now let $k=[2\sqrt{d}]+1$ and use $1\le h(d)=K(d)\sqrt{d}/\ln\delta$. Prove directly if $d\le 5$.

Problems 11

1. $p(10)=42$, $p(20)=627$, $p(30)=5604$.

2. A partition into distinct parts has the form $2^{a_1}+2^{a_2}+\ldots+2^{b_1}\cdot 3+2^{b_2}\cdot 3+\ldots+2^{c_1}\cdot 5+\ldots$ where the a_i, b_i, ... are distinct, and a partition into odd parts has the form $(2^{a_1}+2^{a_2}+\ldots)\cdot 1+(2^{b_1}+2^{b_2}+\ldots)\cdot 3+\ldots=a\cdot 1+b\cdot 3+\ldots$, this defines the correspondence.

3. Use $\prod(1+q^n+q^{2n}+q^{3n})=\prod(1+q^n)(1+q^{2n})=\prod(1+q^{2n})(1-q^{2n-1})^{-1}$. General case corresponds to partitions in which parts divisible by 2^{c-1} are not repeated.

4. Use $\prod(1+q^{2n})(1+q^{3n})=\prod\frac{(1-q^{4n})(1-q^{6n})}{(1-q^{2n})(1-q^{3n})}=\prod(1-q^{4n+2})^{-1}(1-q^{6n+3})^{-1}$. A typical term of the left-hand side is $q^{2(a_1+a_2+\ldots)+3(b_1+b_2+\ldots)}$.

5. Use $\prod(1+q^n)^{-1}=\prod(1-q^{2n-1})$ as in Theorem 1.2 and write the left-hand side as $\sum^*(-1)^{a_1+a_2+\cdots}q^{a_1+2a_2+3a_3+\cdots}$. The parity of each term is even (odd) when the partition with a_1 1's, a_2 2's, ... has an even (odd) number of parts. The right-hand side generates partitions with an odd number of distinct parts (ignore signs).

6. Replace each row of one of the first kind of partition by an L-shaped pattern with equal length arms, now fit together.

7. To partition of $a-c$ add new top row of length c and delete first column (of length b), now take conjugates.

8. Suppose largest part has $2n+1$ dots so smallest part has at least $n+1$ dots. Split dot pattern into a rectangular block A, $n+1\times k$, where k is the number of parts, on the left; and a pattern B on the right. Move B to lie directly below A and take conjugate. The resulting partition is of the second kind. Now the equation relates the generating functions of these partitions for the left-hand side is

$$1+(q^1+q^2+\ldots)+q^3(1+q^2+q^4+\ldots)(1+q^3+q^6+\ldots)+\ldots,$$

and the typical terms are $q^{a_1\cdot 1}$, $q^{3+b_1\cdot 2+b_2\cdot 3}, \ldots$. These exponents are partitions of the first kind, now use a similar argument for partitions of the second kind.

9. Use Theorem 1.3 to characterize (ii) and (iii), then use Theorem 1.1.

10. Formula enumerates partitions of $n-N$ into even parts, note that, for a self-conjugate partition, pieces (ii) and (iii) referred to in Problem 9 are themselves conjugate to each other. Finally use Theorem 1.1.

11. Suppose $n=a_1+a_2+\ldots+a_k$ has minimum difference 2, so $a_{k-i}\geq 2i+1$ and $(a_1-(2k-1))+(a_2-(2k-3))+\ldots+(a_k-1)$ is a partition of $n-k^2$. Proceed as in Problem 10.

12. Put q^8 for q in Theorem 2.3(ii).

13. Put $z=-1$ in Jacobi's identity and use the method of Theorem 1.2.

14. We have (mod 2) $\Sigma\, p(n)q^n \equiv \Pi\,(1+q^n)^{-1} = \Pi\,(1-q^{2n-1}) \equiv \Pi\,(1-q^{4n})^{-1}\Pi\,(1-q^{4n})(1+q^{4n-1})(1+q^{4n-3}) = \Sigma\, p(m)q^{4m}\Sigma\, q^{n(n+1)/2}$. Now compare coefficients of q^{4n}.

15. We have $q^2\prod\limits_{n}^{\infty}(1-q^n)^6 = \sum\limits_{n}\sum\limits_{m}(-1)^{n+m}(2n+1)(2m+1)q^t$ where $t=2+n(n+1)/2+m(m+1)/2$. So $8t=14+(2n+1)^2+(2m+1)^2$ and $7|t$ implies $7|2n+1$ and $7|2m+1$, now proceed as in Theorem 2.4.

16. $49|p(n)$ if $n=49m+u$ where $u=19$, 33, 40, or 47. Also $p(243)=133978259344888$, this is not divisible by $7^3=343$ but $24\cdot 243\equiv 1 \pmod{343}$.

17. Use Theorem 2.3(i). We have $\left[\sum\limits_{n} q^{n^2}\right]^2 = \sum\limits_{n} r(n)q^n$ where $r(n)$ is the number of solutions of $n=x^2+y^2$. Finally expand the right-hand series and compare terms.

18. For fixed Durfee square with r^2 dots there are at most n^{2r} partitions of the remaining pieces, sum over r.

Problems 12

1. By Theorem 1.2 with p_n for n we have $p_n > \frac{1}{6}n \ln p_n > \frac{1}{6}n \ln n$ and

$$\frac{\ln p_n}{\sqrt{p_n}} < \frac{1}{8} < \frac{n \ln p_n}{p_n}$$

if $p_n > 5000$ $(n > 668)$. So $\ln p_n < 2 \ln n$ and $p_n < 16n \ln n$.

2. $\pi(x, r) = [x] - \sum_{i \le r} [x/p_i] + \sum_{i,j \le r} [x/p_ip_j] - \ldots$. So as there are at most 2^r terms in these sums, and

$$\pi(x) < x - \sum x/p_i + \sum x/p_ip_j \ldots + r + 2^r < x\prod_i (1 - 1/p_i) + 2^{r+1}.$$

Now use $\prod_i^x (1 - 1/p_i) \to 0$ as $x \to \infty$.

3. By Theorem 1.2 take $e' = 12$. Now write $T = \prod_{n<p\le 2n} p$ and take n large. We have

$$\binom{2n}{n} < T(2n)^{\sqrt{2n}} \prod_{p \le 2n/3} p,$$

and so $T > 4^{n/3}(2n)^{-(\sqrt{2n})-1} > 2^{nk}$ for some $k > 0$. But also $T < (2n)^{\pi(2n)-\pi(n)}$, and so $nk \ln 2 < (\pi(2n) - \pi(n)) \ln 2n$.

4. Use Bertrand's postulate.

5. (i) If $k < n$ then $m < 1$. If $k \ge n$ there is a prime between n and $n + k$, now consider the largest one. (ii) If k, m, and $n > 1$, every prime in $n!$ occurs at least twice, impossible by Theorem 1.3.

6. If $p_{n+1} - p_n < k$ then $c(n + t) \ln(n + t) < p_{n+t} < c'n \ln n + kt$, this is false for large t.

7. (ii) $\ln \sum_{n \le x} \frac{1}{n} \le \ln \prod_{p \le x} \left(1 - \frac{1}{p}\right)^{-1} = \sum_{p \le x} -\ln\left(1 - \frac{1}{p}\right) < \sum_{p \le x} \frac{2}{p}$.

8. We have $\sum_{n \le x} \frac{\Lambda(n)}{n} = \sum_{p \le x} \frac{\ln p}{p} + \sum_{p^2 \le x} \frac{\ln p}{p^2} + \ldots$, use Theorem 2.1 as the remaining series are convergent.

9. Let $t_n = n$. (i) $z_n = 1$ and $f(t) = 1/t$; (ii) $z_n = 1$ and $f(t) = t^{-1/2}$; (iii) $z_n = 1$ and $f(t) = \ln t$; (iv) $z_n = \ln n$ and $f(t) = \ln t$; (v) $z_n = \ln(x/n)$ and $f(t) = \ln(x/t)$, or use (iii) and (iv); (vi) $z_n = 1/n$ and $f(t) = \ln t$; (vii) $z_n = \Lambda(n)/n$ and $f(t) = \ln t$.

10. (i) Use expansion of $\ln\left(1 - \frac{1}{p}\right)$. (ii) Use Theorem 2.3 and (i), take exponentials.

11. (i) Write $\sum_{pq \le x} \frac{1}{pq} = \sum_{p \le x} \left(\frac{1}{p} \sum_{q \le x/p} \frac{1}{q}\right)$ and note that $\ln \ln x/p - \ln \ln x = \ln\left(1 - \frac{\ln p}{\ln x}\right)$. (ii) In Lemma 2.2 put $t_n = p_n$, $z_n = \frac{\ln p_n}{p_n}$ and $f(t) = \ln t$, and use Theorem 2.1. (iii) This is similar to (i) via Theorem 2.1.

12. (i)

$$\ln \frac{\phi(n)}{n} = \sum_{p|n} \ln\left(1 - \frac{1}{p}\right) = -\sum_{p|n} \frac{1}{p} + \sum_{p|n} \left(\ln\left(1 - \frac{1}{p}\right) + \frac{1}{p}\right) > -\sum_{p|n} \frac{1}{p} - c.$$

Let $p_1, \ldots, p_r$ be the prime factors of n with $p_1, \ldots, p_{r-s} < \ln n$. Let

$$\sum_{p|n} \frac{1}{p} = \sum_{k \le r-s} \frac{1}{p_k} + \sum_{r-s<k\le r} \frac{1}{p_k} = S_1 + S_2,$$

S_2 is bounded and, by Theorem 2.3, $S_1 < \ln \ln p_{r-s} + c < \ln \ln \ln n + c$. (ii) Use (i).

13. (i) (a) Let $n_i = p_i$, (b) let $n_i = 2^i$, and (c) let n_i be the product of the first i primes. (ii) We have

$$\sum_n \omega(n) = \sum_{n \le x} \sum_{p|n} 1 = \sum_{p \le x} [x/p] = x \sum_{p \le x} 1/p + O(\pi(x)),$$

and

$$\sum_n \Omega(n) = \sum_{p \le x} [x/p] + \sum_{m>1} \sum_{p^m \le x} [x/p^m],$$

and this second sum equals $\sum_p 1/(p(p-1)) + o(x) = c'' + o(x)$. (iii) Use formula on p. 18.

14. See proof of Theorem 3.1.

15. We have $S = \sum_{n=m}^{k-1} n(n^{-s} - (n+1)^{-s}) - m^{1-s} + k^{1-s}$ and $n(n^{-s} - (n+1)^{-s}) = -s\int_n^{n+1} [x]x^{-s-1}\,dx$. Collect terms and use $s\int_m^k x^{-s}\,dx = s(k^{1-s} - m^{1-s})/(1-s)$. Secondly let $m=1$ and $k \to \infty$ and note integral is convergent for $\sigma > 0$.

16. Use Problem 15 for the first part. If $\zeta(1+it) = 0$ then $|\zeta(1+\tau+it)| < A\tau$ as $\tau \to 0^+$ for some constant A, secondly by Part One $|\zeta(1+\tau)| < 2/\tau$ for τ near zero, and thirdly $|\zeta(1+\tau+2it)| < C$ (a constant) by Problem 15. Hence $|\theta(\tau, t)| < (2/\tau)^3 (A\tau)^4 C \to 0$ as $\tau \to 0^+$ giving a contradiction.

17.
$$\zeta(0) = \lim_{s\to 0} 2^s \pi^{s-1} \Gamma(1-s)\zeta(1-s)\sin(\pi s/2)$$

$$= \pi^{-1} \lim_{s\to 0} (-s)\zeta(1-s)\sin(\pi s/2)/(-s)$$

$$= -1/2 \lim_{s\to 0} \sin(\pi s/2)/(\pi s/2) = -1/2.$$

18. Substitute $-2n$ in functional equation and note that $\sin(-\pi n) = 0$.

19. (i) $\psi(x) = \sum_{p \le x} k \ln p$ where $p^k \le x$, hence $k = \ln x/\ln p$ and $\psi(x) = O(\pi(x)\ln x)$. (ii) As in Theorem 3.1, if $\sigma = 1 + \varepsilon$, $\varepsilon > 0$, series is uniformly convergent so differentiation term-by-term is valid. (iii) We have

$$\zeta(s)\sum_n \Lambda(n)/n^s = \sum_n \left[n^{-s} \sum_{d|n} \Lambda(d)\right] = \sum_n n^{-s} \ln n.$$

For last part see proof of Theorem 3.3 as $\Lambda(n) = \psi(n) - \psi(n-1)$.

20. $B_8 = -1/30$, $B_{10} = 5/66$, $B_{12} = -691/2730$, $B_{14} = 7/6$, $B_{16} = -3617/510$.

21. Use series for e^{jt}. Now $\sum_{j=0}^{m-1} e^{jt} = \frac{e^{nt} - 1}{t}\frac{t}{e^t - 1} = \left(\sum_k n^k t^{k-1}/k!\right)\left(\sum_j B_j t^j/j!\right)$, equate coefficients and multiply by $(m+1)!$. Also $4S_3(n) = n^4 - 2n^3 + n^2$.

22. (i) and (ii) Use definitions. (iii) Consider series for e^{tx} and $t/(e^t - 1)$. (iv) Use (iii). (v) Use $te^{t(1-x)}/(e^t - 1) = -te^{-xt}/(e^{-t} - 1)$ and (iii). (vi) Equate coefficients in (iv) and (v).

23. Estimate $\zeta(2n)$ and use Theorem 3.6.

24. (i) $f'(t) = \sum a_{n+1}t^n/n!$, etc.; use $\displaystyle\int_0^t \frac{f(x)^{n-1}}{(n-1)!} f'(x)\,dx = f(t)^n/n!$. (ii) Note that $(m-1)! \equiv 0\ (m)$. We have (if $0^0 = 1$)

$$(e^t - 1)^m = \sum_{n=0}^{\infty} \left(\sum_{j=0}^{m} (-1)^{m-j} \binom{m}{j} j^n \right) t^n/n!,$$

so

$$(e^t - 1)^3 \equiv \sum_{n=2}^{\infty} (3 + 3^n) t^n/n! \pmod 4,$$

and $3 + 3^n \equiv 0$ or $2\ (4)$. For last part note coefficients are periodic mod p. (iii) We have

$$t/(e^t - 1) = \sum_{m=0}^{\infty} (-1)^m (e^t - 1)^m/(m+1)$$

$$= U(t) - \frac{1}{2}\sum_{k=1}^{\infty} t^k/k! - \frac{2}{4}\sum_{k=1}^{\infty} t^{2k+1}/(2k+1)! - \sum_{p>2} \frac{1}{p} \sum_{k=1}^{\infty} t^{kp-k}/(kp-k)!,$$

where $U(t)$ is an integer. Now equate coefficients. (iv) Note $2 - 1$ and $3 - 1$ both divide 2.

25. Use Problem 24, check that each p satisfying $p - 1 | 2k$ divides $n(n^{2k} - 1)$.

Problems 13

1. If $cn + a = p_1$, replace c by cp_1 and let $cp_1n_1 + a = p_2$, etc.

2. We have $F_a(x)G_a(x) = x^a - 1 \equiv -1\ (a)$, choose x so that $F_a(x) \neq \pm 1$ [$F_a(x) = \pm 1$ has only finitely many solutions], and let p be a prime divisor of $F_a(x)$. By (ii) $p \nmid G_a(x)$, this gives the first part. Suppose $a \nmid p - 1$ so s, t exist with $(a, p-1) = sa + t(p-1)$ and if $n = (a, p-1)$, $x^n \equiv (x^a)^s(x^{p-1})^t \equiv 1\ (p)$, contradiction. Now use Problem 1.

3. $\ln L(s, \chi) = \ln \prod_p (1 - \chi(p)p^{-s})^{-1} = -\sum_p \ln(1 - \chi(p)p^{-s}) = \sum_p \chi(p)p^{-s} + O(1)$, so $\sum_\chi \chi(a^*) \ln L(s, \chi) = \sum_{p \equiv a(c)} \chi(p)p^{-s} + O(1)$. Now $L(s, \chi_0) \to \infty$ as $s \to 1^+$ and remaining terms are bounded.

4. By product formula consider (if $p \nmid c$)

$$\begin{aligned}
&(|1 - \chi(p)p^{-s}|^2)^2\, |1 - \chi^2(p)p^{-s}|^2 \\
&= (1 - 2p^{-s}\cos\theta_p + p^{-2s})^2(1 - 2p^{-s}\cos 2\theta_p + p^{-2s}) \qquad \text{by (ii)} \\
&= (1 - 2p^{-s}(2\cos\theta_p + \cos 2\theta_p)/3 + p^{-2s})^3 \qquad \text{by (i)} \\
&\leq (1 + p^{-s} + p^{-2s})^3 \text{ [by (iii)]} \\
&\leq (1 - p^{-s})^{-3}. \text{ [If } p | c,\ \chi(p) = 0.]
\end{aligned}$$

Now take product of reciprocals. If $\chi^2 \neq \chi_0$ then $L'(s, \chi)$ is continuous at $s = 1$, and we have if $L(1, \chi) = 0$, $|L(s, \chi)| = |\int_1^s L'(u, \chi)\, du| < A(s-1)$, and so $(s-1)((s-1)L(s, \chi_0))^3\, |L(s, \chi)/(s-1)|^4\, |L(s, \chi^2)|^2 \geq 1$. A contradiction follows as all terms are bounded as $s \to 1^+$.

5. (i)

$$\sum_{\substack{p \leq x \\ p \equiv a(c)}} 1/p = \sum_\chi \chi(a^*) \sum_{p \leq x} \chi(p)/p = \sum_\chi \chi(a^*) \left[\sum_{p \leq x} (\chi(p) \ln p/p)/\ln p \right]$$

$$= \frac{1}{\ln x} \sum_{\substack{p \leq x \\ p \equiv a(c)}} \frac{\ln p}{p} + \int_2^x \sum_{\substack{p \leq t \\ p \equiv a(c)}} \frac{\ln p}{p} \frac{dt}{t \ln^2 t}.$$

Now proceed as in the proof of Theorem 2.3, Chapter 12, using Theorem 1.4.

(ii) We need $\sum_{n \leq x} \dfrac{\chi(n)\mu(n)}{n} = O(1)$ for $\chi \neq \chi_0$. By Theorem 1.3

$$O(1) = \sum_{n \leq x} \frac{\chi(n)\Lambda(n)}{n} = \sum_{m \leq x} \frac{\chi(m)\mu(m)}{m} \sum_{k \leq x/m} \frac{\chi(k) \ln k}{k}.$$

Replace last sum as in proof of Lemma 1.2 and use Problem 2(ii), Chapter 2.

6. If $k \equiv (p_1 - 1)! - 1 \pmod{p_1!}$ then the $p_1 - 2$ integers preceding and following k are composite.

7. As a_i is not square, k_i exist: $(a_i/k_i) = \varepsilon_i$. Use the Chinese remainder theorem and properties of the Jacobi symbol.

8. If $q \neq p$, $f(p + tq^{m+1}) = r^j$ (r prime) for infinitely many t.

9. We have $p | g(\zeta^p) - (g(\zeta))^p$, so if $g(\zeta) = 0$ then $p | g(\zeta^p)$. Now if $p \equiv m$ (n), $\zeta^p = \zeta^m$ and so $p | g(\zeta^m)$ for infinitely many p, impossible if $g(\zeta^m) \neq 0$.

10. Put $t_n = n$, $z_n = 1$ if n is prime and $z_n = 0$ otherwise, and $f(t) = \ln t$ in Lemma 2.2, Chapter 12. Proceed as in the proof on p. 245.

11. Using (i) check that each inequality involving constants can be replaced by an equality of the form $f(x) = c + \varepsilon$ where $\varepsilon \to 0$ as $x \to \infty$.

12. Use Lemma 1.4, Chapter 2. Now left-hand side $= M(x) \ln[x] - \sum_{n \leq x-1} M(n) \ln(1 + 1/n)$, and sum is $O(x)$ as $M(x) < x$. For right-hand side note $\psi(x) = \sum_{n \leq x} \Lambda(n)$. By (iii) given $\varepsilon > 0$ we have $T = |[x/n] - \psi(x/n)| < \varepsilon x/n$ if $x/n > A(\varepsilon)$ and $T < Bx/n$ (all x/n) for constants A and B, so $|\sum T| \leq \sum_{n \leq x/A} \varepsilon x/n + \sum_{x/A < n \leq x} Bx/n \leq 2\varepsilon x \ln x$ for large x.

13. Note $\sum_{n \leq x} [x/n]\mu(n) = 1$ and use partial summation. (vi) gives $|S(x)| = x/A + o(x)$ as in the problem above.

14. First note that $\sum_{n \leq x} \psi(x/n) \ln(x/n) = \ln x \sum_{n \leq x} \psi(x/n) - \sum_{mn \leq x} \Lambda(n)\psi(x/mn)$. Secondly $\sum_{n \leq x} \psi(x/n) = \sum_{mn \leq x} \Lambda(m) = \sum_{m \leq x} \Lambda(m)[x/m]$.

15. We have $\psi(x)\ln x = \theta(x)\ln x + O(x)$ and $\sum' \ln p \ln q = O(x)$, a sum over all p, q: $p^\alpha q^\beta \le x$ with $\alpha > 1$. Secondly $\sum_{pq\le x} \ln p \ln q = \sum_{p\le x} \theta(x/p)\ln p$ and $\sum_{p\le x} \ln^2 p = \theta(x)\ln x + O(x)$.

16. Use $\sum_{n\le x} \mu(n)\ln n = -\sum_{mn\le x} \mu(m)\Lambda(n)$.

17. $\pi((1+\varepsilon)x) - \pi(x) = \varepsilon x/\ln x + o(x/\ln x) > 0$ if x is large.

18. (i) $n = \prod_{p\le x} p$ so $\tau(n) = 2^{\pi(x)} = 2^{(1+o(1))x/\ln x}$. Also $\ln n = \theta(x) = x(1+o(1))$, hence $\ln x = (1+o(1))\ln\ln n$. Note $1 + o(1) > 1 - \varepsilon$ for large n. (ii) Use a similar argument.

19. Mimic the proof given in Section 3.

Problems 14

1. (i) No, modulus 9; (ii) Yes; (iii) Yes.

2. By solubility in $\mathbb{R}$ assume $a > 0$ and $b < 0$. Solution exists in $\mathbb{Q}$ if and only if $-ab$ is a square.

3. As $\sqrt{(ad)}$, $\sqrt{a}$ and $\sqrt{d}$ are irrational the square-free condition is not needed. (i) In the proof on p. 265 (with $a = 1$, $b = -d$, $c = -a$) note that in (8) we have $a_1 = a_4 = 1$ and so, if $(w_1, d) > 1$ in (10), we can replace (10) by $(w_1 \pm a_3) - a_2 y_1 - a_3(z_1 \mp 1) \equiv 0$ (ad) as $(a_3, d) = 1$. (ii) Assume z even and w, y odd. If $a = 1$, let $w = z = 1$, $y = 0$; if $a > 1$, let $a = u^2 - v^2$ then $zu + w$ is odd and $dy^2v^2 = (w^2 - az^2)(u^2 - a) = (wu + az)^2 - a(zu + w)^2$.

4. If Fermat's conjecture is true for p then $(x^n)^p + (y^n)^p = (z^n)^p$ is insoluble for all n.

5. If $(c/d)^2 = (a/b)^4 + 1$ then $(b^2cd)^2 = (ad)^4 + (bd)^4$.

6. Assume $x > 0$, $y > 0$, $z > 0$ with x least (and odd). If y is odd, $x^2 = a^2 + b^2$, $y^2 = a^2 - b^2$, and $z = 2ab$, and so $(xy)^2 = a^4 - b^4$ with $0 < a < x$. If y is even, $x^2 = a^2 + b^2$, $y^2 = 2ab$ with $a > 0$, $b > 0$, and $(a, b) = 1$. We may suppose a even and b odd, this gives $a = 2m^2$, $b = n^2$, n odd, $m > 0$, $n > 0$, and $(m, n) = 1$. So $x^2 = 4m^4 + n^4$, $y = 2mn$. Using result again gives $m^2 = rs$, $n^2 = r^2 - s^2$, $r > s > 0$, and $(r, s) = 1$. So $r = u^2$, $s = v^2$, $u > v > 1$ and $(u, v) = 1$. Thus $n^2 = u^4 - v^4$ with $u = \sqrt{r} \le m < \sqrt{x}$. Secondly x and y are odd, and so $z^4 - (xy)^4 = ((x^4 - y^4)/2)^2$.

7. If $(x+y)(x+\omega y)(x+\omega^2 y) = \varepsilon\lambda^{3n+2}z^3$, then λ divides one of the factors, and so all of them, i.e. $n > 0$. Now we have $x + y = \varepsilon_1\lambda^{3n}\alpha^3$, $x + \omega y = \varepsilon_2\lambda\beta^3$ and $x + \omega^2 y = \varepsilon_3\lambda\gamma^3$, which gives $\beta^3 + \varepsilon_4\gamma^3 + \varepsilon_5\lambda^{3n-1}\alpha^3 = 0$, now proceed as before.

8. $1^3 + 2^3 + 9(-1)^3 = 0$.

9. (i) $-x_1^p = (y_1 + z_1)(y_1^{p-1} - \dots + z_1^{p-1})$; if $p' \ne p$ and p' divides both factors then $py^{p-1} \equiv 0$ (p'), and so $p' | y$ and similarly $p' | x$. (ii) Note $a^{(q-1)/2} \equiv \pm 1$ (q) as q is prime. (iii) Combine equations in (i). (iv) By (i) $y_1 \equiv b^p$ (q), so $t^p \equiv p(b^{p-1})^p$ (q) and this gives $\pm 1 \equiv \pm p$ (q).

10. If $\theta = \sqrt[4]{2}$, $N(x + y\theta^3) = x^4 - 8y^4$. Equate coefficients.

11. If $k = t^3$, $y = 0$, and $x = t$; a vertical tangent. If $k = u^2$, $x = 0$, and $y = \pm u$; points of inflexion with horizontal tangent.

12. If $(k, x) = t$, $ty'^2 = t^2x'^3 + k'$ with $k't = k$.

13. $(y/z^3)^2 = (x/z^2)^3 + k$, etc.

14. If $x \equiv 3 \ (8)$, $x^2 + ax + a^2 \equiv 1 + 3a + a^2 \equiv \pm 3 \ (8)$, also if $x \equiv -1 \ (8)$, $x - a \equiv \pm 3 \ (8)$. $k = -6$ and 34.

15. Follow the method suggested. $k = -43$ and 29.

16. (i) We have $y + \sqrt{-11} = (a + b(1 + \sqrt{-11})/2)^3$ and so $2 = b(3a^2 + 3ab - 2b^2)$ giving solutions $\{3, \pm 4\}$, $\{15, \pm 58\}$. (ii) Here $(y + \sqrt{-3})\omega^\alpha = (a + b(1 + \sqrt{-3})/2)^3$. If $\alpha = 0$, $2 = 3ab(a + b)$ impossible; if $\alpha = 1$, $2 = -a^3 + b^2(3a + b)$ also impossible (try a, b even and odd); $\alpha = 2$ is similar. No solution.

17. Consider $y^2 = x^3 + z^6$ in integers. If $u \mid y - z^3$ and $u \mid y + z^3$ then $u \mid 2z^3$, but $(u, z) = 1$ so $u \mid 2$. If $u = 1$, $y - z^3 = a^3$ and $y + z^3 = b^3$ so $2z^3 = b^3 - a^3$. $u = 2$ is similar. Tangent at $(2, 3)$ passes through $(0, -1)$; $(0, \pm 1)$ are points of inflexion, etc.

18. $4(r^3 + s^3) = c^3 + 3a^2c = c^3(1 + 3(y/36)^2) = 4b^3$.

Problems 15

1. (i) $(0\colon 1\colon 0)$, $(0\colon 0\colon 1)$, $(2\colon 0\colon 1)$. (ii) $(0\colon 1\colon 0)$, $(1\colon 0\colon 1)$. (iii) None. (iv) $(0\colon 0\colon 1)$, $(0\colon 1\colon 0)$, $(1\colon 0\colon 0)$.

2. (i) Two intersections both with intersection number 2. (ii) Intersection $(0\colon 0\colon 1)$ has intersection number 6 (think of a cusp as the limit of a loop), and $(0\colon 1\colon 0)$ has intersection number 3. (iii) One intersection with $I_{(0:\,0:\,1)} = 3$.

3. (i) $g = 3$; (ii) and (iii) $g = 1$; (iv) $g = 0$.

4. All curves have genus zero.

5. Zero, there is a singularity at $(0\colon 0\colon 1)$.

6. $(0\colon 1\colon 0) \in C$ and there is a singularity at $(0\colon 0\colon 1)$ iff $4a^3 = -27b^2$.

7. It is a flex point.

8. Use Lemmas 3.1 and 3.2; $(-3\colon \pm 3\colon 1)$, $(0\colon \pm 6\colon 1)$, $(4\colon \pm 10\colon 1)$.

9. (i) Any curve whose group is infinite cyclic, e.g. $y^2 = x^3 - 4$; (ii) $y^2 = x^3 + 4$ has two points of order 3, so $W_C = 2$ and $R_C = 2$.

10. $\Delta = -\Delta^*/4^3$ where

$$\Delta^* = \det\begin{pmatrix} 0 & 0 & 1 & 0 & 0 & 0 & 4 \\ 0 & 1 & 0 & 0 & 0 & 4 & 0 \\ 1 & 0 & -2a & 0 & 4 & 0 & 4a \\ 0 & -2a & -8b & 4 & 0 & 4a & 4b \\ -2a & -8b & a^2 & 0 & 4a & 4b & 0 \\ -8b & a^2 & 0 & 4a & 4b & 0 & 0 \\ a^2 & 0 & 0 & 4b & 0 & 0 & 0 \end{pmatrix} = -4^4(4a^3 + 27b^2)^2.$$

$g_1(x, z)/4 = (4a^3 + 27b^2)x^3 - a^2bx^2z + a(3a^3 + 22b^2)xz^2 + 3b(a^3 + 8b^2)z^3$,

$g_2(x, z) = a^2bx^3z + a(5a^3 + 32b^2)x^2z^2 + 2b(13a^3 + 96b^2)xz^3 - 3a^2(a^3 + 8b^2)z^4$,

$g_3(x, z)/4z = 3x^2 + 4az^2$, and $g_4(x, z)/z = -3x^3 + 5axz^2 + 27bz^3$.

11. Let order of $C_t(\mathbb{Q})$ be T. $T = 1$ for $y^2 = x^3 - 3$, $T = 2$ for $y^2 = x^3 - 8$, $T = 3$ for $y^2 = x^3 + 4$, $T = 4$ (non-cyclic) for $y^2 = x(x^2 - a^2)$, $T = 6$ for $y^2 = x^3 + 1$, and $T = 4$ (cyclic) for $y^2 = x^3 + 4x$ [tangent at $(2, 4)$ intersects curve again at $(0, 0)$].

12. $N(y^2 \equiv x^3 + b(p)) = \sum_{t-u=b} N(y^2 \equiv t\,(p))N(x^3 \equiv u\,(p))$. $N(x^3 \equiv u\,(p)) = 1$, and $N(y^2 \equiv t\,(p)) = 0$, 1, or 2, now add up.

13. Use the methods of Section 2 and Lemma 3.1.

14. If $r =$ order of $C_t(\mathbb{Q})$, $r \mid p + 1$ by Problem 12 and assumption, i.e. $p \equiv -1$ (r). Density of primes satisfying this congruence is $1/\phi(r)$, but density of primes $\equiv 2$ (3) is 1/2, and so $\phi(r) \leq 2$.

15. (ii) Use Lemmas 3.2 and 3.1 as $3\mathsf{P}_1 = \mathsf{P}_1 + 2\mathsf{P}_1$, note $y_1y_{11} = (x_1^6 + 20x_1^3k - 8k^2)/8(x_1^3 + k)$. The second part of (i) shows that $C_2(\mathbb{Q}) \subset C_1(\mathbb{Q})$, so use formula for $3\mathsf{P}_1$. (iii) and (iv) Use the methods of proof of Lemmas 3.1 and 3.2.

16. Use $x_1^2 + y_1^2 = z_1^2$, $2n = x_1y_1$. Rational point is $(z_1^2/4, (x_1^2 - y_1^2)z_1/8)$. Conversely let $u = \sqrt{x}$, $v = y/u$, then $v^2 + n^2 = x^2$. Finally note 'right-angled triangle' points P can be written in the form $\mathsf{P} = 2\mathsf{P}_1$ for $\mathsf{P}_1 \in C(\mathbb{Q})$, that is they belong to the identity element of $C(\mathbb{Q})/2C(\mathbb{Q})$.

17. (i) We have $f_2f_1((x, y)) = ((x^2 - a)^2/4y^2, (x^2 - a)(y^4 - (c^2 - 4a)x^4)/8x^2y^3)$ so consider tangent at (x, y). (ii) Note: first coordinate of $f_1((x, y))$ is a square. (iii) Use $(y_1/x_1)^2 = x_1 + c + x_2$ etc. to show $f_1((x_1, y_1)) = (x, y)$ and $f_1((x_2, y_2)) = (x, -y)$. (iv) If (x_i, y_i), $i = 1, 2, 3$, lie on the line $y = ux + v$ and on $C_1(\mathbb{Q})$ then x_i are roots of $(ux + v)^2 = x^3 + cx^2 + ax$, that is $x_1x_2x_3 = v^2$ and so $\theta_1((x_1, y_1))\theta_1((x_2, y_2))\theta_1((x_3, y_3)) = \mathbb{Q}^{*2}$. θ_1 is a homomorphism follows using $(0, 0) + (x, y) = (a/x, -ay/x^2)$. Now suppose $x = m/e^2$, $y = n/e^3$ with $(m, e^2) = (n, e^3) = 1$, if $(x, y) \in C_1(\mathbb{Q})$ then $n^2 = m(m^2 + cme^2 + ae^4)$, consider prime factorizations. (v) We have isomorphisms $C_1(\mathbb{Q})/f_2C_2(\mathbb{Q}) \approx \text{image}(\theta_1)$, $C_2(\mathbb{Q})/f_1C_1(\mathbb{Q}) \approx \text{image}(\theta_2)$ and $C_2(\mathbb{Q})/f_1C_1(\mathbb{Q}) \approx f_2C_2(\mathbb{Q})/f_2f_1C_1(\mathbb{Q}) \approx f_2C_2(\mathbb{Q})/2C_1(\mathbb{Q})$, now use (iv).

TABLES

Prime numbers and primitive roots

Table 1 lists all primes less than 2000, and for those less than 750 the least positive primitive root is also given.

TABLE 1

2	—	197	2	461	2	751	1051	1381	1697
3	2	199	3	463	3	757	1061	1399	1699
5	2	211	2	467	2	761	1063	1409	1709
7	3	223	3	479	13	769	1069	1423	1721
11	2	227	2	487	3	773	1087	1427	1723
13	2	229	6	491	2	787	1091	1429	1733
17	3	233	3	499	7	797	1093	1433	1741
19	2	239	7	503	5	809	1097	1439	1747
23	5	241	7	509	2	811	1103	1447	1753
29	2	251	6	521	3	821	1109	1451	1759
31	3	257	3	523	2	823	1117	1453	1777
37	2	263	5	541	2	827	1123	1459	1783
41	6	269	2	547	2	829	1129	1471	1787
43	3	271	6	557	2	839	1151	1481	1789
47	5	277	5	563	2	853	1153	1483	1801
53	2	281	3	569	3	857	1163	1487	1811
59	2	283	3	571	3	859	1171	1489	1823
61	2	293	2	577	5	863	1181	1493	1831
67	2	307	5	587	2	877	1187	1499	1847
71	7	311	17	593	3	881	1193	1511	1861
73	5	313	10	599	7	883	1201	1523	1867
79	3	317	2	601	7	887	1213	1531	1871
83	2	331	3	607	3	907	1217	1543	1873
89	3	337	10	613	2	911	1223	1549	1877
97	5	347	2	617	3	919	1229	1553	1879
101	2	349	2	619	2	929	1231	1559	1889
103	5	353	3	631	3	937	1237	1567	1901
107	2	359	7	641	3	941	1249	1571	1907
109	6	367	6	643	11	947	1259	1579	1913
113	3	373	2	647	5	953	1277	1583	1931
127	3	379	2	653	2	967	1279	1597	1933
131	2	383	5	659	2	971	1283	1601	1949
137	3	389	2	661	2	977	1289	1607	1951
139	2	397	5	673	5	983	1291	1609	1973
149	2	401	3	677	2	991	1297	1613	1979
151	6	409	21	683	5	997	1301	1619	1987
157	5	419	2	691	3	1009	1303	1621	1993
163	2	421	2	701	2	1013	1307	1627	1997
167	5	431	7	709	2	1019	1319	1637	1999
173	2	433	5	719	11	1021	1321	1657	
179	2	439	15	727	5	1031	1327	1663	
181	2	443	2	733	6	1033	1361	1667	
191	19	449	3	739	3	1039	1367	1669	
193	5	457	13	743	5	1049	1373	1693	

Continued fraction representations of surds

Tables 2 and 3 give the continued fraction representation of numbers of the form $\sqrt{d}$ for $d \leq 200$ except where $d = n^2 - 1$, $n^2 + 1$ or $n^2 + 2$. For these exceptions we have (see Problems 20 and 21 on p. 134)

$$\sqrt{(n^2-1)} = [n-1, \overset{*}{1}, \overset{*}{2n-2}],$$
$$\sqrt{(n^2+1)} = [n, \overset{*}{2n}],$$
$$\sqrt{(n^2+2)} = [n, \overset{*}{n}, \overset{*}{2n}].$$

Using the symmetry properties of the representations (see the problems mentioned above) we list in Table 2 those integers whose square roots have an odd period, and give the *first half* of the period only. For example 13–3; 1, 1 stands for $\sqrt{13} = [3, \overset{*}{1}, 1, 1, 1, \overset{*}{6}]$; that is the second half of the period is the mirror image of the first half with twice the first entry added at the end. The surds with period one are given above.

TABLE 2. Odd periods

13	3; 1, 1	109	10; 2, 3, 1, 2, 4, 1, 6
29	5; 2, 1	113	10; 1, 1, 1, 2
41	6; 2	125	11; 5, 1
53	7; 3, 1	130	11; 2
58	7; 1, 1, 1	137	11; 1, 2, 2, 1
61	7; 1, 4, 3, 1, 2	149	12; 4, 1, 5, 3
73	8; 1, 1, 5	157	12; 1, 1, 7, 1, 5, 2, 1, 1
74	8; 1, 1	173	13; 6, 1
85	9; 4, 1	181	13; 2, 4, 1, 8, 6, 1, 1, 1, 1, 2
89	9; 2, 3	185	13; 1, 1
97	9; 1, 5, 1, 1, 1	193	13; 1, 8, 3, 2, 1, 3
106	10; 3, 2, 1, 1		

Table 3 lists the representations of those integers whose square roots have an even period and gives, as above, the first half of the period only. For example 19–4; 1, 2, 3 stands for $\sqrt{19} = [4, \overset{*}{1}, 2, 3, 2, 1, \overset{*}{8}]$; that is the second half of the period is the mirror image, except for the central element, of the first half with twice the first entry added at the end. The surds $\sqrt{(n^2-1)}$ and $\sqrt{(n^2+2)}$ are given above.

TABLE 3. Even periods

7	2; 1, 1	84	9; 6	142	11; 1, 10
12	3; 2	86	9; 3, 1, 1, 1, 8	147	12; 8
14	3; 1, 2	87	9; 3	148	12; 6
19	4; 2, 1, 3	88	9; 2, 1, 1	150	12; 4
20	4; 2	90	9; 2	151	12; 3, 2, 7, 1, 3, 4, 1, 1, 1, 11
21	4; 1, 1, 2	91	9; 1, 1, 5, 1	152	12; 3
22	4; 1, 2, 4	92	9; 1, 1, 2, 4	153	12; 2, 1, 2, 2
23	4; 1, 3	93	9; 1, 1, 1, 4, 6	154	12; 2, 2, 3, 1, 2
28	5; 3, 2	94	9; 1, 2, 3, 1, 1, 5, 1, 8	155	12; 2, 4
30	5; 2	95	9; 1, 2	156	12; 2
31	5; 1, 1, 3, 5	96	9; 1, 3	158	12; 1, 1, 3, 12
32	5; 1, 1	98	9; 1, 8	159	12; 1, 1, 1, 1, 3
33	5; 1, 2	103	10; 6, 1, 2, 1, 1, 9	160	12; 1, 1, 1, 5
34	5; 1, 4	104	10; 5	161	12; 1, 2, 4, 1, 2
39	6; 4	105	10; 4	162	12; 1, 2, 1, 2, 12
40	6; 3	107	10; 2, 1, 9	163	12; 1, 3, 3, 2, 1, 1, 7, 1, 11
42	6; 2	108	10; 2, 1, 1, 4	164	12; 1, 4, 6
43	6; 1, 1, 3, 1, 5	110	10; 2	165	12; 1, 5, 2
44	6; 1, 1, 1, 2	111	10; 1, 1, 6	166	12; 1, 7, 1, 1, 1, 2, 4, 1, 3, 2, 12
45	6; 1, 2, 2	112	10; 1, 1, 2	167	12; 1, 11
46	6; 1, 3, 1, 1, 2, 6	114	10; 1, 2, 10	172	13; 8, 1, 2, 2, 1, 1, 3, 6
47	6; 1, 5	115	10; 1, 2, 1, 1, 1	174	13; 5, 4
52	7; 4, 1, 2	116	10; 1, 3, 2, 1, 4	175	13; 4, 2, 1
54	7; 2, 1, 6	117	10; 1, 4, 2	176	13; 3, 1
55	7; 2, 2	118	10; 1, 6, 3, 2, 10	177	13; 3, 3, 2, 8
56	7; 2	119	10; 1, 9	178	13; 2, 1, 12
57	7; 1, 1, 4	124	11; 7, 2, 1, 1, 1, 3, 1, 4	179	13; 2, 1, 1, 1, 3, 5, 13
59	7; 1, 2, 7	126	11; 4, 2	180	13; 2, 2
60	7; 1, 2	127	11; 3, 1, 2, 2, 7, 11	182	13; 2
62	7; 1, 6	128	11; 3, 5	183	13; 1, 1, 8
67	8; 5, 2, 1, 1, 7	129	11; 2, 1, 3, 1, 6	184	13; 1, 1, 3, 2, 1, 2
68	8; 4	131	11; 2, 4, 11	186	13; 1, 1, 1, 3, 4
69	8; 3, 3, 1, 4	132	11; 2	187	13; 1, 2, 13
70	8; 2, 1, 2	133	11; 1, 1, 7, 5, 1, 1, 1, 2	188	13; 1, 2, 2, 6
71	8; 2, 2, 1, 7	134	11; 1, 1, 2, 1, 3, 1, 10	189	13; 1, 2
72	8; 2	135	11; 1, 1, 1, 1	190	13; 1, 3, 1, 1, 1, 2, 2
75	8; 1, 1	136	11; 1, 1	191	13; 1, 4, 1, 1, 3, 2, 2, 13
76	8; 1, 2, 1, 1, 5, 4	138	11; 1, 2	192	13; 1, 5
77	8; 1, 3, 2	139	11; 1, 3, 1, 3, 7, 1, 1, 2, 11	194	13; 1, 12
78	8; 1, 4	140	11; 1, 4	199	14; 9, 2, 1, 2, 2, 5, 4, 1, 1, 13
79	8; 1, 7	141	11; 1, 6	200	14; 7

Binary quadratic forms

Tables 4 and 5 list all primitive reduced forms whose discriminants d satisfy $|d| \leq 200$ as given by the results of Section 4, Chapter 9. The genera (see Section 1, Chapter 10) are indicated by a vertical bar on the left; if no bar is given then each form is in a distinct genus. $a, \pm b, c$ stands for the two non-equivalent forms $\langle a, b, c\rangle$ and $\langle a, -b, c\rangle$, Gn denotes the number n of genera, and Im denotes the number m of (unlisted) imprimitive reduced forms (this is omitted if $m = 0$).

TABLE 4. Positive definite forms

(Each row below is one genus; forms sharing a vertical bar on the page are given in one cell.)

d	G, I	Forms
−3	G1	1, 1, 1
−4	G1	1, 0, 1
−7	G1	1, 1, 2
−8	G1	1, 0, 2
−11	G1	1, 1, 3
−12	G1, I1	1, 0, 3
−15	G2	1, 1, 4
		2, 1, 2
−16	G1, I1	1, 0, 4
−19	G1	1, 1, 5
−20	G2	1, 0, 5
		2, 2, 3
−23	G1	1, 1, 6 2, ±1, 3
−24	G2	1, 0, 6
		2, 0, 3
−27	G1, I1	1, 1, 7
−28	G1, I1	1, 0, 7
−31	G1	1, 1, 8 2, ±1, 4
−32	G2, I1	1, 0, 8
		3, 2, 3
−35	G2	1, 1, 9
		3, 1, 3
−36	G2, I1	1, 0, 9
		2, 2, 5
−39	G2	1, 1, 10 3, 3, 4
		2, ±1, 5
−40	G2	1, 0, 10
		2, 0, 5
−43	G1	1, 1, 11
−44	G1, I1	1, 0, 11 3, ±2, 4
−47	G1	1, 1, 12 2, ±1, 6 3, ±1, 4
−48	G2, I2	1, 0, 12
		3, 0, 4
−51	G2	1, 1, 13
		3, 3, 5
−52	G2	1, 0, 13
		2, 2, 7
−55	G2	1, 1, 14 4, 3, 4
		2, ±1, 7
−56	G2	1, 0, 14 2, 0, 7
		3, ±2, 5
−59	G1	1, 1, 15 3, ±1, 5
−60	G2, I2	1, 0, 15
		3, 0, 5
−63	G2, I1	1, 1, 16 4, 1, 4
		2, ±1, 8
−64	G2, I2	1, 0, 16
		4, 4, 5
−67	G1	1, 1, 17
−68	G2	1, 0, 17 2, 2, 9
		3, ±2, 6
−71	G1	1, 1, 18 2, ±1, 9 3, ±1, 6 4, ±3, 5
−72	G2, I1	1, 0, 18
		2, 0, 9
−75	G2, I1	1, 1, 19
		3, 3, 7
−76	G1, I1	1, 0, 19 4, ±2, 5
−79	G1	1, 1, 20 2, ±1, 10 4, ±1, 5
−80	G2, I2	1, 0, 20 4, 0, 5
		3, ±2, 7
−83	G1	1, 1, 21 3, ±1, 7
−84	G4	1, 0, 21
		2, 2, 11
		3, 0, 7
		5, 4, 5
−87	G2	1, 1, 22 4, ±3, 6
		2, ±1, 11 3, 3, 8
−88	G2	1, 0, 22
		2, 0, 11
−91	G2	1, 1, 23
		5, 3, 5
−92	G1, I3	1, 0, 23 3, ±2, 8
−95	G2	1, 1, 24 4, ±1, 6 5, 5, 6
		2, ±1, 12 3, ±1, 8
−96	G4, I2	1, 0, 24
		3, 0, 8
		4, 4, 7
		5, 2, 5
−99	G2, I1	1, 1, 25
		5, 1, 5
−100	G2, I1	1, 0, 25
		2, 2, 13

TABLE 4. (*Continued*)

D	Type	Forms
−103	G1	1, 1, 26 2, ±1, 13 4, ±3, 7
−104	G2	1, 0, 26 3, ±2, 9 2, 0, 13 5, ±4, 6
−107	G1	1, 1, 27 3, ±1, 9
−108	G1, I3	1, 0, 27 4, ±2, 7
−111	G2	1, 1, 28 3, 3, 10 4, ±1, 7 2, ±1, 14 5, ±3, 6
−112	G2, I2	1, 0, 28 4, 0, 7
−115	G2	1, 1, 29 5, 5, 7
−116	G2	1, 0, 29 5, ±2, 6 2, 2, 15 3, ±2, 10
−119	G2	1, 1, 30 2, ±1, 15 4, ±3, 8 3, ±1, 10 5, ±1, 6 6, 5, 6
−120	G4	1, 0, 30 2, 0, 15 3, 0, 10 5, 0, 6
−123	G2	1, 1, 31 3, 3, 11
−124	G1, I3	1, 0, 31 5, ±4, 7
−127	G1	1, 1, 32 2, ±1, 16 4, ±1, 8
−128	G2, I3	1, 0, 32 4, 4, 9 3, ±2, 11
−131	G1	1, 1, 33 3, ±1, 11 5, ±3, 7
−132	G4	1, 0, 33 2, 2, 17 3, 0, 11 6, 6, 7
−135	G2, I2	1, 1, 34 4, ±3, 9 2, ±1, 17 5, 5, 8
−136	G2	1, 0, 34 2, 0, 17 5, ±2, 7
−139	G1	1, 1, 35 5, ±1, 7
−140	G2, I2	1, 0, 35 4, ±2, 9 3, ±2, 12 5, 0, 7
−143	G2	1, 1, 36 3, ±1, 12 4, ±1, 9 2, ±1, 18 6, 1, 6 6, ±5, 7
−144	G2, I4	1, 0, 36 4, 0, 9 5, ±4, 8
−147	G2, I1	1, 1, 37 3, 3, 13
−148	G2	1, 0, 37 2, 2, 19
−151	G1	1, 1, 38 2, ±1, 19 4, ±3, 10 5, ±3, 8
−152	G2	1, 0, 38 6, ±4, 7 2, 0, 19 3, ±2, 13
−155	G2	1, 1, 39 5, 5, 9 3, ±1, 13
−156	G2, I4	1, 0, 39 3, 0, 13 5, ±2, 8
−159	G2	1, 1, 40 4, ±1, 10 6, ±3, 7 2, ±1, 20 3, 3, 14 5, ±1, 8
−160	G4, I2	1, 0, 40 4, 4, 11 5, 0, 8 7, 6, 7
−163	G1	1, 1, 41
−164	G2	1, 0, 41 2, 2, 21 5, ±4, 9 3, ±2, 14 6, ±2, 7
−167	G1	1, 1, 42 2, ±1, 21 3, ±1, 14 4, ±3, 11 6, ±1, 7 6, ±5, 8
−168	G4	1, 0, 42 2, 0, 21 3, 0, 14 6, 0, 7
−171	G2, I1	1, 1, 43 7, 5, 7 5, ±3, 9
−172	G1, I1	1, 0, 43 4, ±2, 11
−175	G2, I1	1, 1, 44 4, ±1, 11 2, ±1, 22 7, 7, 8
−176	G2, I4	1, 0, 44 5, ±2, 9 3, ±2, 15 4, 0, 11
−179	G1	1, 1, 45 3, ±1, 15 5, ±1, 9
−180	G4, I2	1, 0, 45 2, 2, 23 5, 0, 9 7, 4, 7
−183	G2	1, 1, 46 3, 3, 16 4, ±3, 12 2, ±1, 23 6, ±3, 8
−184	G2	1, 0, 46 2, 0, 23 5, ±4, 10
−187	G2	1, 1, 47 7, 3, 7
−188	G1, I5	1, 0, 47 3, ±2, 16 7, ±6, 8
−191	G1	1, 1, 48 2, ±1, 24 3, ±1, 16 4, ±1, 12 5, ±3, 10 6, ±1, 8 6, ±5, 9
−192	G4, I4	1, 0, 48 3, 0, 16 4, 4, 13 7, 2, 7
−195	G4	1, 1, 49 3, 3, 17 5, 5, 11 7, 1, 7
−196	G2, I1	1, 0, 49 2, 2, 25 5, ±2, 10
−199	G1	1, 1, 50 2, ±1, 25 4, ±3, 13 5, ±1, 10 7, ±5, 8
−200	G2, I1	1, 0, 50 6, ±4, 9 2, 0, 25 3, ±2, 17

TABLE 5. Indefinite forms

D	Class	Forms
5	G1	1, 1, −1
8	G1	1, 0, −2
12	G2	1, 0, −3 −1, 0, 3
13	G1	1, 1, −3
17	G1	1, 1, −4
20	G1, I1	1, 0, −5
21	G2	1, 1, −5 −1, 1, 5
24	G2	1, 0, −6 −1, 0, 6
28	G2	1, 0, −7 −1, 0, 7
29	G1	1, 1, −7
32	G2, I1	1, 0, −8 −1, 0, 8
33	G2	1, 1, −8 −1, 1, 8
37	G1	1, 1, −9
40	G2	1, 0, −10 2, 0, −5
41	G1	1, 1, −10
44	G2	1, 0, −11 −1, 0, 11
45	G2	1, 1, −11 −1, 1, 11
48	G2, I2	1, 0, −12 −1, 0, 12
52	G1, I1	1, 0, −13
53	G1	1, 1, −13
56	G2	1, 0, −14 −1, 0, 14
57	G2	1, 1, −14 −1, 1, 14
60	G4	1, 0, −15 −1, 0, 15 3, 0, −5 −3, 0, 5
61	G1	1, 1, −15
65	G2	1, 1, −16 2, 1, −8
68	G1, I1	1, 0, −17
69	G2	1, 1, −17 −1, 1, 17
72	G2, I1	1, 0, −18 −1, 0, 18
73	G1	1, 1, −18
76	G2	1, 0, −19 −1, 0, 19
77	G2	1, 1, −19 −1, 1, 19
80	G2, I2	1, 0, −20 −1, 0, 20
84	G2, I2	1, 0, −21 −1, 0, 21
85	G2	1, 1, −21 3, 1, −7
88	G2	1, 0, −22 −1, 0, 22
89	G1	1, 1, −22
92	G2	1, 0, −23 −1, 0, 23
93	G2	1, 1, −23 −1, 1, 23
96	G4, I2	1, 0, −24 −1, 0, 24 3, 0, −8 −3, 0, 8
97	G1	1, 1, −24
101	G1	1, 1, −25
104	G2	1, 0, −26 2, 0, −13
105	G4	1, 1, −26 −1, 1, 26 2, 1, −13 −2, 1, 13
108	G2, I2	1, 0, −27 −1, 0, 27
109	G1	1, 1, −27
112	G2, I2	1, 0, −28 −1, 0, 28
113	G1	1, 1, −28
116	G1, I1	1, 0, −29
117	G2, I1	1, 1, −29 −1, 1, 29
120	G4	1, 0, −30 −1, 0, 30 2, 0, −15 −2, 0, 15
124	G2	1, 0, −31 −1, 0, 31
125	G1, I1	1, 1, −31
128	G2, I3	1, 0, −32 −1, 0, 32
129	G2	1, 1, −32 −1, 1, 32
132	G2, I2	1, 0, −33 −1, 0, 33
133	G2	1, 1, −33 −1, 1, 33
136	G2	1, 0, −34 −1, 0, 34 3, 2, −11 −3, 2, 11
137	G1	1, 1, −34
140	G4	1, 0, −35 −1, 0, 35 2, 2, −17 −2, 2, 17
141	G2	1, 1, −35 −1, 1, 35
145	G2	1, 1, −36 4, 1, −9 2, 1, −18 2, −1, −18
148	G1, I1	1, 0, −37 3, 2, −12 3, −2, −12
149	G1	1, 1, −37
152	G2	1, 0, −38 −1, 0, 38
153	G2, I1	1, 1, −38 −1, 1, 38
156	G4	1, 0, −39 −1, 0, 39 2, 2, −19 −2, 2 19
157	G1	1, 1, −39
160	G4, I2	1, 0, −40 −1, 0, 40 3, 2, −13 −3, 2, 13

TABLE 5. (*Continued*)

Discriminant	Forms
161 G2	1, 1, −40 −1, 1, 40
164 G1, I1	1, 0, −41
165 G4	1, 1, −41 −1, 1, 41 3, 3, −13 −3, 3, 13
168 G4	1, 0, −42 −1, 0, 42 2, 0, −21 −2, 0, 21
172 G2	1, 0, −43 −1, 0, 43
173 G1	1, 1, −43
176 G2, I2	1, 0, −44 −1, 0, 44
177 G2	1, 1, −44 −1, 1, 44
180 G2, I4	1, 0, −45 −1, 0, 45
181 G1	1, 1, −45
184 G2	1, 0, −46 −1, 0, 46
185 G2	1, 1, −46 −1, 1, 46; 2, 1, −23 2, −1, −23
188 G2	1, 0, −47 −1, 0, 47
189 G2, I2	1, 1, −47 −1, 1, 47
192 G4, I4	1, 0, −48 −1, 0, 48 3, 0, −16 −3, 0, 16
193 G1	1, 1, −48
197 G1	1, 1, −49
200 G2, I1	1, 0, −50 2, 0, −25

Mordell's equation

Table 6 lists those values of k in the range $|k| \leq 50$ such that Mordell's equation $y^2 = x^3 + k$ has at least one integer solution. The table also lists the x coordinates of the generators of the group of rational points; if a

TABLE 6. Integer solutions

k	x	k	x
1	2 (6)	−1	1 (2)
2	−1	−2	3
3	1	−4	2
4	0 (3)	−7	2
5	−1	−8	2 (2)
8	−2 (2), 2	−11	3, 15
9	0 (3), −2	−13	17
10	−1	−15	4
12	−2	−18	3
15	1, 109	−19	7
16	0 (3)	−20	6
17	−2, −1	−23	3
18	7	−25	5
19	5	−26	3, 35
22	3	−27	3 (2)
24	−2, 1	−28	4
25	0 (3)	−35	11
26	−1	−39	4, 10
27	−3 (2)	−40	14
28	2	−44	5
30	19	−45	21
31	−3	−47	12, 63
33	−2	−48	4
35	1	−49	65
36	0 (3), −3		
37	−1, 3		
38	11		
40	6		
41	2		
43	−3, 57/49		
44	−2		
48	1		
49	0 (3)		
50	−1		

generator x has finite order m this is indicated by $x(m)$. Note that the second generator for $k = 43$ is not integral. Also note that, for some k, a number of integer solutions exist; for example if $k = -28$ the solutions are given by $x = 4$, 8, and 37 (see Ellison *et al.*, 1972).

Table 7 lists those values of k in the range $|k| \leq 50$ such that Mordell's equation has rational, but not integer, solutions; as above the x coordinate of the generator of the group of rational points is also given. In this range each group is cyclic of infinite order.

TABLE 7. Rational number solutions

k	x
11	−7/4
39	217/4
46	−7/4
47	17/4
−21	37/9
−22	71/25
−29	3133/9
−30	31/9
−38	4447/441
−43	1177/36
−50	211/9

Note that for the values of k ($|k| \leq 50$) not listed in either of these tables Mordell's equation has no rational affine solution. Tables 6 and 7 have been taken from Cassels (1950), by courtesy of the author.

BIBLIOGRAPHY

Andrews, G. E. (1976). *The theory of partitions*. Addison-Wesley, Reading, Mass.

Apostol, T. M. (1976). *Introduction to analytic number theory*. Springer, New York.

Ax, J. and Kochen, S. (1966). Diophantine problems over local fields III, decidable fields. *Ann. Math.* **83** (2), 437–56.

Bachmann, P. (1968). *Niedere Zahlentheorie* (reprint of 1910 edn). Chelsea, New York.

Baker, A. (1975). *Transcendental number theory*. Cambridge University Press, London.

—— (1984). *A concise introduction to the theory of numbers*. Cambridge University Press, London.

Birch, B. and Swinnerton-Dyer, H. P. F. (1963; 1965). Notes on elliptic curves I and II. *J. Reine Angew. Math.* **212,** 7–25: **218,** 79–108.

Borevich, Z. I. and Shafarevich, I. R. (1966). *Number theory* (trans. from Russian by N. Greenleaf). Academic Press, New York.

Burgess, D. A. (1957). The distribution of quadratic residues and non-residues. *Mathematika* **4,** 106–12.

Burton, D. M. (1980). *Elementary number theory*. Allyn and Bacon, Boston, Mass.

Cassels, J. W. S. (1950). The rational solutions of the diophantine equation $y^2 = x^3 - D$. *Acta Math.* **82,** 243–73.

—— (1978). *Rational quadratic forms* (LMS monograph 13). Academic Press, London.

—— (1986). Mordell's finite basis theorem revisited. *Math. Proc. Cambridge Philosophical Soc.* **100,** 31–41.

Chowla, A. (1965). *The Riemann hypothesis and Hilbert's tenth problem*. Gordon and Breach, New York.

Cohen, H. (1962). *A second course in number theory*. Wiley, New York.

Davenport, H. (1962). *The higher arithmetic*. Hutchinson, London.

—— (1967). *Multiplicative number theory*. Markham, Chicago, Ill.

Davis, C. S. (1945). On some simple continued fractions connected with e. *J. London Math. Soc.* **20,** 194–8.

Diamond, H. G. (1982) Elementary methods in the study of the distribution of prime numbers. *Bull. Am. Math. Soc.* **7,** 553–89.

Dickson, L. E. (1919–23). *A history of the theory of numbers* (3 vols). Carnegie Institution, Washington, DC.

Edwards, H. M. (1974). *Riemann's zeta function*. Academic Press, New York.

—— (1977). *Fermat's last theorem*. Springer, New York.

Ellison, W. J. (1971). Waring's problem. *Am. Math. Monthly* **78,** 10–36.

—— *et al.* (1972). The diophantine equation $y^2 + k = x^3$. *J. Number Theory* **4,** 107–17.

Fulton, W. (1969). *Algebraic curves*. Benjamin Cummings, Reading, Mass.

Gauss, C. F. (1863). *Werke, volume* 2. Koniglichen Gesellschaft der Wissenschaften, Gottingen.

—— (1966). *Disquisitiones arithmeticae* (trans. 1801 edn by A. A. Clarke). Yale University Press, New Haven, Conn.

Gelfond, A. O. and Linnik, Ju. V. (1965). *Elementary methods in analytic number theory* (trans. from Russian by A. Feinstein). Rand McNally, Chicago, Ill.

Goldfeld, D. (1985). Gauss's class number problem for imaginary quadratic fields. *Bull. Am. Math. Soc.* **13,** 23–37.

Grosswald, E. (1984). *Topics from the theory of numbers*. Birkhäuser, Boston, Mass.

Guy, R. K. (1981). *Unsolved problems in number theory*. Springer, New York.

—— (editor) (1984). *Reviews in number theory* 1973–83 (6 vols). American Mathematical Society, Providence, RI.

Halberstam, H. and Richert, H. E. (1974). *Sieve methods* (LMS monograph 4). Academic Press, London.

Hardy, G. H. and Wright, E. M. (1954). *An introduction to the theory of numbers* (3rd edn). Clarandon Press, Oxford.

Heath, T. L. (1910). *Diophantus of Alexandra*. Cambridge University Press, London.

Heath–Brown, D. R. (1978; 1979). The difference between consecutive primes I, II, and III. *J. London Math. Soc.* **18,** 7–13; **19,** 207–20; **20,** 177–8.

Hooley, C. (1967). On Artin's conjecture. *J. Reine Angew. Math.* **225,** 209–20.

Hua, L.-K. (1982). *Introduction to number theory* (trans. from Chinese by P. Shiu). Springer, Berlin.

Husemöller, D. (1987). *Elliptic curves*. Springer, New York.

Ireland, K. F. and Rosen, M. I. (1982). *A classical introduction to modern number theory*. Springer, New York.

Jacobi, C. G. J. (1956). *Canon arithmeticus* (reprint of the 1839 edn). Akademie Verlag, Berlin.

Jones, B. W. (1950). *The arithmetic theory of quadratic forms,* Mathematical Association of America monograph. Wiley, New York.

Khinchin, A. Ya. (1964). *Continued fractions*. University of Chicago Press, Chicago, Ill.

Knopp, M. I. (1970). *Modular functions in analytic number theory*. Markham, Chicago, Ill.

Koblitz, N. I. (1984). *Introduction to elliptic curves and modular functions*. Springer, New York.

Landau, E. (1927). *Vorlesungen uber Zahlentheorie* (3 vols). Hirzel, Leipzig.

—— (1958). *Elementary number theory* (trans. of part of Vol. 1 of the above). Chelsea, New York.

Lang, S. (1978). *Elliptic curves; diophantine analysis*. Springer, Berlin.

Lenstra, H. W. (1986). Factoring integers with elliptic curves. *Report no.* 68-18, Math. Inst., University of Amsterdam.

LeVeque, W. J. (1955). *Topics in number theory* (2 vols). Addison-Wesley, Reading, Mass.

—— (editor) (1974). *Reviews of number theory* (1940–72). American Mathematical Society, Providence, RI.

—— (1977). *Fundamentals of number theory*. Addison-Wesley, Reading, Mass.

Lewy, H. (1946). Waves on beaches. *Bull. Am. Math. Soc.* **52,** 737–75.

Marcus, D. A. (1977). *Number fields*. Springer, New York.

Mazur, B. (1977). Rational points on modular curves. *Modular functions in one variable V*. Lecture Notes 601. pp. 107–48. Springer, Berlin.

—— (1986). Arithmetic on curves. *Bull. Am. Math. Soc.* **14,** 207–60.

Miller, G. L. (1975). *Riemann's hypothesis and tests for primality*. Seventh Annual ACM Symposium on the Theory of Computing, Albuquerque. Association of Computing Machinery, New York.

Mordell, L. J. (1966). The infinity of rational solutions of $y^2 = x^3 + k$. *J. London Math. Soc.* **41,** 523–5.

—— (1969). *Diophantine equations*. Academic Press, London.

Newman, D. J. (1980). Simple analytic proof of the prime number theorem. *Am. Math. Monthly* **87,** 693–6.

Niven, I. (1956). *Irrational numbers*. Mathematical Association of America monograph. Wiley, New York.

—— and Zuckerman, H. S. (1980). *An introduction to the theory of numbers* (4th edn). Wiley, New York.

Odylzko, A. M. and te Riele, H. J. J. (1985). Disproof of Merten's conjecture. *J. Reine Angew. Math.* **357,** (138–60).

Olsen, L. D. (1974). Points of finite order on elliptic curves with complex multiplication. *Manuscr. Math.* **14,** 195–205.

Ore, O. (1948). *Number theory and its history*. McGraw-Hill, New York.

Perron, O. (1950). *Die Lehre von der Kettenbruchen* (reprint of the 1929 edn). Chelsea, New York.

Pollard, H. (1950). *Theory of algebraic numbers*. Mathematical Association of America monograph. Wiley, New York.

Pollard, J. M. (1971). An algorithm for testing the primality of any integer. *Bull. Am. Math.Soc.* **3,** 337–40.

Rademacher, H. (1964). *Lectures on number theory*. Blaisdell, New York.

Ribenboim, P. (1972). *Algebraic numbers*. Wiley-Interscience, New York.

—— (1979). 13 *lectures on Fermat's last theorem*. Springer, New York.

Riesel, H. (1985). *Prime numbers and computer methods of factorization*. Birkhäuser, Boston, Mass.

Robinson, J. (1949). Definability and decision problems in arithmetic. *J. Symbolic Logic* **14,** 98–114.

Rose, H. E. (1984). *Subrecursion, functions and hierarchies*. Clarendon Press, Oxford.

Salié, H. (1949). Über den kleinsten positiven quadratischen Nichtrest nach einen Primzahl. *Math. Nachr.* **3,** 7–8.

Scharlau, W. and Opolka, H. (1984). *From Fermat to Minkowski*. Springer, New York.

Schmidt, W. M. (1980). *Diophantine approximation*. Lecture notes 785. Springer, Berlin.

Scott, W. R. (1964). *Group Theory*. Prentice-Hall, Englewood Cliffs, NJ.

Serre, J.-P. (1973). *A course in arithmetic* (trans. from French). Springer, New York.

Shanks, D. (1978) *Solved and unsolved problems in number theory* (2nd edn). Chelsea, New York.

Shapiro, H. N. (1983). *Introduction to the theory of numbers*. Wiley–Interscience, New York.

Siegel, C. L. (1949) *Transcendental numbers. Annals of Mathematical Studies,* Vol. 16. Princeton University Press, Princeton, NJ.

Silverman, J. (1986). *The arithmetic of elliptic curves*. Springer, New York.

Stark, H. (1970). *An introduction to number theory*. Markham, Chicago, Ill.

Stewart, I. N. and Tall, D. O. (1979). *Algebraic number theory*. Chapman and Hall, London.

Taussky, O. (editor) (1982). *Ternary quadratic forms and norms*. Marcel Dekker, New York.

teRiele (1986) *On the sign of the difference* $\pi(x) - \text{li}(x)$. NM-R8609 Centre for Mathematics and Computer Science, Amsterdam.

Titchmarsh, E. C. (1951). *Theory of the Riemann zeta function*. Clarendon Press, Oxford.

Venkov, B. A. (1970). *Elementary number theory* (trans. by H. Alderson of the 1937 edn). Wolters-Noordhoff, Groningen.

Watson, G. L. (1960). *Integral quadratic forms*. Cambridge University Press, London.

Weil, A. (1949). Number of solutions of equations in finite fields. *Bull. Am. Math. Soc.* **55,** 497–508.

—— (1983). *Number theory, an approach through history*. Birkhäuser, Boston, Mass.

INDEX OF NOTATION

GENERAL INDEX